100

TE PAPA

PRESS

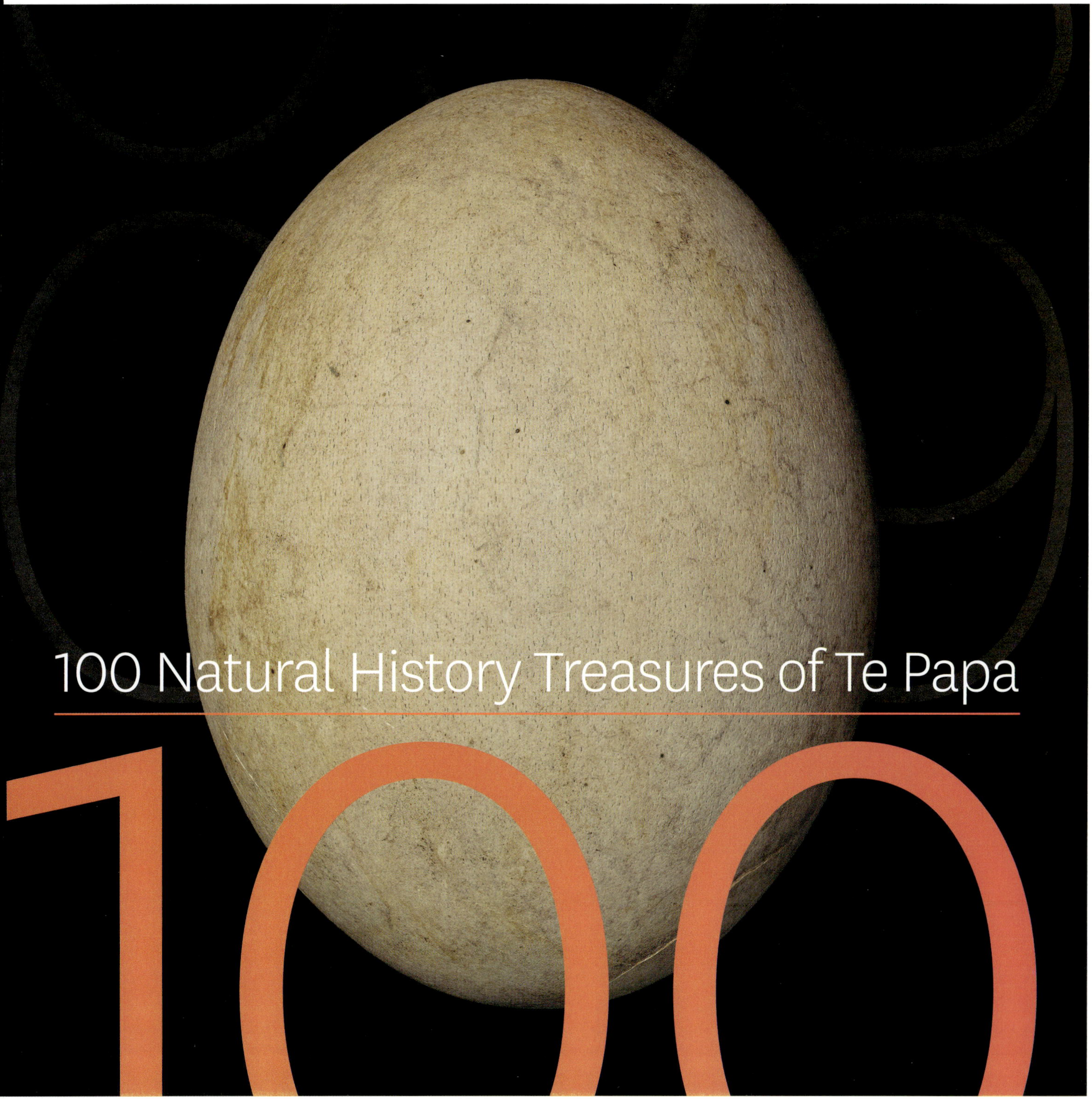

100 Natural History Treasures of Te Papa

Introduction:
Susan Waugh

The publication of *100 Natural History Treasures of Te Papa* coincides with, and we hope complements, two events that are significant to Te Papa's natural history collections – an opening and an anniversary. 2019 sees the launch of Te Taiao | Nature, the renewed natural environment exhibition in the museum, while October 2019 marks the 250th anniversary of the meeting of two peoples in this country – iwi and hapū Māori and the (mostly) European members of James Cook's *Endeavour* expedition – and the encounter of their distinctive world views.

That anniversary commemorates the landing here of the first European naturalists, Joseph Banks and Daniel Solander, with their shipmate, artist and scientific illustrator Sydney Parkinson. They were the first eager collectors of the local flora and fauna in this (from their perspective) virgin territory for Western science, introducing to it their rational and systematic understanding of its nature. In Te Papa's collections, we have physical links with that early scientific expedition: some of the specimens they collected; priceless taonga that make for us a special connection with the anniversary.

As a bicultural institution, we also use that anniversary to acknowledge the long-established presence of Māori understanding of the natural world in this land. For as much as science is about describing relationships in the web of life, mātauranga Māori – Māori knowledge – with its emphasis on whakapapa, or interconnectedness, offers a coherent and systematic universe of described relationships. We mark at least seven centuries of detailed observations and discoveries prior to Western contact – the original scholarship about nature in New Zealand, whose application enabled communities to live and thrive within their environment. After twenty years of display since the museum's opening at its current site, Te Papa's natural history exhibits have been rebuilt with a new focus on the public's connection to nature. We integrate narrative from mātauranga Māori into the new exhibits, exploring the intimate connection of New Zealanders to the whenua, the land, and the power of the elements, and the importance of the natural world in our existence as dwellers in Aotearoa New Zealand.

Te Papa has had a long association with science, having recently commemorated 150 years since the Colonial Museum's opening in 1865 under Director James Hector, later Sir James. The museum's major focus then was on providing an inventory of natural resources around the nation, with geology being a key feature, and to map the potential for the extractive industries. But Hector also had a penchant for making collections and describing biological life, including

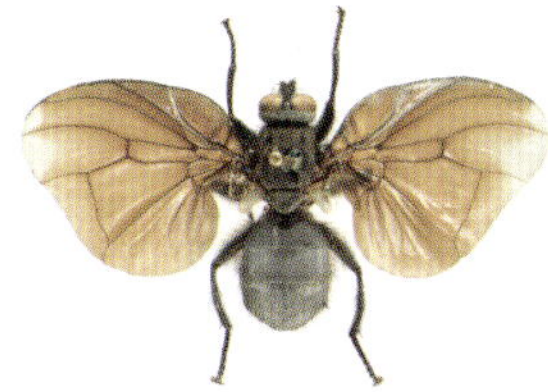

Bat-winged fly (see page 34)

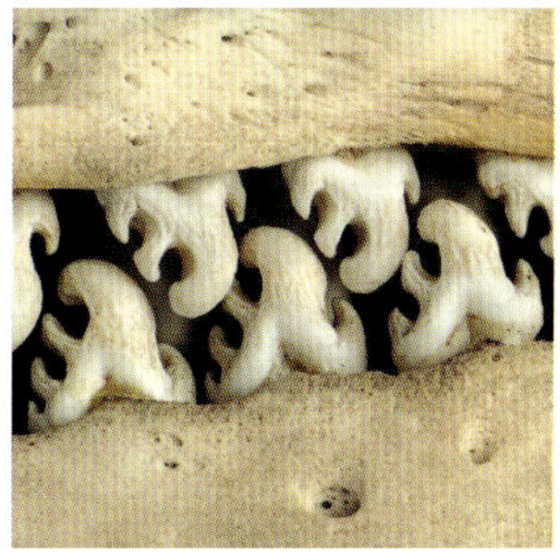

Crabeater seal skull (see page 60)

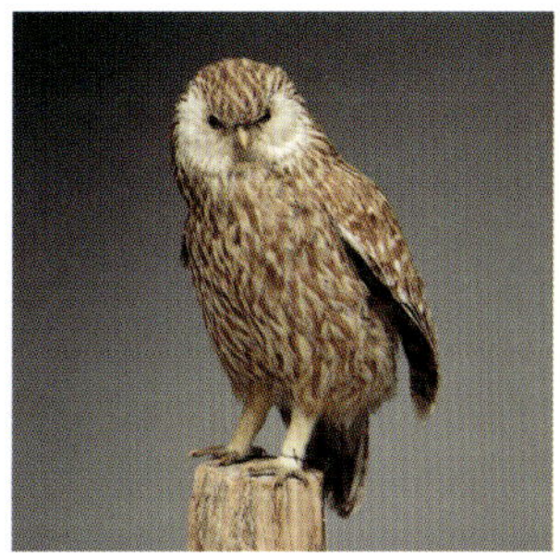

Laughing owl (see page 128)

describing several taxa new to science such as twenty-one species of fish, six species of mollusc, one plant and ten fossil reptiles, including the genus *Taniwhasaurus* that he named after the magical monsters of Aotearoa New Zealand.

In the museum's early years, despite the interests of scientists like Hector, there was a tendency to defer to established European museums as recipients of natural history collecting. Many important specimens, including type specimens – chosen to represent a newly described taxon – were sent overseas to be held in trust by those museums, with relatively few being retained in New Zealand. Today, researchers who wish to compare the type specimens for many endemic species of this country's fauna and flora must therefore work within the extensive global network of collecting institutions. These include the Rothschild collection in New York; the Musée National d'Histoire Naturelle in Paris; the Imperial Natural History Museum at Vienna, where many collections of the avid collector Andreas Reischek are held; Kew Gardens in London, the repository for the botanical collections made by Joseph Hooker and William Colenso; and the Natural History Museum in London, founded with the collections of Joseph Banks. But as soon as the Colonial Museum was established, scientists in Aotearoa New Zealand began to deposit their type specimens there. Te Papa's collections have since become the broadest of any New Zealand natural history collection, with specialist areas that are both deep and comprehensive.

Early on, the New Zealand Institute (later to become the Royal Society of New Zealand) and the Geological Survey (now GNS Science) were by and large run by the same people, all housed under the roof of the Colonial Museum. Over the decades they have become increasingly differentiated as institutions, but in the tiny community that comprises the New Zealand science sector we remain part of the same whānau. The museum forms part of a network of some twenty taxonomic collection institutions across the nation, each with its own focus. These include NIWA, Landcare Research, Scion, GNS Science and the Cawthron Institute – all research institutions funded in part by the Crown. New Zealand universities and museums in the major metropolitan centres (Auckland, Christchurch and Dunedin) also hold nationally important collections of fauna and flora.

In 2015, after many decades of collaborative work and shared research programmes, these agencies came together, under the leadership of Professor Wendy Nelson of NIWA, with the Royal Society of New Zealand to document the challenges for the sector. Their work and their report to government resulted in the establishment of the National Systematics and Taxonomic Collections Working

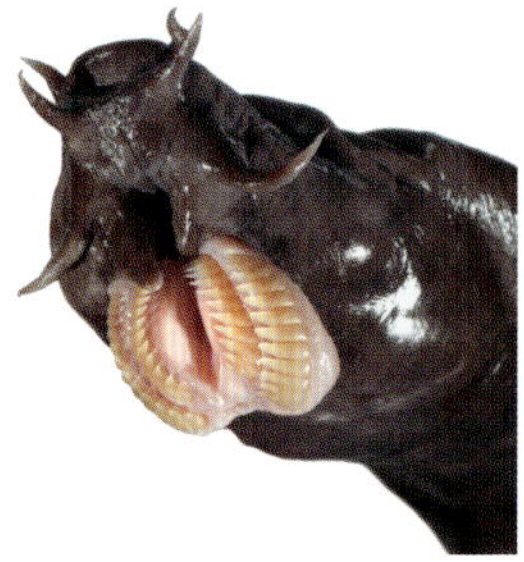

Hagfish (see page 110)

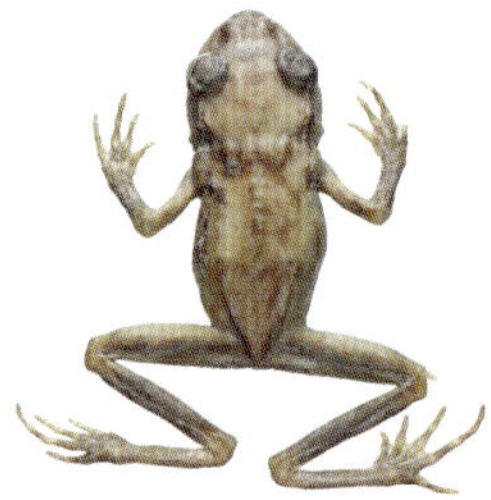

Hamilton's frog (see page 112)

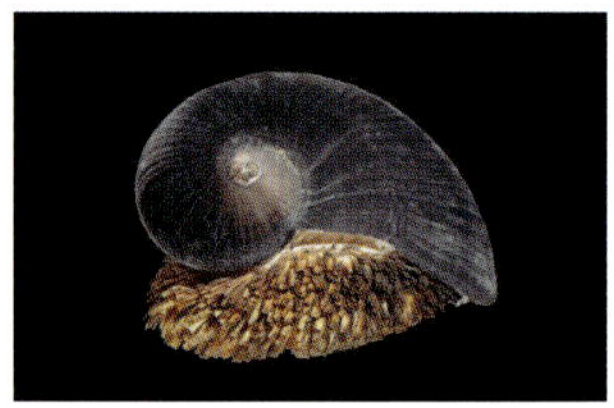

Scaly-foot gastropod (see page 136)

Group. This group now coordinates the natural history collections sector and its development. In a digital millennium, the work of our taxonomic collecting institutions increasingly lies in sharing information and standards of collection care, with the challenge of making the enormous numbers and diversity of specimens widely available across Aotearoa New Zealand and to international audiences.

Why develop natural history collections? People have many motives for collecting in this area: some may be driven by the wonder of nature, or the desire to understand how evolution has shaped the biological material at hand, or how species adapt to their environment. For others, the impetus may be in the spirit summed up by Everest climbing pioneer George Mallory: simply 'because it's there'. But whatever inspires the collecting, Te Papa today does it with clear purpose – to describe the biodiversity of our region, and to define which species influence our existence here in Aotearoa New Zealand.

We do this as a modern research institution, using current technology and methods such as scanning electron microscopy, molecular ecology and genetics work, and the collections give us the enviable advantage of a ready-made store brimming with the riot of life that makes up the biodiversity of Aotearoa New Zealand. Our work assists the conservation of species undertaken by the New Zealand government. It helps define which species are naturally occurring in Aotearoa New Zealand and which are recent arrivals: something that is of enormous benefit to the agencies ensuring the biosecurity of the country. There are many outside users of the collections and the associated databases, including schools, researchers and resource management agencies such as the Department of Conservation, the Ministry for Primary Industries, the Ministry of Health and the Environmental Protection Authority. We have begun work on projects with the wānanga of our region, to develop programmes exploring the collections from a mātauranga Māori perspective.

Many collectors have shaped the Te Papa natural history collections over the institution's 150-year history. Their particular scientific interests and expertise have made an indelible impression on the nature of the collecting and scientific description work carried out at Te Papa. In addition to such 'self-generated collections', many other major collections have been donated to the museum over time. They include those of William Colenso, Donald Petrie, George Hudson, Charles Fleming and many other notable contributors to our understanding of the flora and fauna of Aotearoa New Zealand – passionate amateurs as well as professionals in

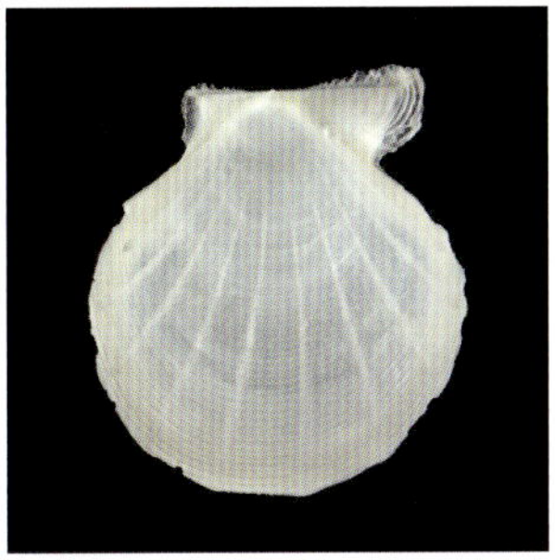

Parvamussium maorium **(see page 140)**

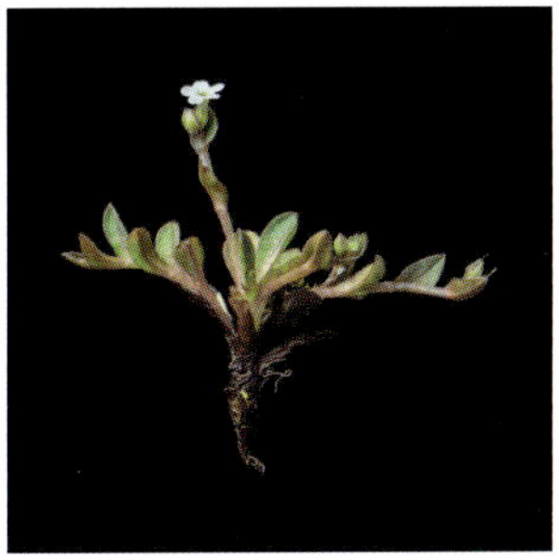

Moss-dwelling forget-me-not (see page 144)

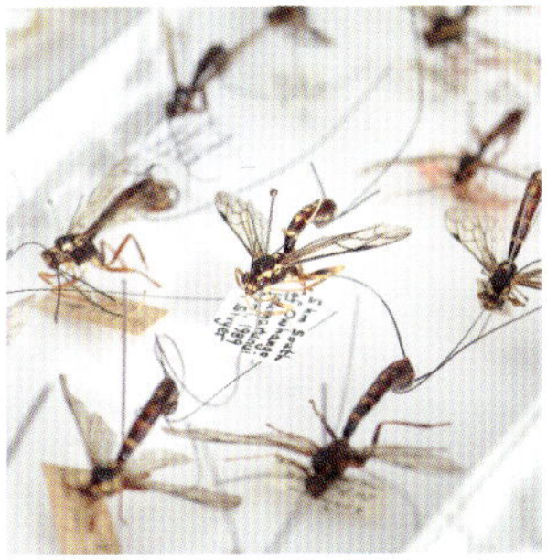

Parasitic wasp (see page 168)

specific biological fields. Te Papa holds the national collections of land snails, fishes, lizards and whales, and the strongest collection of birds and fossil vertebrates of any New Zealand collection agency. Some collection items have been 'repatriated' to Aotearoa New Zealand after being held in overseas museums for many years. For example, the British Museum sent duplicate specimens from the Banks and Solander botanical collections to Aotearoa New Zealand in 1895 and again in the 1980s, and, as mentioned earlier, these form some of the jewels in Te Papa's crown of natural history.

What does the museum do with its collections? And why have more than one example of a particular species? Our collections serve to define the biodiversity of Aotearoa New Zealand and its region. A specimen in the collections can be re-examined over time, enabling a high level of certainty about what it represents, even as the scholarship around the biota changes. Something called a *Hebe* in the 1960s is now part of the global genus of *Veronica*, thanks to our ability to re-examine the characteristics of specimens and compare them with related species across the planet. The specimens serve as a record of what occurs naturally or has been introduced into Aotearoa New Zealand over time, and they are a benchmark for determining what is native and what is not. Because the peopling of Aotearoa New Zealand occurred so late, in the thirteenth century, it is possible to separate native from naturalised species with much more certainty than in many other parts of the world.

The collections function rather like a library (including the loan of specimens to national and international institutions). Old favourites can be re-read and new meanings found from them. New scientific techniques allow us to use the species' minute structures and biochemical properties to explore their uniqueness and relatedness – a sort of forensic study of nature. The 'duplicates' in the collections serve to define the geographical spread of particular species across the nation, but also to document the natural variation within a species that may occur throughout its range. Sometimes, with new techniques or careful re-examination, this variation leads to a new species being defined.

Today, Te Papa's team of scientists covers the fields of botany and zoology, with specialists in ferns, mosses, flowering plants, seaweeds, molluscs, mosquitoes and other flies, spiders, birds and lizards, fossil vertebrates, fish and crustaceans. Some important faunal groups are under-represented in the current staff – the invertebrate collections are far broader than our institute can study,

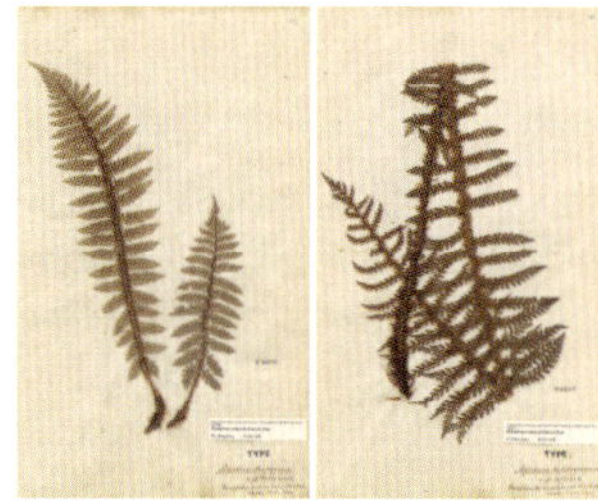

Prickly shield fern (see page 174)

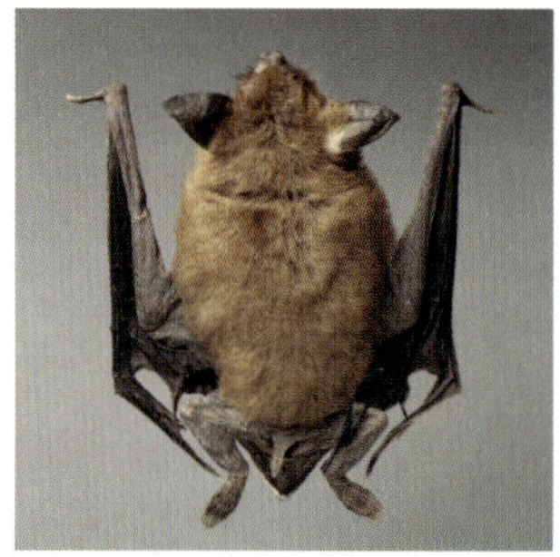

Short-tailed bat (see page 196)

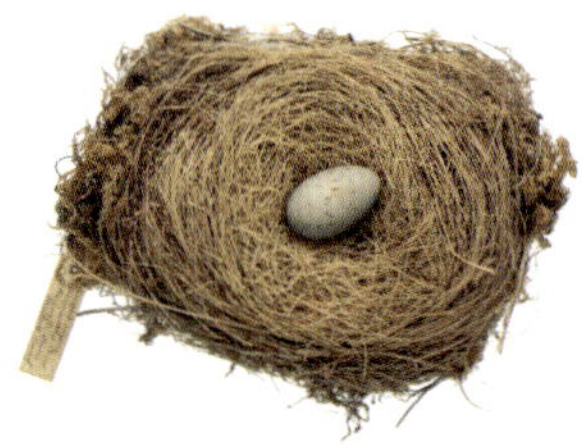

South Island kōkako nest (see page 202)

although invertebrate fauna make up 70 percent of the world's biodiversity. For example, with over 350,000 described species of beetle alone on the planet, we simply have to accept that we can't do all of it all of the time. However, the balance within Te Papa's science staff enables the museum to service research and support collection activity across the whole biota of Aotearoa New Zealand species groups. The single exception is the non-lichenised fungi, which are uniquely managed by Landcare Research in Auckland. This breadth of activity is no mean task, and our staff have an extraordinary range of skills to enable them to manage the collections, undertake field collecting activity, write scientific articles to document their work, and engage the public with science and their research specialities.

The items described in this book comprise a selection made by the scientists in the natural history team, with input from our expert photography team and Te Papa Press editors. We chose a set of objects that told the stories of Te Papa's natural history collections – a celebration of our natural world, portrayed through the lens of Western scientific methodology. It's impossible to 'representatively' document the over 1.5 million specimen lots in the collection, so we focused on some of the choice items – those for which there was a real connection for the scientists – and also on some quirky ones. This gives the book the feel of a 'behind the scenes' tour for the reader, revealing some of the hidden gems along with old favourites, each with an interesting narrative. We have endeavoured to showcase items from some key historical collectors as well as those that illustrate current research undertaken by the natural history team.

The work links to the renewal of Te Papa's natural history exhibits. We hope that readers will find interest and enjoyment in this book to complement their experience of the renewed context for Te Papa's natural history collections 'on the floor'.

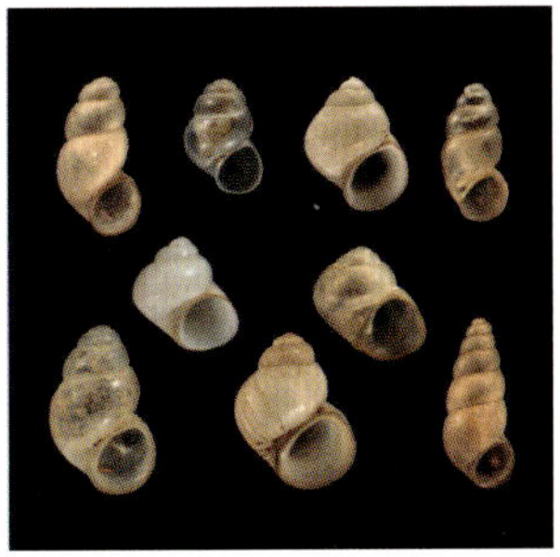

Tateid snails (see page 218)

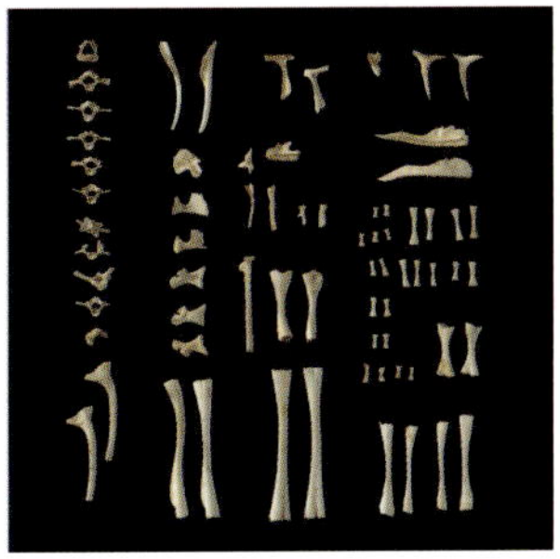

Waitomo frog bones (see page 234)

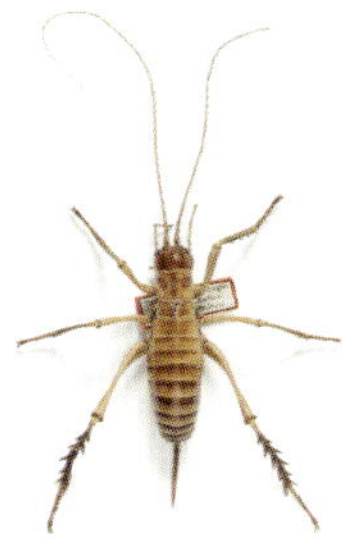

West Coast bush wētā (see page 238)

Objects
A—B

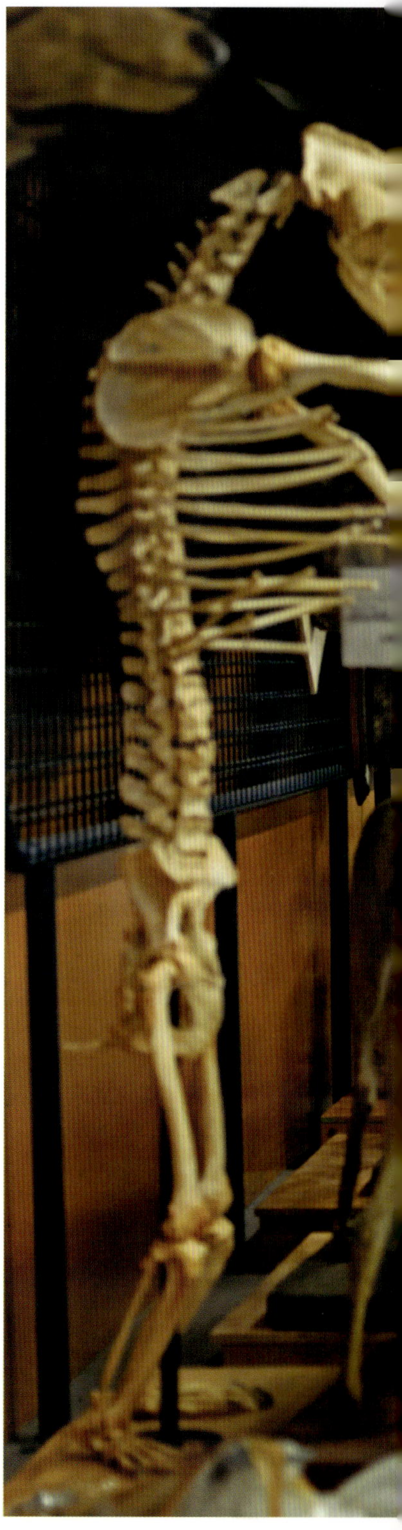

Photographer Jean-Claude Stahl, Kaitiaki Taonga Collection Manager for Vertebrates Tom Schultz and Head of Science Susan Waugh in one of Te Papa's collection storage rooms at its Tory Street, Wellington, collections facility.

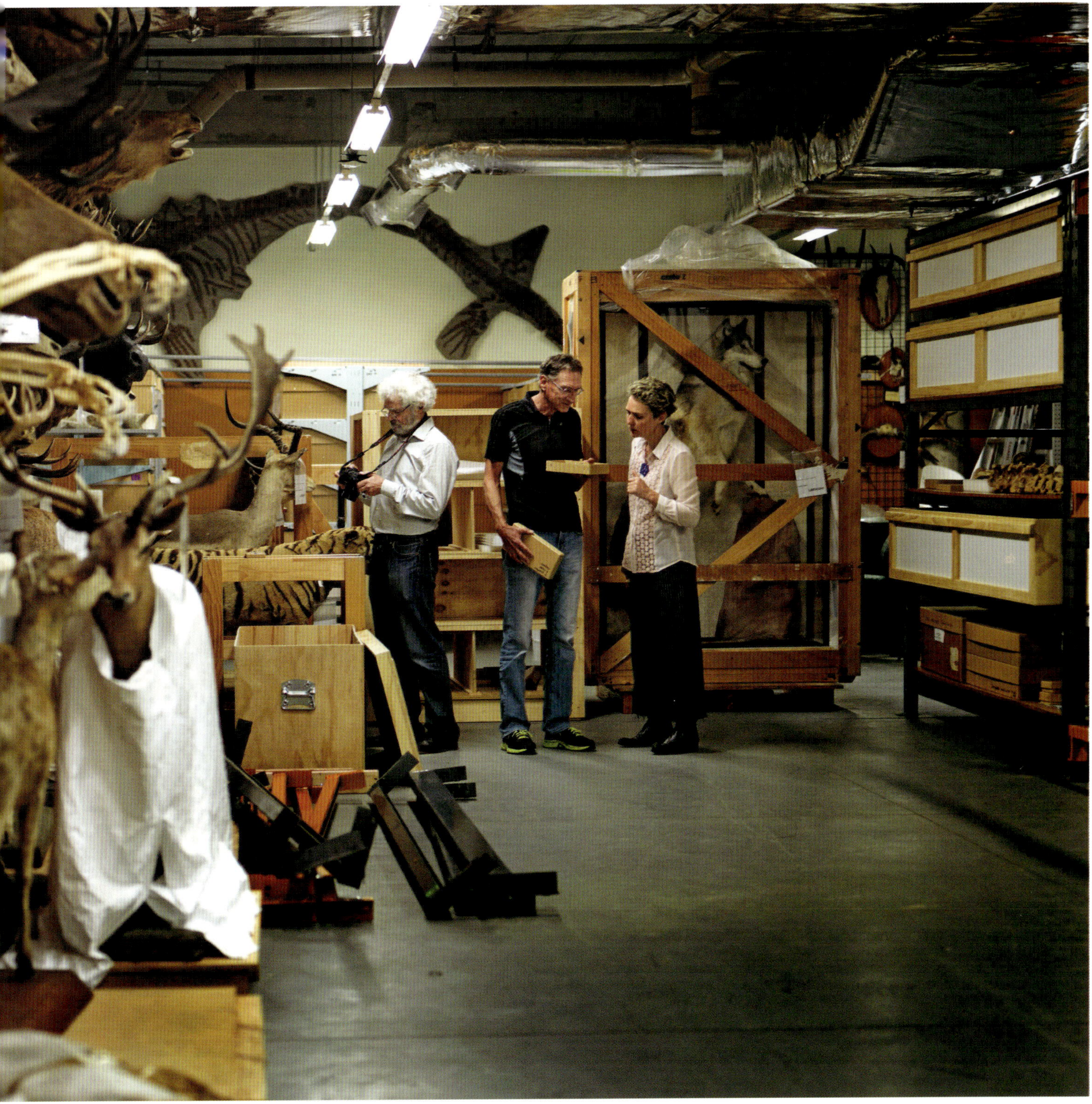

001 002 003

007 008 00

013 014 015

019 020 021

025 026 027

004 005 00

010 011 012

016 017 018

022 023 02

028 029 03

A Handbook of the New Zealand Mosses

GOK Sainsbury

When I first became interested in mosses, in the 1980s, there was only one reliable and comprehensive source of information – GOK Sainsbury's *A Handbook of the New Zealand Mosses*. Published in 1955, after more than thirty years of collecting, researching and collaboration, it is a remarkable compilation of descriptions and identification keys, augmented by seventy-six pages of illustrations by Nancy Adams, for almost 500 species of New Zealand mosses. Surprisingly, Sainsbury was not a trained botanist; he was a lawyer, based in Wairoa, Hawke's Bay.

While always interested in botany, it was not until Sainsbury turned forty that he took up bryology (the study of mosses and liverworts), quickly realising that he also needed to master German to understand the earlier literature on mosses. With close associates Amy Hodgson and KW Allison, he collected extensively in inland Hawke's Bay and the forests of the Urewera ranges. He called on mountaineering friends to collect from areas he could no longer access himself. Exchanging widely with overseas bryologists, Sainsbury amassed over 15,000 specimens of mosses; these underpin his book and are now held in the herbarium at Te Papa.

Sixty years later, there are other books, electronic keys and vastly more illustrations to help with identification. But when it gets difficult, or a problematic moss is found, we still resort to the clear, concise descriptions in our now dog-eared copies of 'Sainsbury'. Sainsbury described about twenty new species; some of these, like the one shown here, have been skilfully photographed by Jean-Claude Stahl, revealing the intricacies and beauty of even dead and dried specimens that first captured the imagination of a lawyer almost a century ago. PB

Moss (*Fifea aciphylla*), M005678/A, holotype, collected by GB Rawlings, Conical Hill, near Tapanui, Otago, November 1936.

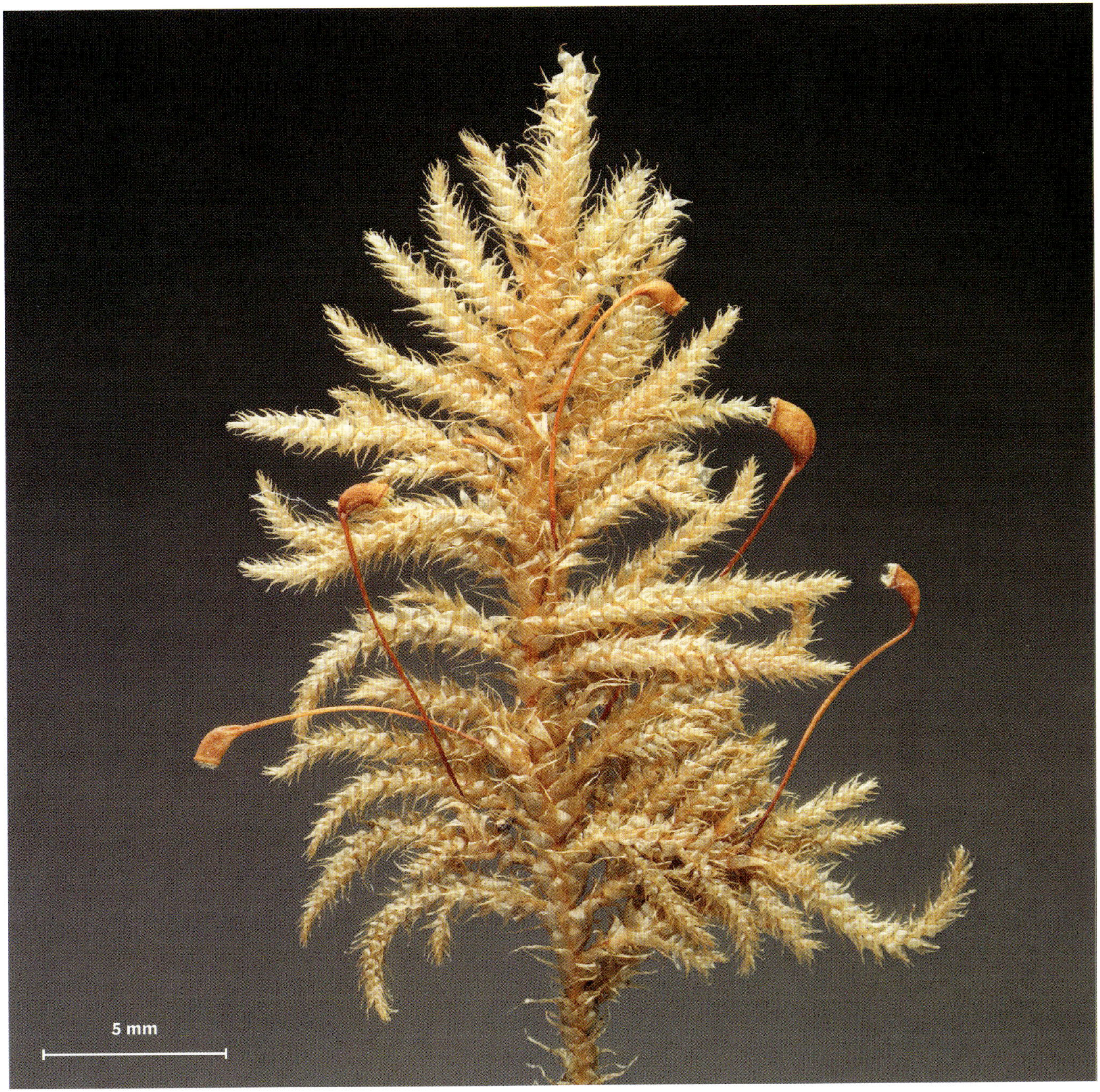
5 mm

Antarctic lichen

Usnea antarctica

The International Geophysical Year, 1957–1958, inspired many countries to send research teams to Antarctica, including the Commonwealth Trans-Antarctic Expedition. New Zealand's contribution is best remembered for Edmund Hillary's cheeky run to the South Pole. His team, driving converted Massey-Ferguson farm tractors, was setting up supply depots for Fuchs' party who were making the actual continental crossing – but the Pole was within reach, so Hillary and four men made the two-week dash.

Less is remembered of other, equally important, expeditions to study the aurora, sunspot activity, geomagnetism, ice movement and seismology. The New Zealand Geological Survey sent a team of eight geologists to the Cape Hallett area. Led by Larry Harrington, Ed Hillary's brother-in-law (they were married to sisters), the team man-hauled sledges for nearly two months, mapping and surveying the mountains around the Tucker Glacier. All New Zealand expeditions were also instructed to collect lichens and mosses. These specimens, sent back to New Zealand for identification and study, are now tucked into packets and boxes in Te Papa's herbarium.

Antarctica presents not only a formidable challenge to human explorers but also a difficult environment for the survival of any life. Only two flowering plant species live in the Antarctic Peninsula, a grass and a cushion-forming plant, in the warmest parts. Surprisingly, however, Antarctica is home to more than 250 lichen species and 125 mosses and liverworts. Lichens grow wherever bare rocks emerge from the ice and are adapted to the harsh conditions, some even growing below the rock surface between the grains and crystals.

This specimen of *Usnea antarctica* was collected from rocks above the Tucker Glacier, but is also found on subantarctic islands and in the New Zealand Southern Alps. BH

Antarctic shrub lichen (*Usnea antarctica*), L000067, unknown collector, Crater Hill, Hallett Base, Antarctica, 11 January 1958.

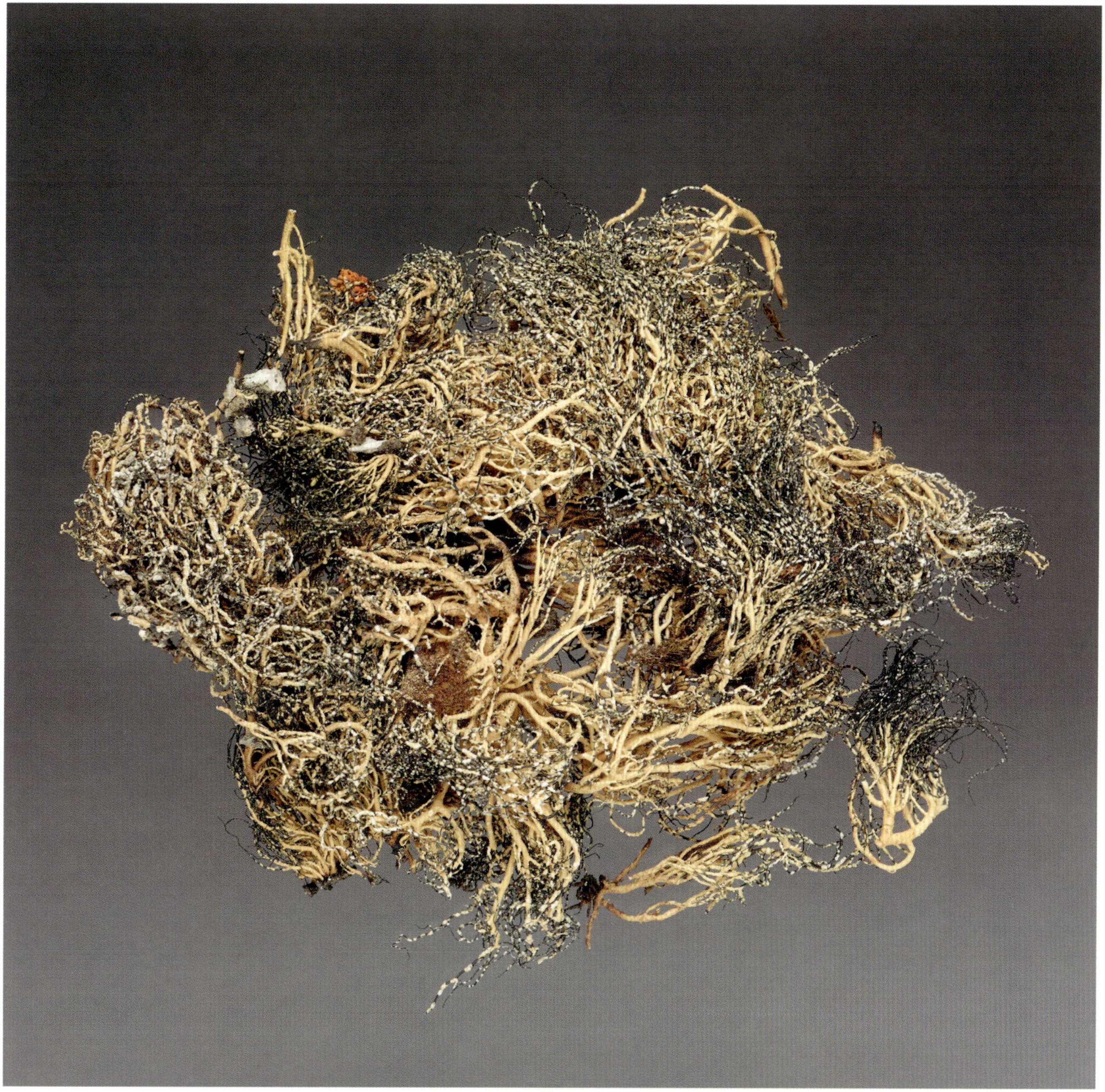

003

100

Antipodean *Durvillaea*

Durvillaea species

One of the first cabinet drawers I opened during my first days working with Te Papa's botany collection revealed an impressive specimen: a *Durvillaea*, a massive brown seaweed more commonly known as bull kelp. This specimen was collected in 1985 from one of the remote corners of New Zealand, the Antipodes Islands.

While out on a marine survey at this time, museum collector Cameron Hay and Director John Yaldwyn discovered the large plant at depth and collected it, knowing that it was different. They had the large specimen flown by helicopter from the island, from where it was brought on board a navy vessel, frozen, and transported all the way back to Wellington.

Te Papa's botany curator Nancy Adams was waiting for it. She carefully processed and meticulously hand-pressed it over a number of very large sheets, as though to also preserve every bit of effort that went into collecting it. The specimen is now part of the incredible library of seaweeds held by Te Papa. Many of the specimens are well understood, while many more await study and even a name, like this *Durvillaea*.

The species was known to have been new to science since it was first collected in 1969. With only a few specimens to work with, it's taken time – and perhaps advancements in tools and techniques, such as DNA identification – to really understand it. Nearly forty years later, the *Durvillaea* is nearing the end of its long wait for a formal name.

There are many such examples in Te Papa's natural history collections: what we know to be unknown is pretty extraordinary. This *Durvillaea* has always been a conspicuous, and my favourite, reminder of this. JD

Manager Science and Humanities Collection Jenn Dalen and Kaitiaki Taonga Collection Manager of Botany Antony Kusabs examine the *Durvillaea* specimen laid out in the Botany storeroom.

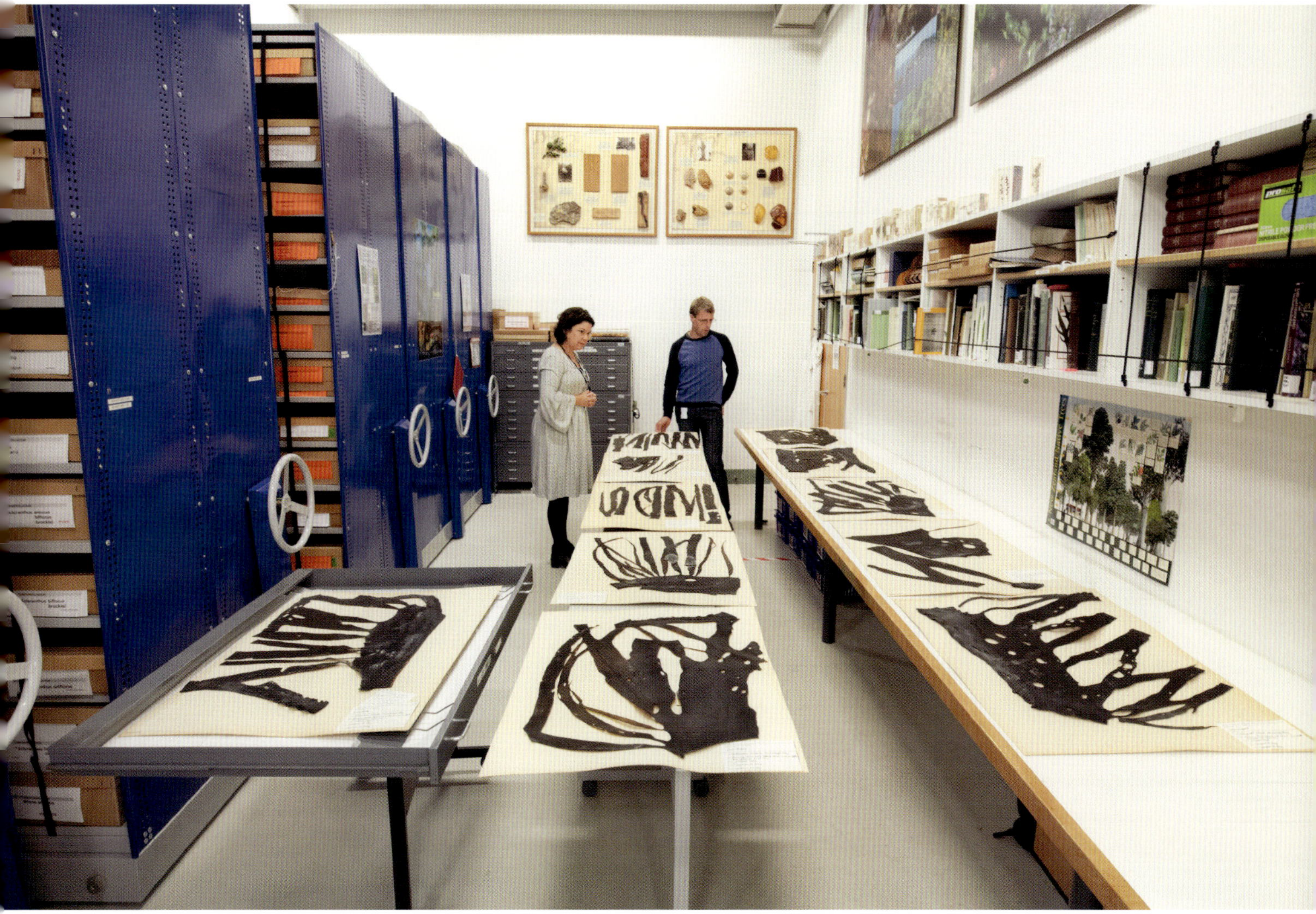

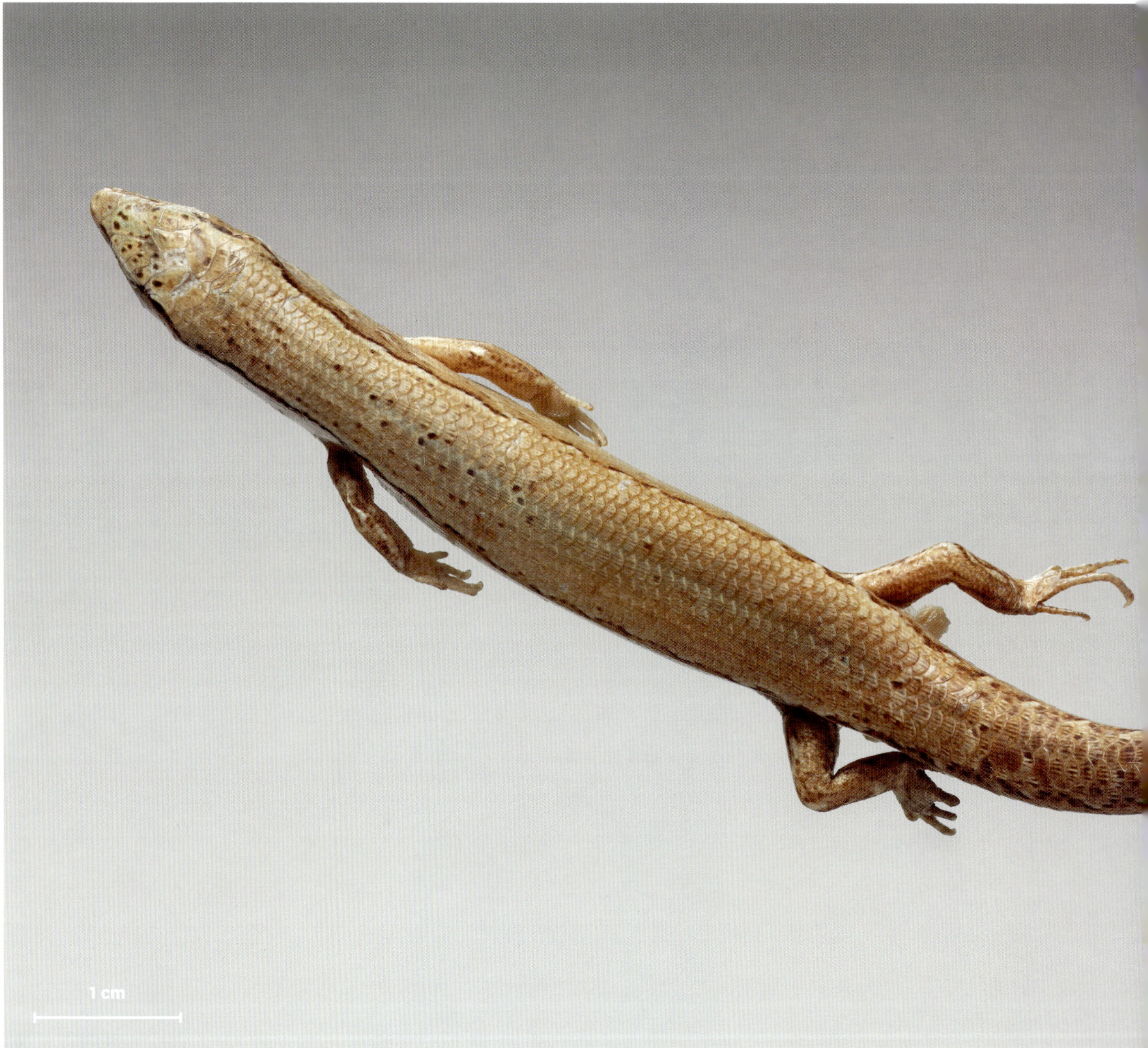
1 cm

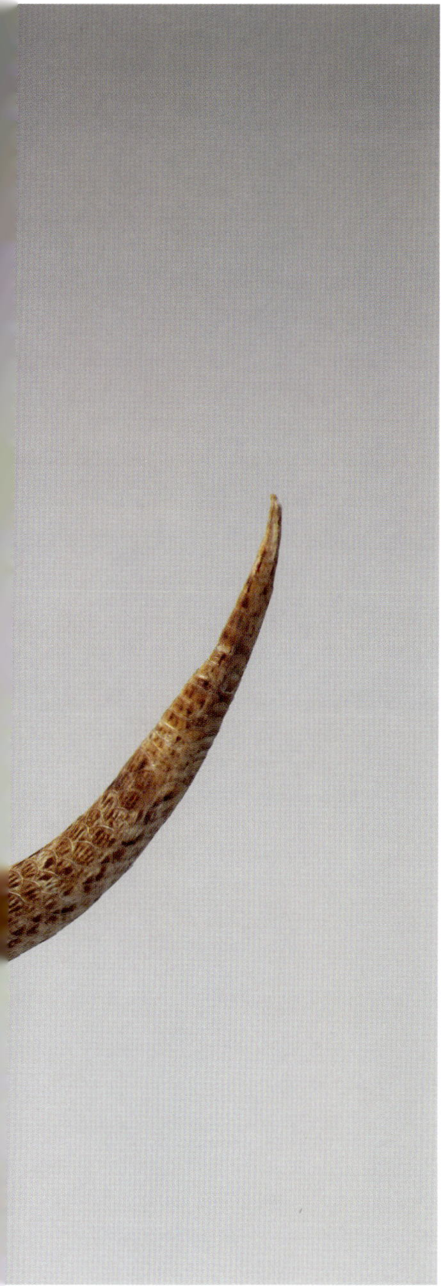

Aorangi skink

Oligosoma roimata

People know New Zealand as a land of birds, but few are aware that it is equally a land of lizards. When the first waka touched our shores, there were about one hundred species of native land or freshwater bird on the three main islands, but at least 110 species of lizard. So why are so few New Zealanders aware that we share our islands with so many endemic lizards?

Part of the reason is that predation by introduced rats, mice, stoats and cats (in particular) has led to most lizard species becoming very rare or confined to tiny portions of their original range. Also, only recently has the true diversity of New Zealand lizards been recognised.

When I first took an interest in New Zealand lizards, in the late 1970s, there were about twenty known species of skink and fifteen known species of gecko. Why has the known number of lizard species increased more than threefold over the past forty years? Part of the reason is that many new species with restricted ranges have been discovered, but the main reason is awareness that many of the 'species' recognised before 1990 were actually clusters of similar-looking species. Recognition of these 'cryptic species' has been mainly undertaken in the laboratory, using a combination of microscopy and genetic techniques.

One of the new species (described in 2013) is the Aorangi skink, known only from the Poor Knights Islands off the Northland coast. Closely related to the more widespread ornate skink, the Aorangi skink has been separated from it for at least 6.2 million years – much longer than the one to two million years since this island group was last connected to the nearby mainland. CM

Aorangi skink (*Oligosoma roimata*), RE.001626/32, snout to vent length 57 mm, collected G Hardy and L Moran, Aorangi Island, Poor Knights Islands, November 1973.

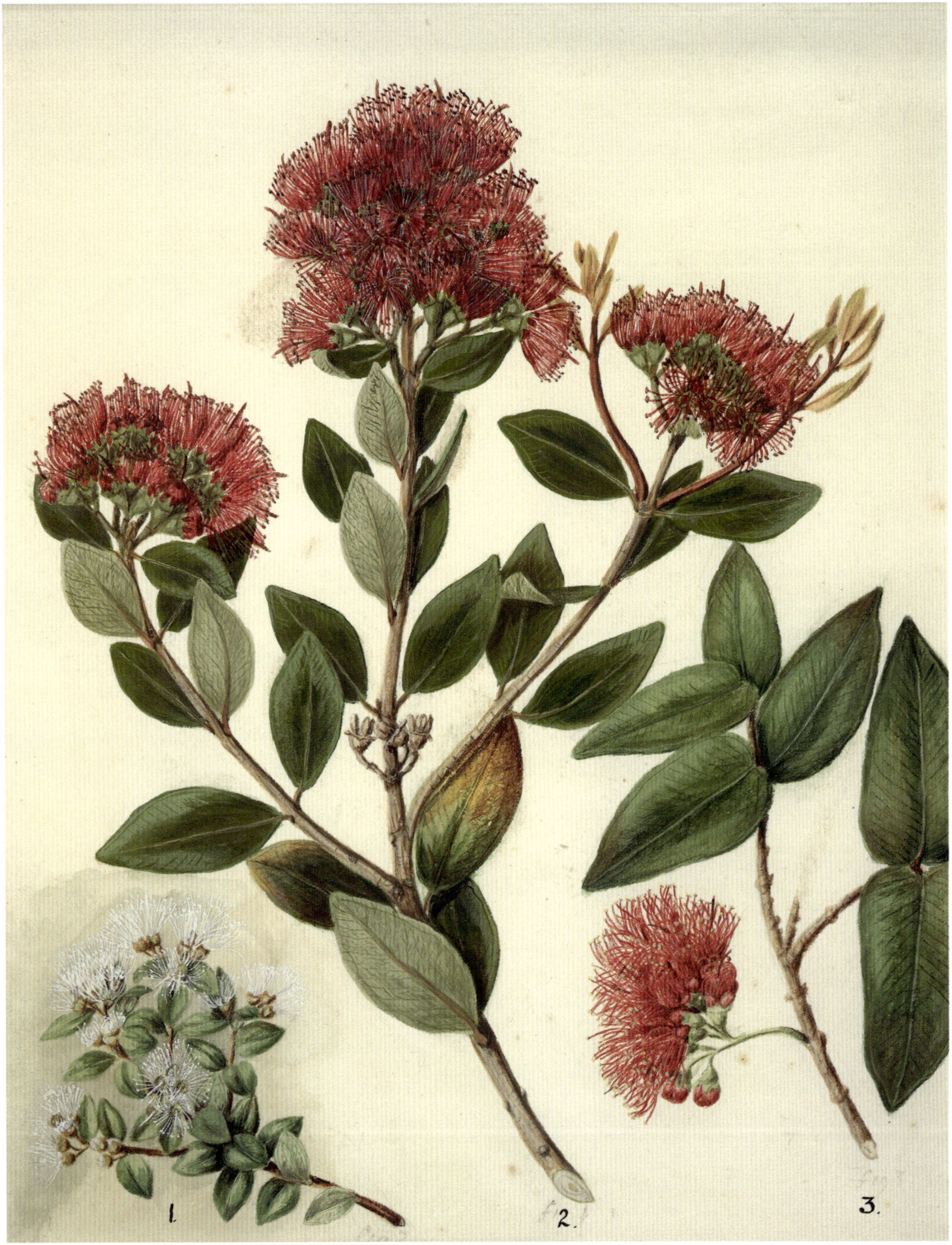
1.
2.
3.

The Art Album of New Zealand Flora

Edward and Sarah Featon

This delightful book was published in three parts between 1887 and 1889. The authors, Edward and Sarah Featon, were keen to disprove the commonly held belief of the time that New Zealand had no flowers of any note. The Featons asserted that 'the flowers of our adopted land ... display a beauty of form and character which in many instances are peculiarly their own'. Native plants are held in much higher esteem these days, but in the late nineteenth century the Featons' views were radical.

The book was ground-breaking in other ways. It was the first volume containing full-colour illustrations to be produced and published entirely in New Zealand. The text was written by Edward Featon, and the forty plates were printed by the exacting technique of chromolithography from original watercolours painted by Sarah Featon, including this one showing three species of *Metrosideros*. It was produced by the Wellington firm of William Bock and Alfred Cousins, who were renowned for their skill in designing and engraving postage stamps.

Unfortunately, the book was expensive to produce and published at a time of economic depression. It was not a commercial success, and the firm was dissolved soon afterwards. Two further volumes had been planned, and Te Papa still holds another ninety-three unpublished plates, but these volumes were never completed.

Despite its lack of commercial success, the book was well received by the public, and Sarah's bright colours opened many people's eyes to the beauty and diversity of the New Zealand flora. The government even presented a copy, in a specially made casket of native timbers, to Queen Victoria at the time of her Diamond Jubilee in 1897. **PB**

Aka, Rata, **watercolour by Sarah Featon, c. 1885, 1992-0035-2277/77, 264 mm x 214 mm. Published as Plate 36, *Metrosideros scandens, Metrosideros robusta, Metrosideros parkinsonii,* in Edward and Sarah Featon, *The Art Album of New Zealand Flora* (Bock and Cousins, Wellington, 1889).**

006 / 100

The Baker and Thompson Herbarium

John G Baker, Bridget Thompson and Silvanus Thompson

Museum collections have the unique power to transport you in space and time. They are a tangible link between you and someone who lived hundreds of years ago, sometimes just down your street and sometimes on the other side of the world.

The Baker and Thompson Herbarium is one of these collections. About 140 years ago, 28,000 specimens of dried plants were purchased from the British Museum by James D Hector (Director of the Colonial Museum, Te Papa's predecessor) to create a reference collection to assist with the identification of naturalised plants (weeds) and the study of indigenous species. It combined the collections of three people with an abiding interest in botany: Silvanus Thompson (a schoolmaster), his wife Bridget Thompson and British botanist John G Baker, who was taught by Silvanus.

Because of poor storage conditions locally and a lack of botanical staff, most of the collection remained in its original zinc-lined packing cases for almost a century after its arrival in 1877. Thanks to the help of volunteers, we have now learnt more about this collection. It includes plants from Asia, Europe, India, the Middle East and North America. Some specimens were collected as early as 1765, four years before Captain Cook first sighted New Zealand!

The pristine state of the collection makes it extremely special. Specimens are still wrapped up in bundles, as they were when shipped here. Some samples remain in between the same paper sheets or newspapers used to press-dry them. And, most astoundingly, some flowers have kept their colours, the same colours that no doubt enticed these three people to collect them over a century ago. CL

Dwarf morning-glory (*Convolvulus tricolor*), SP106548, collected by unknown collector from a botanical garden in Europe, August 1803.

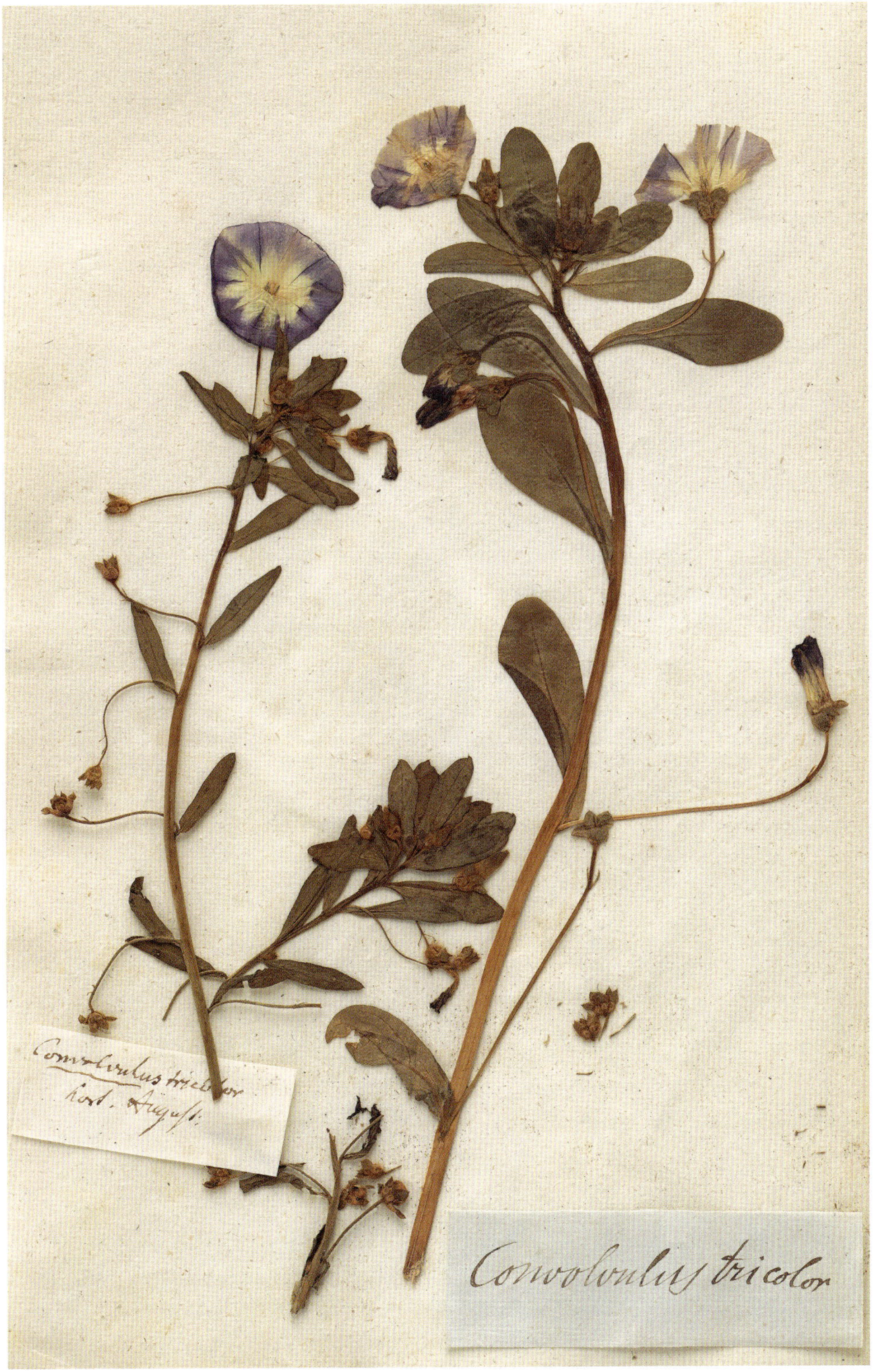
Convolvulus tricolor
hort. August.
Convolvulus tricolor

The Banks and Solander specimens

Joseph Banks and Daniel Solander

Museum visitors light up whenever I show them any of the 541 sheets in this collection. They are the oldest New Zealand plant specimens in existence. Collected between 1769 and 1770 by Joseph Banks and Daniel Solander, the senior naturalists on board Captain James Cook's first voyage on the *Endeavour*, they mark the beginning of modern taxonomy in this country.

007 / 100

The collection at Te Papa represents about 316 of the approximately 350 vascular plant species Banks and Solander collected. Importantly, they show us what plants were growing in lowland habitats when Pākehā first arrived in New Zealand, and amount to approximately 14 percent of our present-day indigenous flora. It was this fact that prompted botanist Thomas Kirk, with help from Colonial Museum Director James Hector, to write to the British Museum to request duplicate specimens from Cook's first voyage. As the first New Zealand scientist tasked with writing a complete Flora of New Zealand, Kirk knew the Banks and Solander specimens would be vital. When the specimens arrived at the Colonial Museum in 1895, Kirk carefully studied them, as attested by his handwritten identification labels. Sadly, he died in 1898 before completing the Flora. An abridged version, *The Students' Flora of New Zealand*, was published posthumously in 1899.

The mixture of historical and modern identification labels associated with these specimens provides a clue to their ongoing use for taxonomic study. Some specimens are currently being studied as part of the revision of all fern and seed plant groups for an updated New Zealand Flora, published online (www.nzflora.info). The collection is truly nationally significant, and will continue to be a key focus for researchers and visitors at the Te Papa herbarium.→

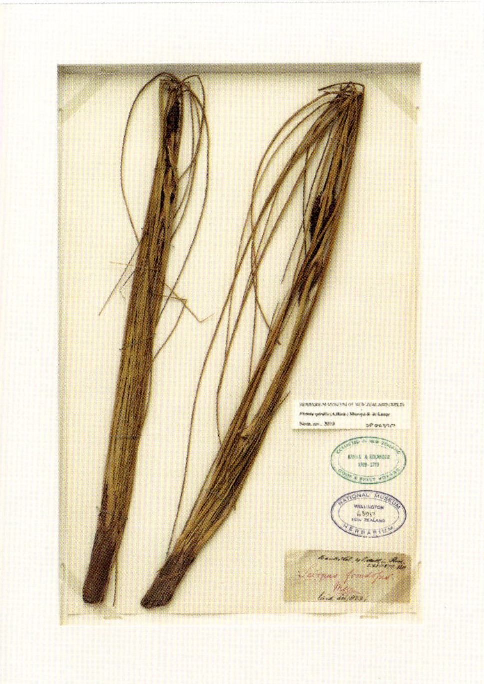
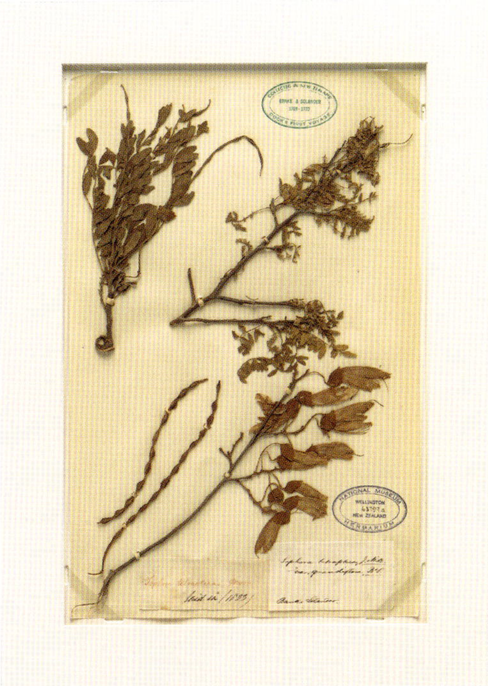
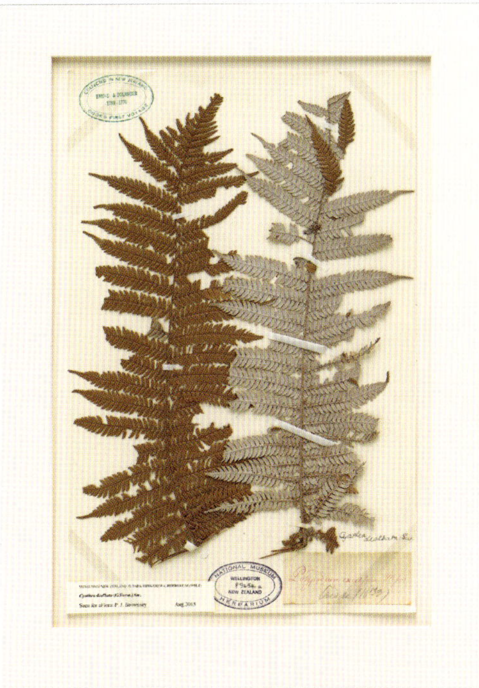
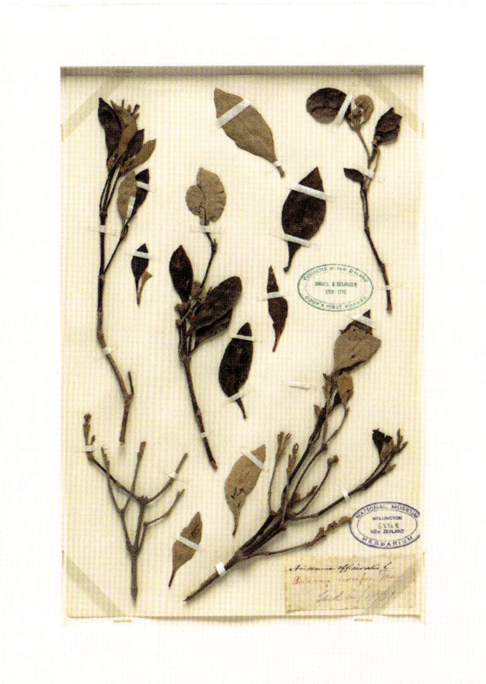
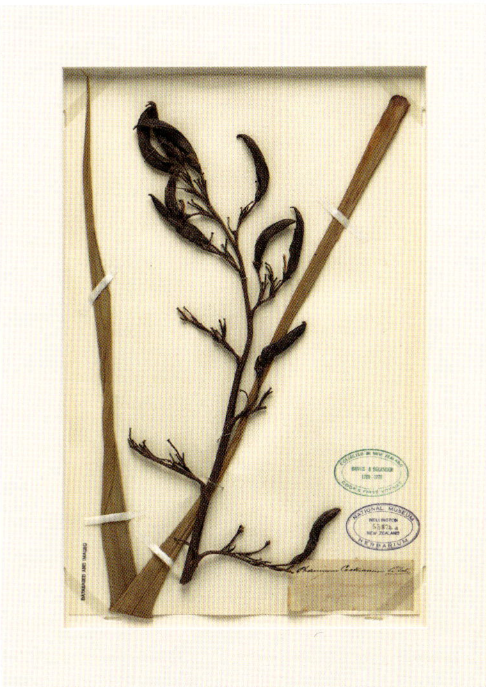
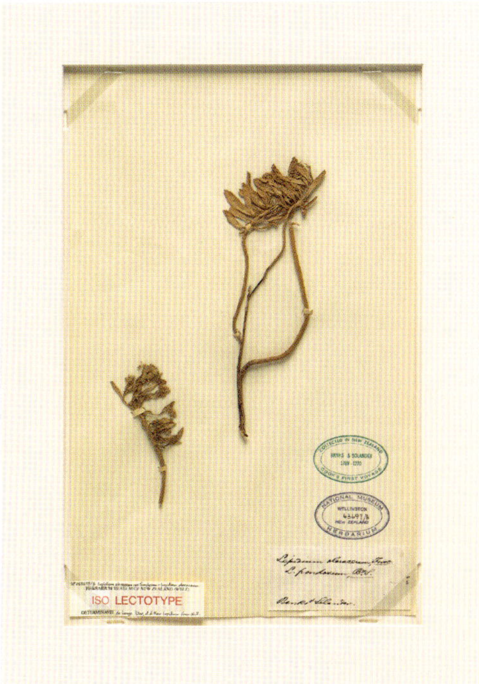
ISO LECTOTYPE

In the early 2000s, Te Papa staff realised that, given their significance, improved storage of the Banks and Solander specimens was needed. Collection managers, curators and conservation staff continually introduce new practices, materials and environmental control systems to improve the storage conditions of our natural history collections. We do this because we are passionate about caring for the collections entrusted to us on behalf of all New Zealanders. We are also obliged, under the Act of Parliament that established Te Papa, to 'develop, conserve, and house securely the collections of art and items relating to history and the natural environment'.

Most plant specimens in the Te Papa herbarium are stored on sheets, and stacked in boxes within folders. This means that some plant material is subjected to the weight of specimens above it. For the Banks and Solander specimens, new customised, mat-board frames were constructed, one for each specimen. The edges of these frames are slightly higher than the individual specimens, so that they bear the weight of specimens stacked above. The frames are housed in robust, acid-free boxes that exclude both dust and light, protecting the plant material and label ink from fading. These new practices, together with strict humidity controls, will enable the specimens to maintain their good condition for centuries to come. AK

Page 29: Specimens collected by D Solander and J Banks: Pīngao/golden sand sedge (*Ficinia spiralis*), SP063917, length 545 mm, width 385 mm, depth 10 mm, New Zealand, 1769. Kōwhai (*Sophora tetraptera*), SP063797/A, length 545 mm, width 385 mm, depth 10 mm, New Zealand, 8 October 1769. Ponga/silver fern (*Cyathea dealbata*), P009454/A, length 545 mm, width 385 mm, depth 10 mm, Mercury Bay, Cook's Bay, Cook's Beach, New Zealand (36°50.30'S, 175°45.38'E), 5 November 1769. Mānawa/mangrove (*Avicennia marina* var. *resinifera*), SP063768, length 545 mm, width 385 mm, depth 10 mm, 'prope Opuragi' [Mercury Bay, Cook's Bay, Cook's Beach] (36°50.30'S, 175°45.38'E), 5 November 1769. Wharariki/mountain flax (*Phormium cookianum*), SP063874/A, length 545 mm, width 385 mm, depth 10 mm, New Zealand, 1769. Nau/Cook's scurvy grass (*Lepidium oleraceum*), SP063697/B, length 545 mm, width 385 mm, depth 10 mm, New Zealand, 1769–1770.

Facing: A Te Papa framer puts the finishing touches on the mat-board framing of a Banks and Solander specimen.

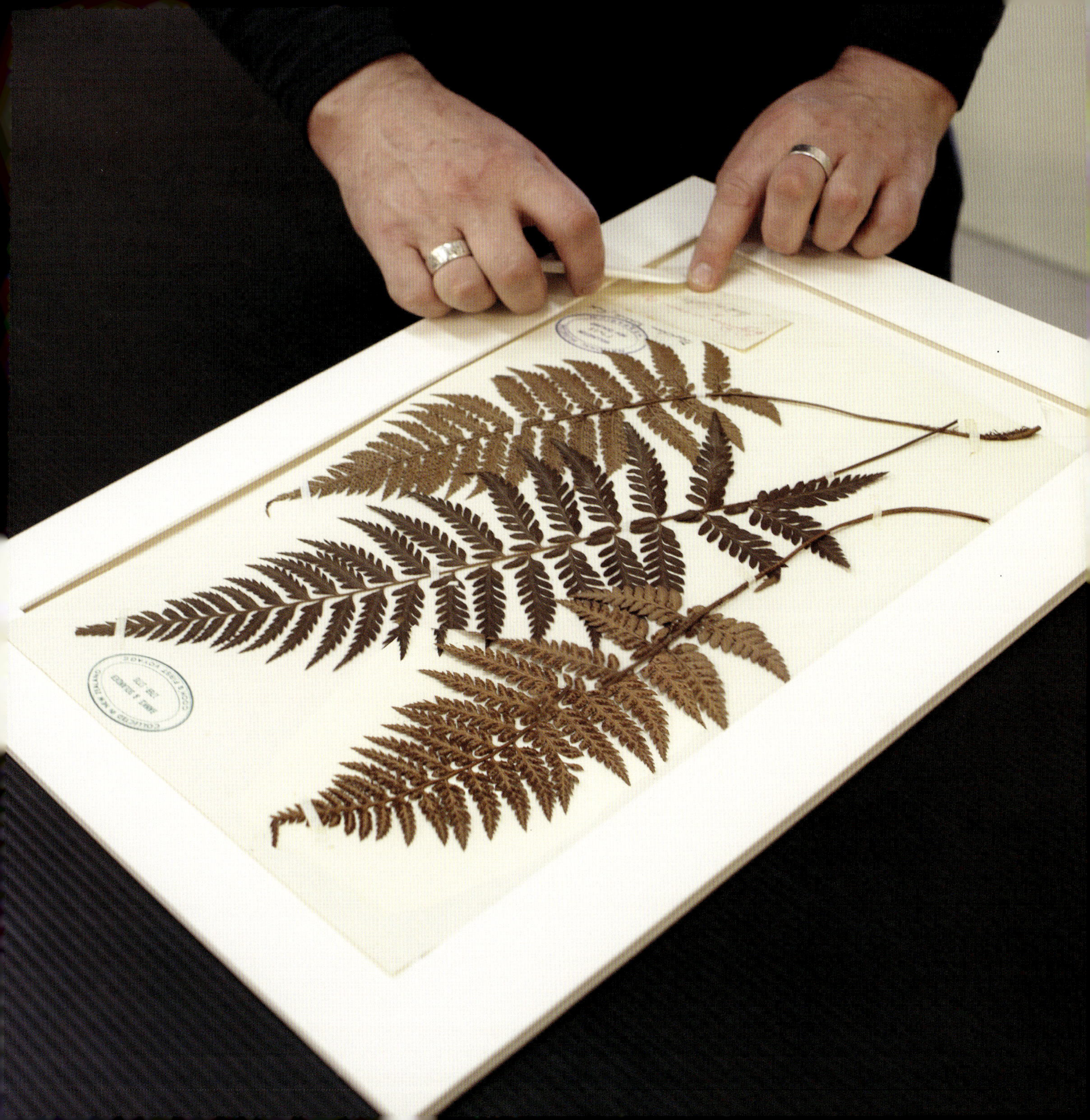

Basket star

Astrodendrum elingamita

This unusual, plant-like tangle is an animal: a basket star. It was trawled from deep water and has rolled up tightly in defence; in their natural bottom habitats, basket stars are not normally tightly bunched up. Named *Astrodendrum elingamita*, it belongs to a large class known as brittle stars. The brittle stars' nearest relatives are sea stars or 'starfish', and you can see the connection in the five-part central disc.

008

100

A. elingamita was described in 1974 by the late Alan Baker of the National Museum (the forerunner of Te Papa) from the photographed specimen, found at the Three Kings Islands. He named it after the passenger steamer *Elingamite*, which was wrecked at these islands in 1902 with much loss of life.

Most brittle star species have five unbranched, snake-like arms, but basket stars have evolved arms that branch repeatedly. Some look like fern fronds while others appear more random in their branching, but all develop to form a basket-like net. This net is held up in the currents above to catch organic particles and tiny animals which are snared in mucus and passed to the mouth beneath the disc. The arms can also capture prey on corals or sponges living nearby.

In *A. elingamita* the arm branching is relatively even, and Dr Baker described the arms as having 'forked approximately ten times'. That might not sound particularly impressive – but if you divide one arm ten times you get 1024; multiply this by five and you end up with 5120 hair-thin arms! Most animals won't have precisely that number, however, because the outer branches are often broken off by accident or by predators; new arms will eventually grow in their place. RW

Basket star (*Astrodendrum elingamita*), EC.001191, average diameter of photographed specimen approx. 70 mm, collected by AN Baker at depth

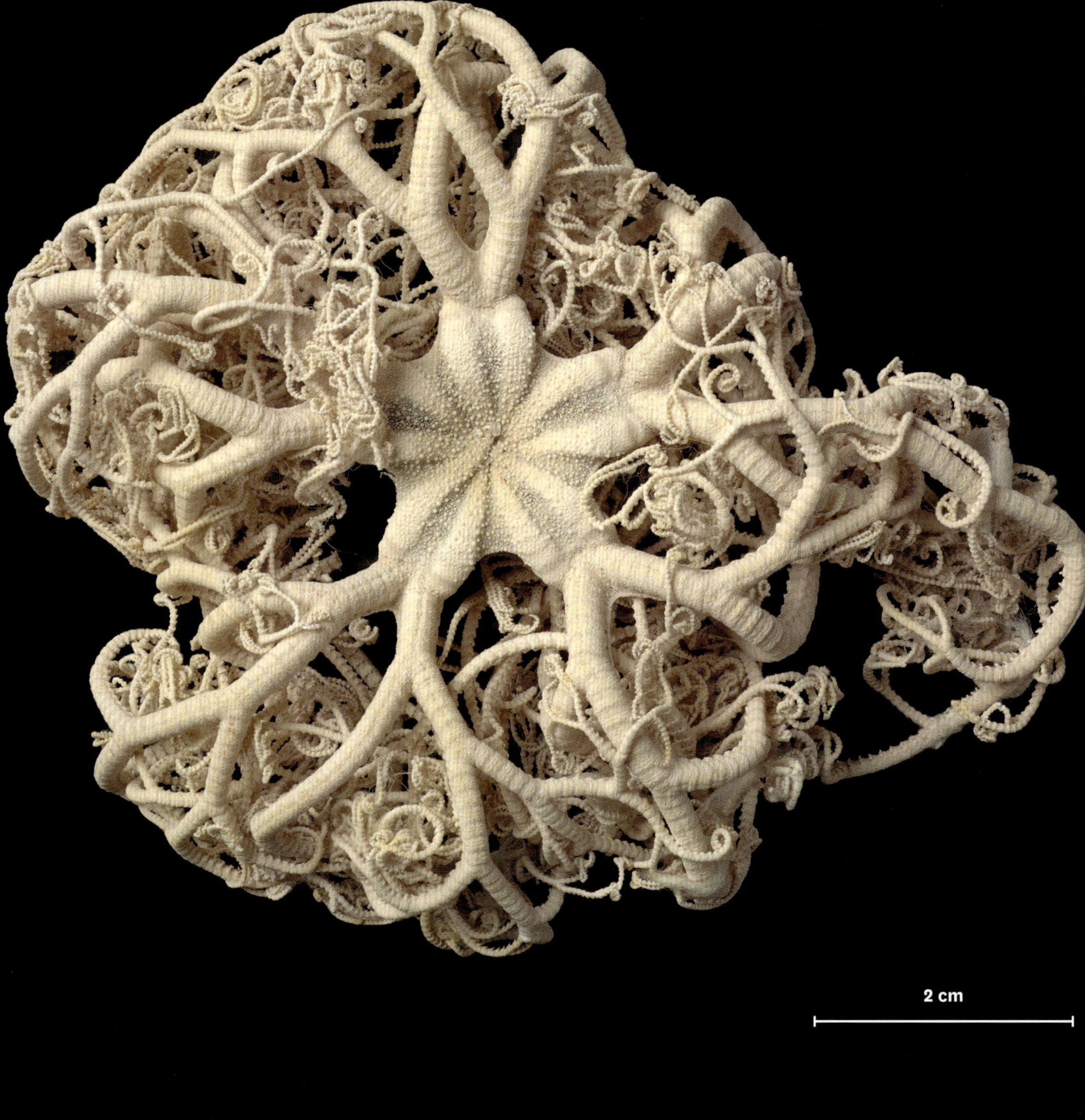
2 cm

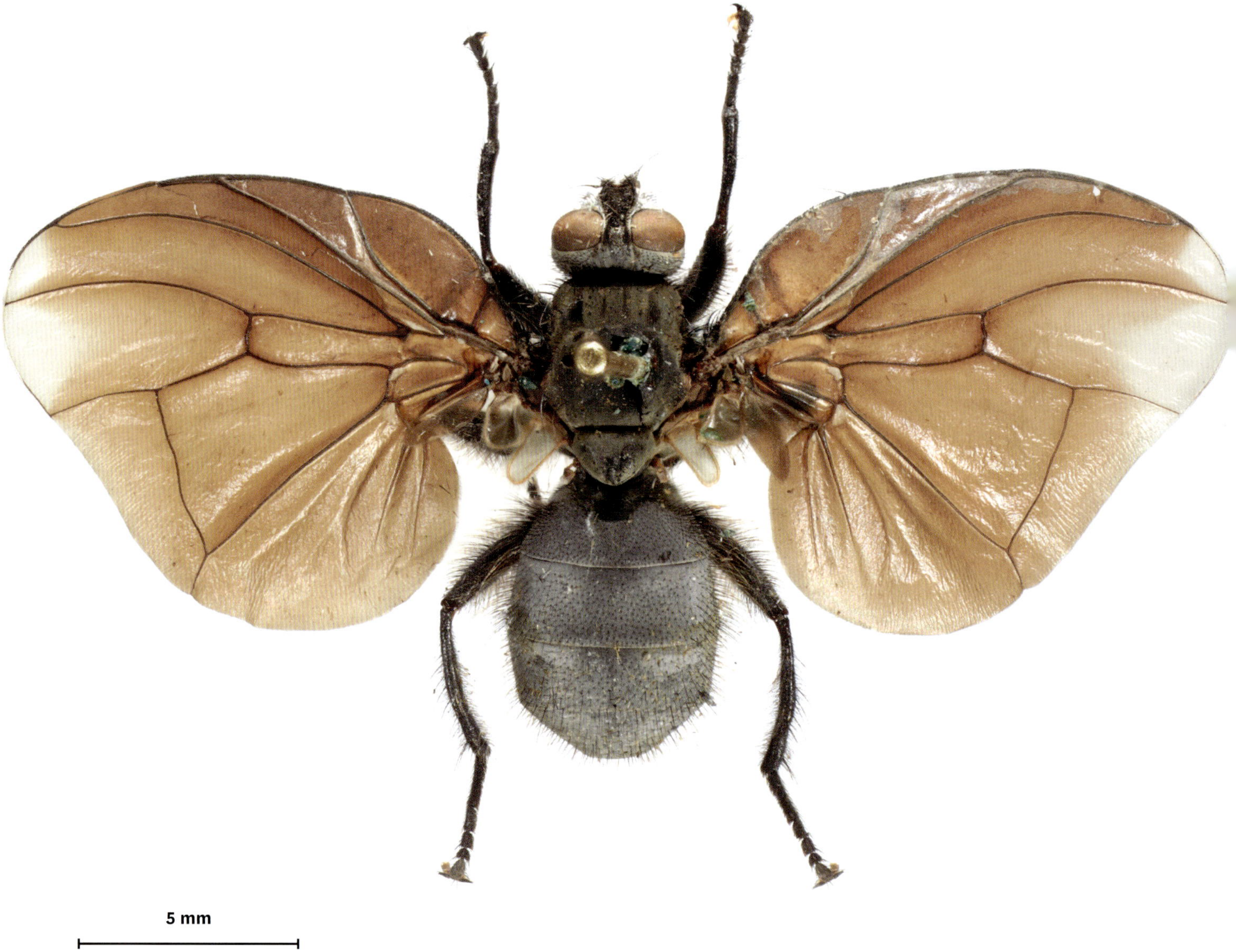
5 mm

Bat-winged fly

Exsul singularis

The bat-winged fly used to be considered one of the rarest and most unusual flies in the world, with only a few records and specimens existing. Now there are more than seventy specimens in the museums of New Zealand – thanks to a kid who liked to be outdoors rather than inside playing video games.

Eight-year-old Jamie Morris from Dunedin, son of wildlife photographer Rod Morris, was camping with his family in the Fiordland mountains late in 1988 when he saw a strange-looking insect, with a black body and a pair of large, dark, rounded wings, almost like a butterfly. After shooting some photos, his dad brought the specimen to the Otago Museum. Jamie's story went global and was published in local, national and international newspapers – it even was featured in *New Scientist* magazine!

Since then, scientists have learned much more about *Exsul singularis*. Found from the Paparoa Range south to Fiordland, it is probably carnivorous. Groups of up to one hundred individuals gather on rocks and, like other alpine insects, bask in the sun. It is likely that the flies absorb heat from the sunlight through their big wings. Dark colouring is typical of alpine insects, allowing them to maximise the amount of warmth absorbed from the sun to combat changeable, often very low, temperatures. The large wings are more prominent in males and also probably play a role in attracting females.

Although no longer among the world's rarest insects, the bat-winged fly remains extremely under-researched. Perhaps we need more children out there in nature helping us to understand the world. JK

Bat-winged fly (*Exsul singularis*), AI.034828, collected by GV Hudson, Te Anau track, Fiordland, December 1923.

Beaver flea

Hystrichopsylla schefferi

Te Papa holds one of the largest flea collections in the southern hemisphere – possibly the largest. It was donated by Professor Robert Pilgrim just before its former storage facility in Christchurch was badly damaged by the 2011 earthquake. When I sorted through the fleas in the collection, I began to realise just how many species there are worldwide. Cat, dog, human, rat and chicken fleas are well known, but there are also fleas that occur on kea, vultures, shags, echidnas, badgers and grizzly bears; some even burrow into the soles of your feet if you walk barefoot on South American beaches.

010

100

Despite their common names, fleas aren't normally too picky about their hosts, but some species can be very host-specific. My favourite flea, *Hystrichopsylla schefferi*, occurs solely on the mountain beaver (*Aplodontia rufa*) of the American Northwest. It is the largest living flea in the world, with a monstrous body size of 13 millimetres; its head alone is the size of an entire cat flea. It is unclear why this should be so. Bergmann's rule suggests that body sizes tend to be large in colder climates; likewise, organisms on islands have a tendency towards gigantism. But the American Northwest has a mild climate, and this flea happily inhabits the mainland.

The mountain beaver is considered to be the most ancient rodent, and it's therefore unsurprising that it should carry the most ancient of rodent fleas. But the oldest and largest flea ever discovered turned up in Jurassic rocks in China: a 165-million-year-old fossil that measures 20.6 millimetres. That beast was so large that it may have lived on feathered dinosaurs. However, not long afterward its descendants must have discovered their taste for mammal blood. JK

(A) Beaver flea (*Hystrichopsylla schefferi*), AI.026168, collected by D Gettinger, 3 miles south of Burnt Woods, Lincoln County, Oregon, USA, 16 December 1981.
(B) Human flea (*Pulex irritans*), AI.028590, collected by G Rawnsley, Wellington, December 1954.

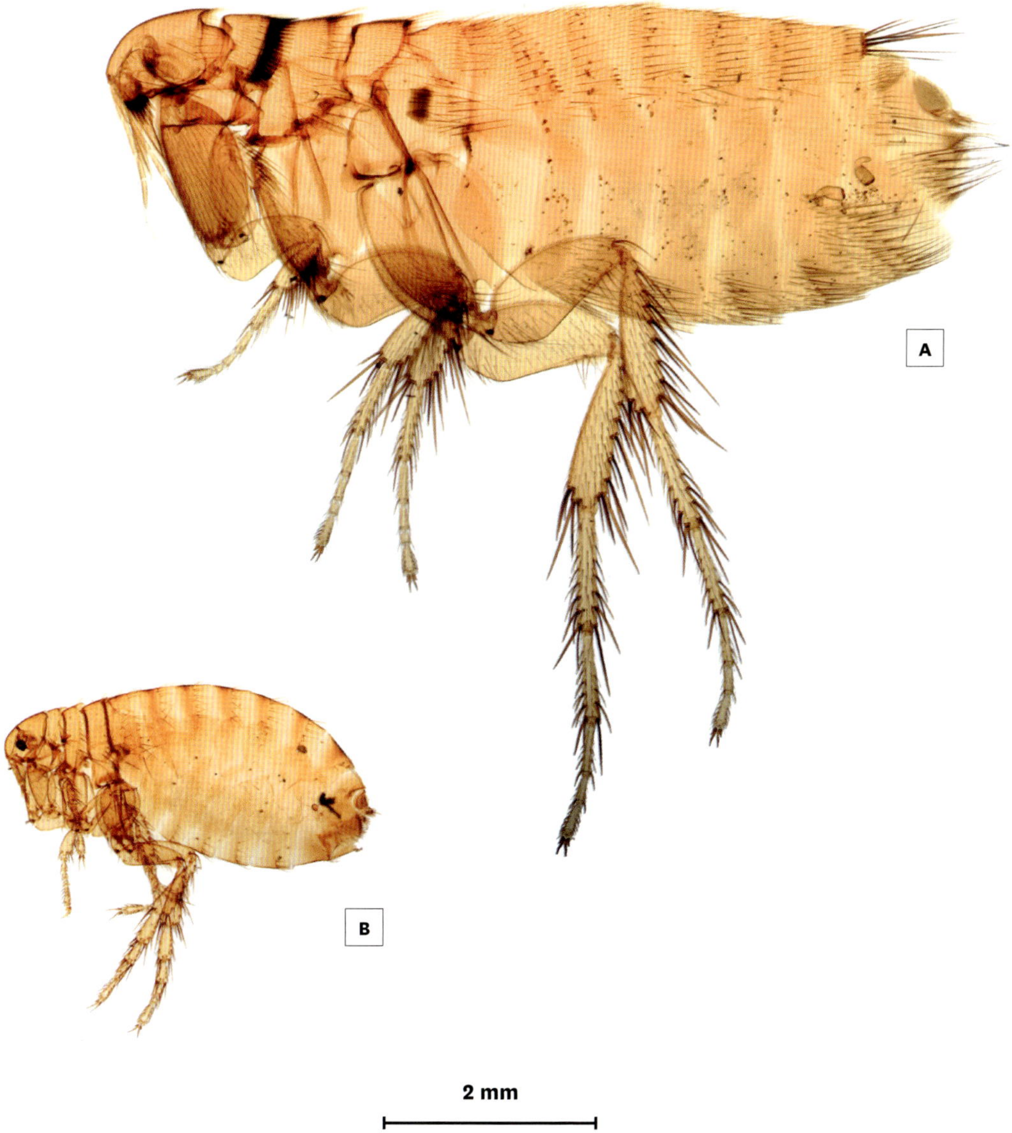
A
B
2 mm

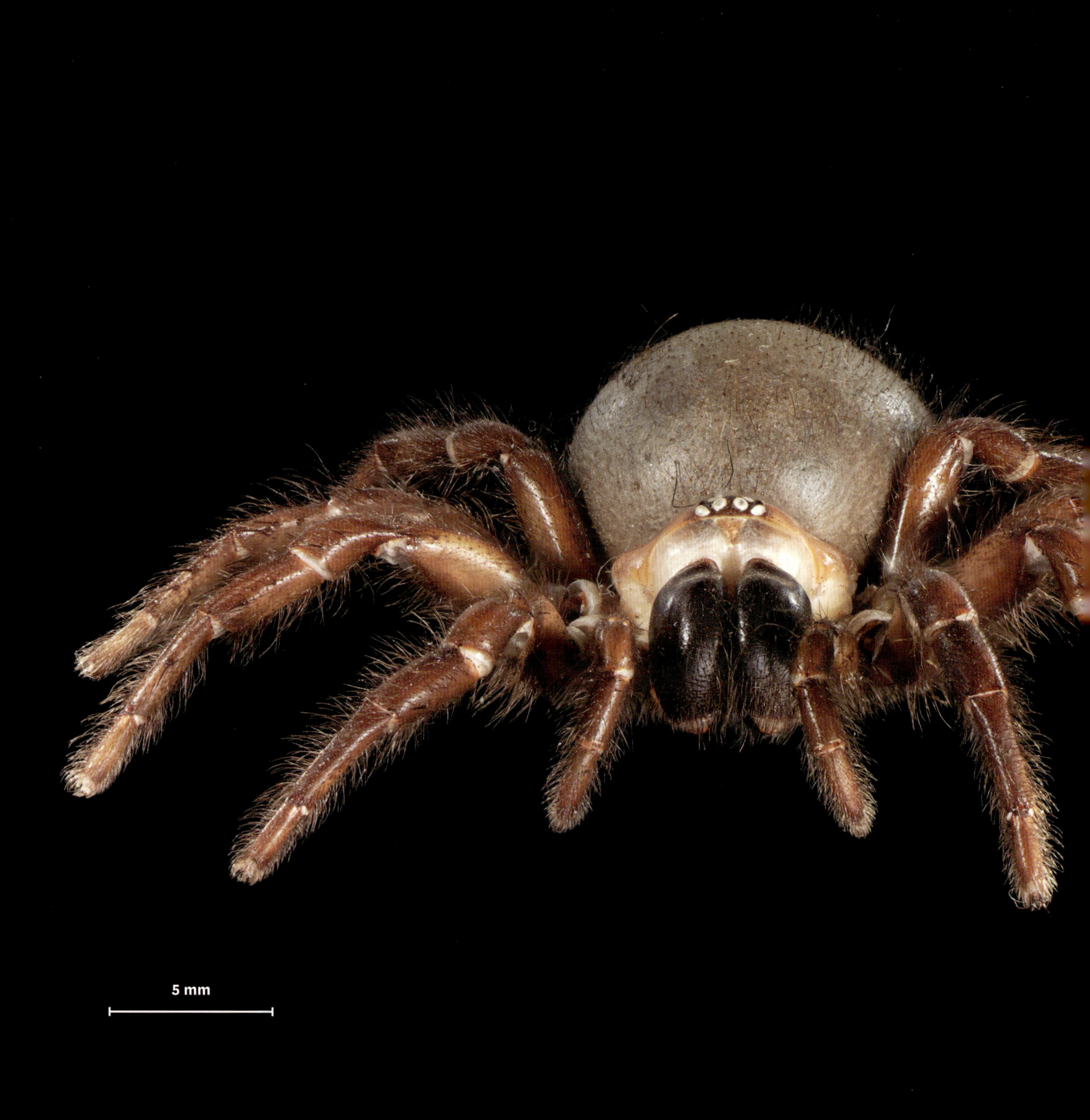
5 mm

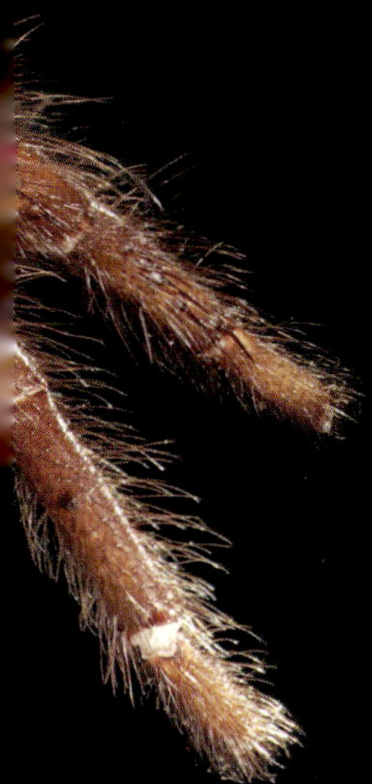

Black tunnelweb spider

Porrhothele antipodiana

The inclusion of the black tunnelweb spider *(Porrhothele antipodiana)* in this book has a rather personal story behind it. Spiders are my favourite animals, but it wasn't always that way. When I was four, one of these climbed up my leg and I was horrified to the point of having nightmares about it. I grew up with a love of animals of all sorts, but until I reached my teens I was prepared to make an exception for spiders.

011 / 100

This species is endemic to New Zealand, as is the family it belongs to (Porrhothelidae). Tunnelwebs resemble their cousins the tarantulas, but are very much smaller. Black tunnelwebs are found across a large part of New Zealand and have been accidentally introduced to the Chatham Islands. They vary a lot in size and other characteristics, so further study may show that what we think is only one species might be a group of several different ones.

As the name suggests, tunnelwebs build tubular, silk-lined retreats. These are often constructed under logs or stones, or even in holes in trees or a damp corner in your garage. The spiders catch prey that wander too close to the tunnel entrance. Normally this will be insects such as beetles, but there are records of more unusual victims such as mice and snails.

Tunnelweb spiders were also part of the design inspiration for Shelob, the giant spider-like creature in the movie version of JRR Tolkien's *The Return of the King*. Although the film-makers took some liberties in how they portrayed spider anatomy, there are moments where the influence of this spider on the design shines through. PS

Black tunnelweb spider, *Porrhothele antipodiana*, TMP.032729, freeze-dried display specimen, no collection data.

Blobfish

Psychrolutes microporos

The blobfish *Psychrolutes microporos*, family Psychrolutidae, occurs widely in the New Zealand region, from the southern Kermadec Ridge and southern Lord Howe Plateau to the Campbell Plateau. It lives on or close to the bottom at depths of 780–1780 metres. As with a number of deep-water fishes, it has loose skin and watery muscles, which are adaptations to life at depth. Out of water, the fish's body is limp and flabby, sagging out of shape because it can't support its own weight.

012

100

A photograph of a saggy-baggy specimen looking sad and wistful, affectionately called 'Mr Blobby' by scientists and crew on board the RV *Tangaroa*, was published on the website of the 2003 NORFANZ Expedition, a joint Australian–New Zealand biodiversity research voyage. The image quickly went viral; 'Mr Blobby' had arrived, gaining immediate international fame.

His achievements are many: his image has appeared on an album cover; he has his own Facebook page and Twitter account; he was voted the World's Ugliest Species by the Ugly Animal Preservation Society and became their official mascot; he was deemed huggable enough to promote plush toys; was a model for slippers; appeared on T-shirts; and inspired an ocean of poems, apps and smartphone games with tag lines like 'Build Up Your Hero and EVOLVE!' and 'Go Home Evolution: You're Drunk'. He also inspired a song by a children's book author; and even appeared in a cameo role in the Hollywood movie *Men in Black 3*!

Science exploration can be sensational – sometimes in the strangest of ways. CR + AS

'Mr Blobby' (*Psychrolutes microporos*), AMS I.42771-001, length 285 mm, trawled by RV *Tangaroa*, NORFANZ Expedition, at depth of 1013–1340 m, Norfolk Ridge, north Tasman Sea, June 2003.

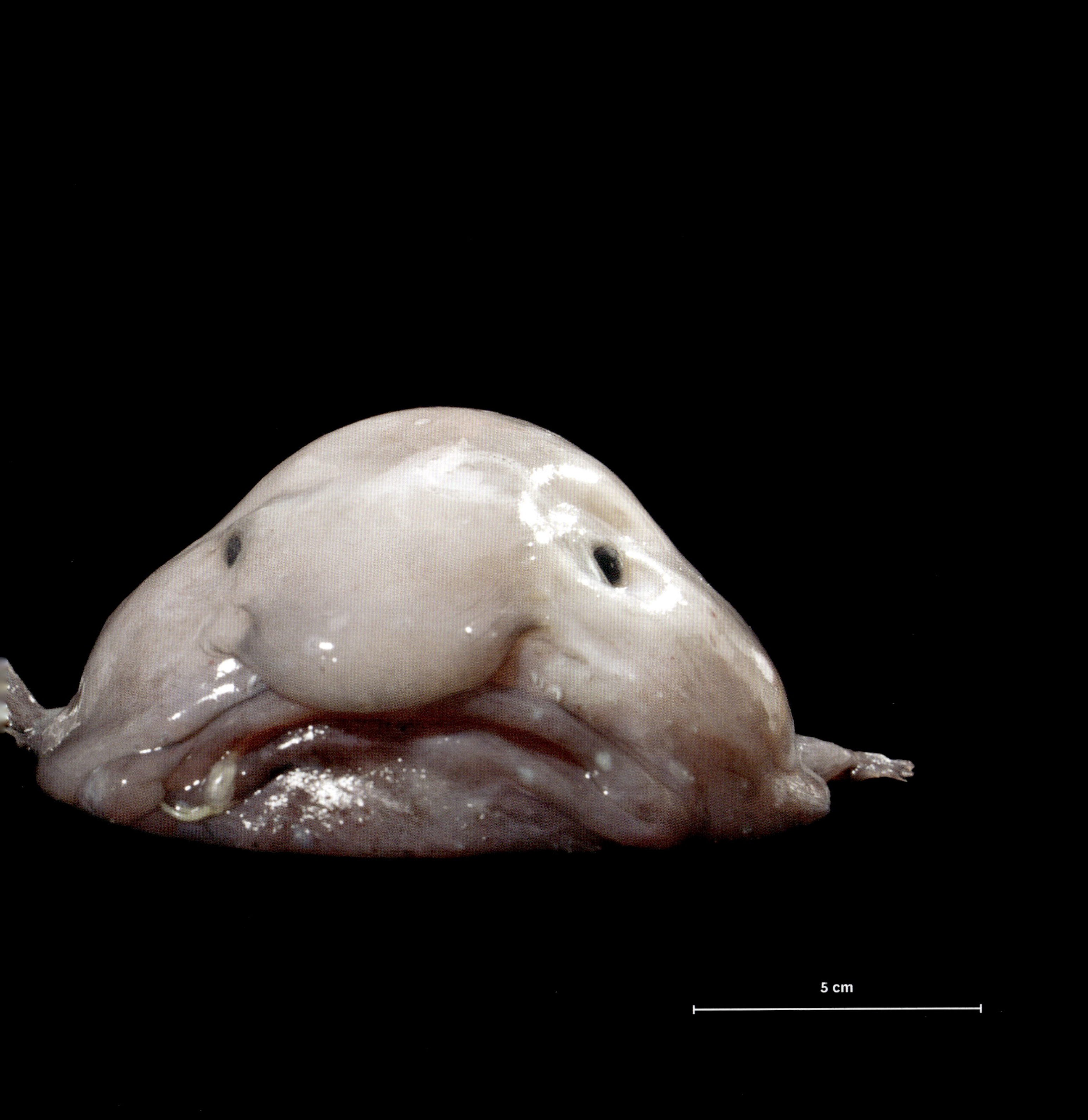
5 cm

Buller collection

Sir Walter Buller

More than a century on from his death in 1906, Sir Walter Buller remains perhaps New Zealand's best-known bird expert, although his reputation has become somewhat sullied. In 1871 the Colonial Museum (now Te Papa) purchased 310 bird specimens from him, comprising its first large and representative collection of New Zealand birds and containing many important specimens, including twenty-three specimens of now-extinct species. Unfortunately, more than three-quarters of the collection was lost or destroyed as a result of inadequate resourcing during the nineteenth century.

Buller continued to collect birds, many of which are still held in the American Museum of Natural History (New York), the Carnegie Museum of Natural History (Pittsburgh) and the Canterbury Museum (Christchurch). However, it was his lavish books, with their outstanding artwork by Dutchman Johannes Keulemans, that made Buller a household name and publicised the plight of New Zealand's native bird fauna.

Buller's standing as a conservation advocate has been eroded by stories of his hunting and trading large numbers of birds and his supposedly direct role in the extinction of birds such as the huia. However, there is no evidence that his collecting activities contributed substantially to the extinction of any species, which can be blamed firmly on introduced predators such as rats, cats and mustelids and on prehistoric hunting; the impact of museum collectors was minimal or limited to a handful of species already at risk of extinction for other reasons. The damage to Buller's reputation came about from his continuing to collect huia while publicly advocating their transfer to island sanctuaries. Buller was in fact a passionate advocate for New Zealand birds, and his enduring legacy was in showing their beauty to the world and documenting the human-caused devastation of this unique fauna. AT

Left to right: North Island piopio (*Turnagra tanagra*), OR.000217, South Island piopio (*Turnagra capensis*), OR.1460, South Island kōkako (*Callaeas cinerea*), OR.000172 and South Island bush wren mātuhituhi (*Xenicus longipes longipes*), OR.005086, study skins of now extinct birds collected in the 19th century and acquired from Walter Buller in 1871.

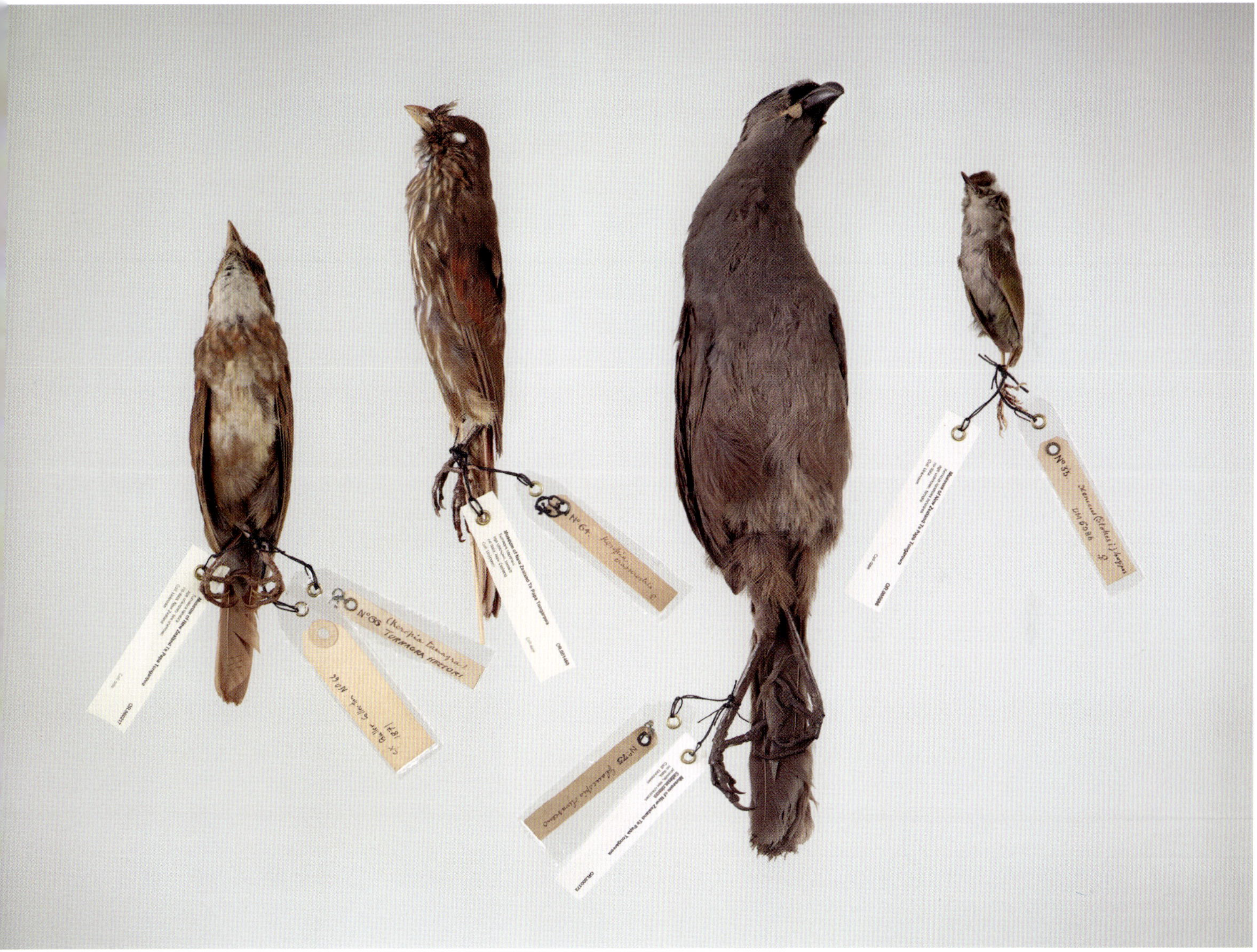

Objects

C—E

Kaitiaki Taonga Collection Manager for Entomology Phil Sirvid and Curator of Invertebrates Julia Kasper in the entomology storage area of Te Papa's Tory Street facility.

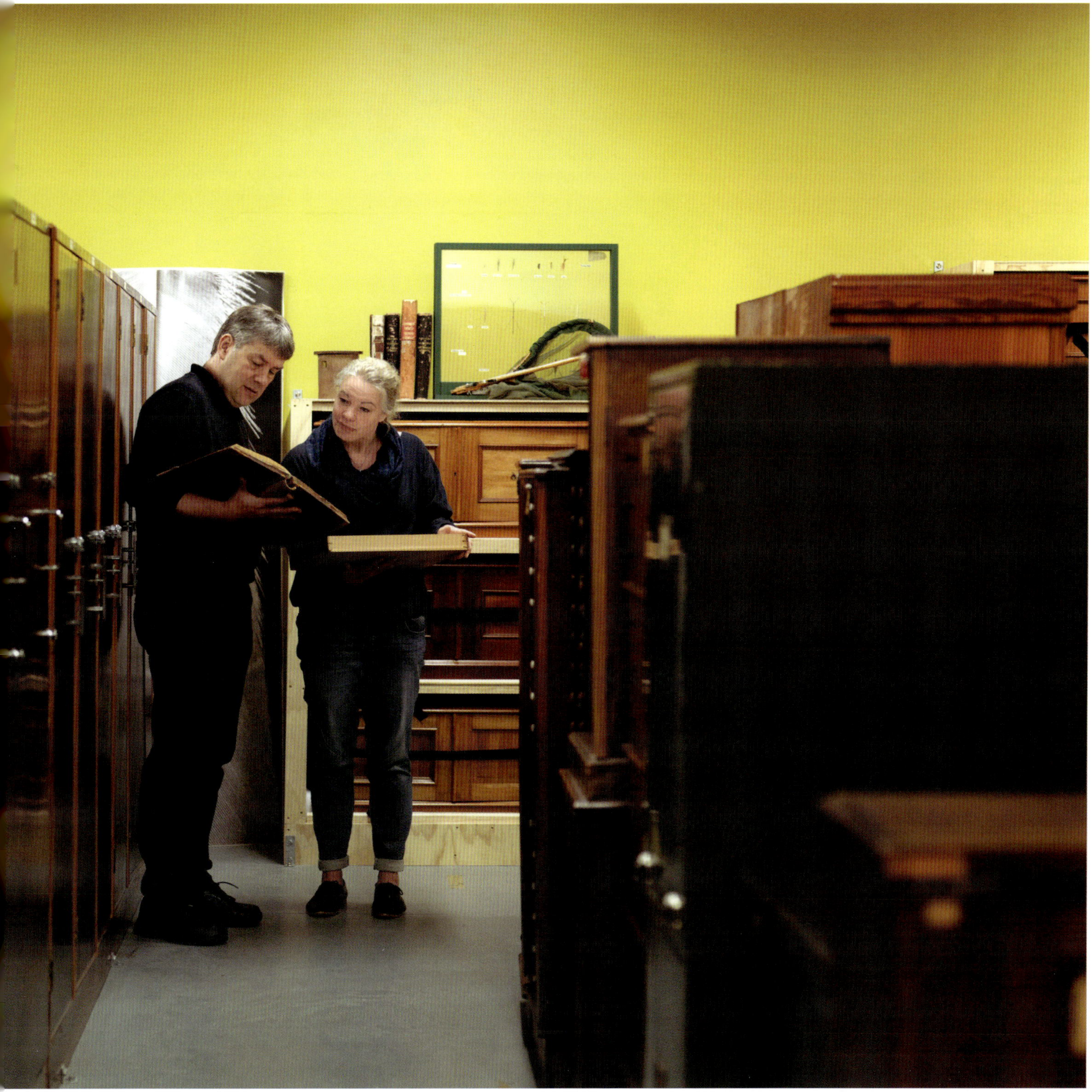

008 009 010

015 016 017

021 022 023

027 028 029

033 034 035

039 040 041

011 012 013 014

018 019 020

024 025 026

030 031 032

036 037 038

042 043 044

The Charles Knight lichen herbarium

Charles Knight

Te Papa holds an amazing collection of more than 15,000 lichens assembled by Charles Knight in the 1800s. He collected many in New Zealand, but also traded with overseas lichenologists. The lichens are beautifully preserved in handmade books, carefully stored in archival boxes in the herbarium.

As a volunteer, I was tasked with imaging each specimen and adding all data to our database so that the information could be readily searched. This proved to be fascinating detective work. Each specimen has an accompanying label. Often these were hand-written, and so I had to decipher, translate (usually from Latin or German) and work out whether a word was the name of the collector, the place it was collected or perhaps the type of tree it was growing on.

014/100

As part of this work, I read diaries of plant collectors in the Amazon jungles or Himalayan mountains, suffering terrible hardships, and I traced the journeys of others crossing America or the Antarctic Ocean, trying to work out which peaks they climbed or which islands they visited. The lichen illustrated here came from Greenland. In 1857, when Captain Francis McClintock sailed the *Fox* into the Arctic Sea to search for the remains of the failed Franklin Expedition, ship's surgeon Dr Walker took advantage of a brief stopover to do some botanising.

Charles Knight spent his retirement describing and illustrating many new species. His herbarium would have been a wonderful reference collection in the days before a Flora of New Zealand lichens was available. My current efforts to make his lichens accessible online is the next step in the journey for these well-travelled little specimens. BH

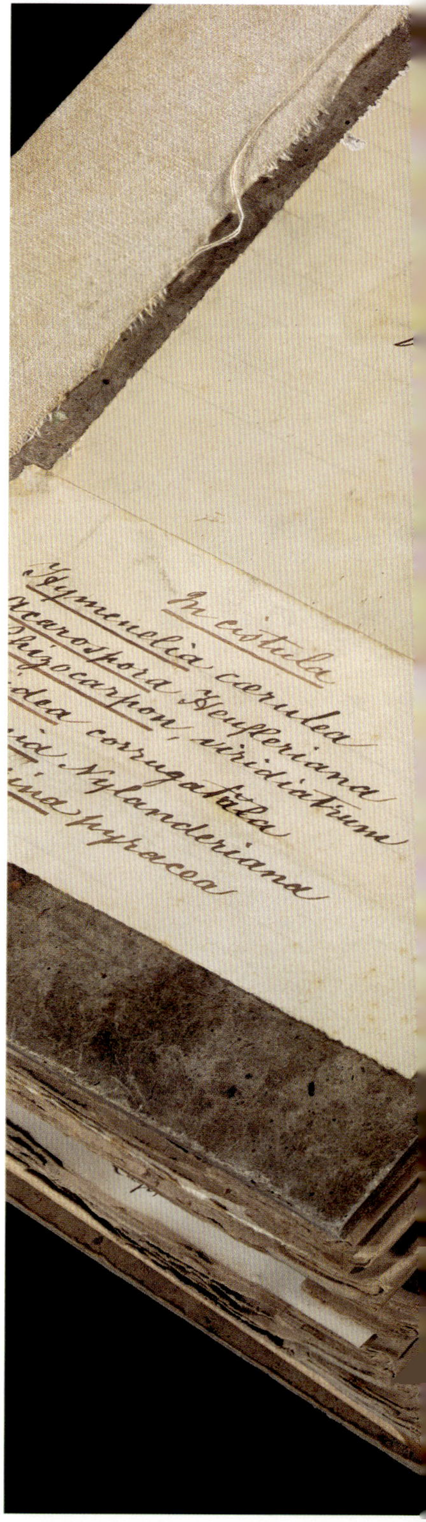

Two of 300 volumes handmade by Charles Knight, c. 1880, showing a specimen of reindeer lichen (L008912, collected by D Walker, Greenland, 1857) pinned onto the page. The books range in size from 205 x 170 x 45 mm to 270 x 180 x 130 mm.

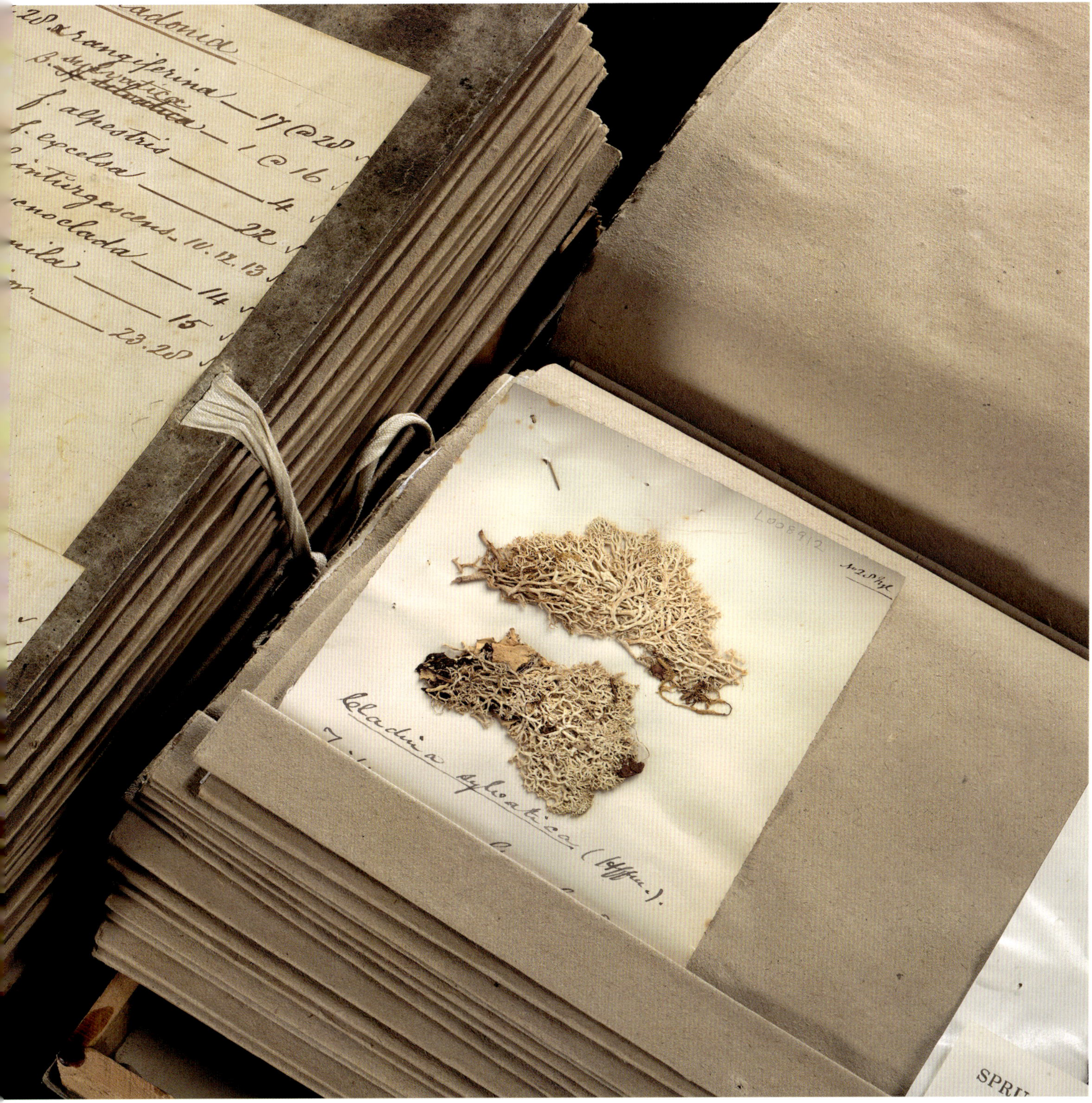

L008912
Cladina sylvatica (Hoffm.)

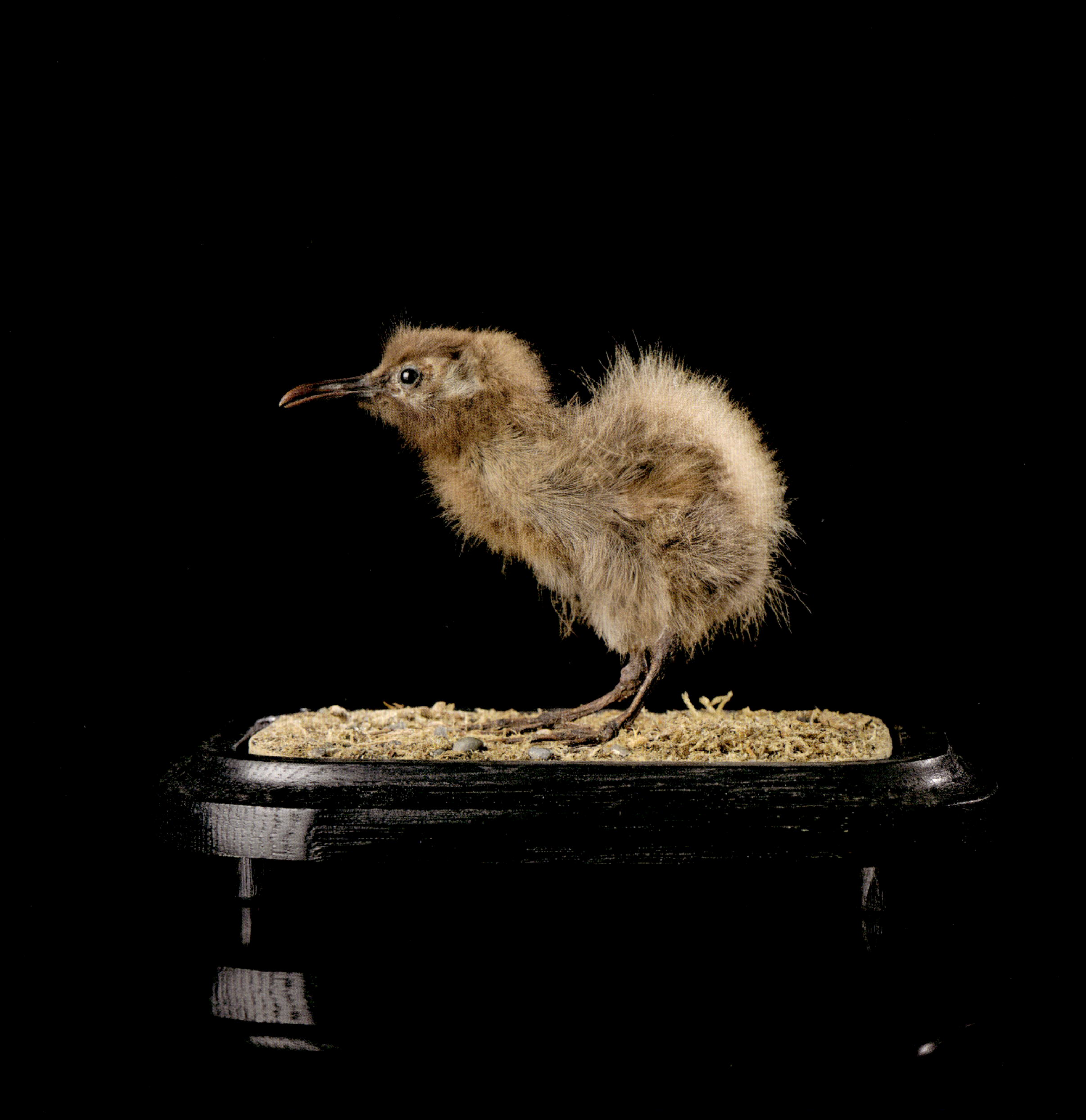

Chatham Island rail

Cabalus modestus

On 5 January 1872, bird collector Henry Travers captured a ground-dwelling bird and its fluffy chick on remote, windswept Māngere Island in the Chatham Islands. Like a miniature weka, weighing only 60 grams, this delicate species was also flightless. It was described as nesting 'in holes in the ground, and when the young are hatched they get into fallen hollow trees. They live on insects, principally the sand-hoppers.' On arrival at the Colonial Museum (now Te Papa), museum Director James Hector asked natural historian Captain Frederick Hutton to document the specimens. Hutton recognised that Travers' rails, already known to Moriori as mātirakahu, were new to the scientific world and formally named them *Rallus modestus* (later changed to *Cabalus modestus*).

A mere twenty years later, the rail met a sad and dramatic end when Māngere Island was ravaged by cats, introduced to control rabbits. The last rails were gone by 1894. Worse, the demise of the little rail was just part of a greater disaster for nature. The island had become the last global refuge of three other birds, the Chatham Island fernbird and bellbird and Imber's petrel, and these perished at the same time along with another ten or so bird species that persisted elsewhere.

Nearly 150 years later, the cats are long gone and Māngere Island has been the subject of a decades-long restoration programme, becoming a key ecological reserve for many threatened Chatham Island plants and animals. More than thirty skinned and mounted specimens of the Chatham Island rail survive in museums around the world, but the only one of Travers' birds still identifiable is his chick, the only chick of the species ever collected. AT

Chatham Island rail (*Cabalus modestus*) mounted chick, OR.001371, syntype, collected by H Travers, Māngere Island, Chatham Islands, 5 January 1872.

Cicadas

Cicadidae family

The late Sir Charles Fleming (1916–1987) was a noted expert in a number of fields, including geology, palaeontology, malacology (the study of molluscs) and ornithology. He had a keen interest in cicadas; he wrote or co-wrote a dozen papers on the subject and built up a fine collection that is held as part of Te Papa's entomology collection based at Tory Street.

With nearly 7000 specimens, Te Papa holds the largest collection of New Zealand cicadas in the country. Fleming's material makes up around half of that figure. As well as thousands of individual cicadas, the collection also includes recordings and sonograms (graphical representations) of cicada songs. For many of us, cicada song is part of the sound of summer, but we wouldn't know one call from another. Fleming was quite sure that we could learn something from cicada songs and used his recordings to show that different species could indeed be distinguished by their calls. This work has been built on by a number of cicada experts, most notably from the Simon Lab at the University of Connecticut.

The collection has also supported research into the evolutionary history of New Zealand cicadas. While all New Zealand cicadas are endemic, they are descended from colonising ancestors that arrived from New Caledonia and Australia and have subsequently diversified into the more than forty species we have today. Much of that diversification is relatively recent. For example, molecular data indicate that most of the green foliage cicadas (members of the genus *Kikihia*) are descended from a species radiation (rapid diversification often resulting from habitat change) no more than five million years ago. PS

Clapping cicadas (*Amphipsalta cingulata*): (A) AI.000904, collected by R Blick, Tauranga, 31 January 1965. (B) AI.000905, unknown collector, near Wainuiomata Valley, Wellington, 11 January 1969. (C) AI.000968, unknown collector, Havelock North, no date. (D) AI.000907, collected by C Fleming, Russell, Northland, 5 February 1971.

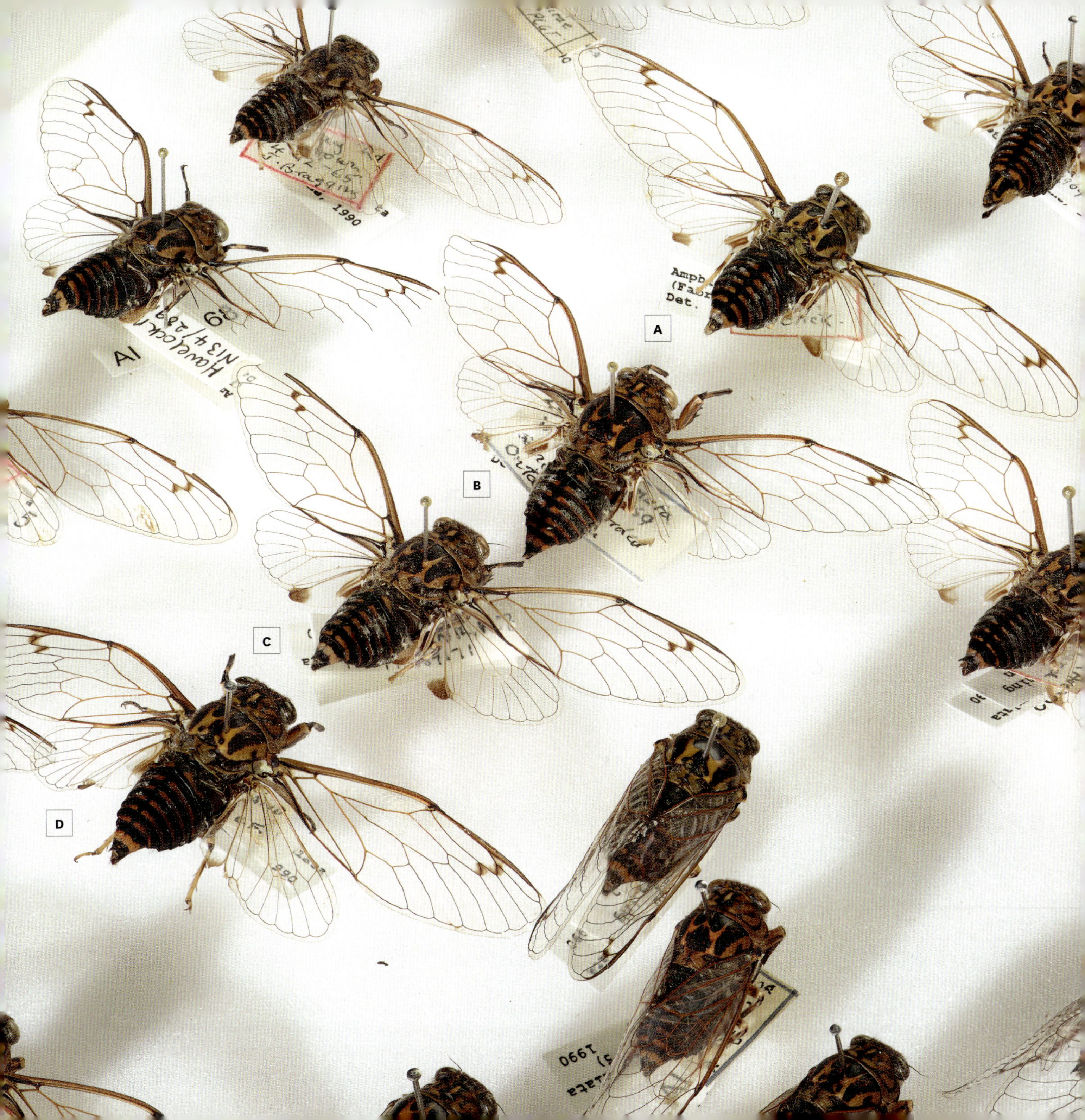
A
B
C
D

017/100

Coconut

Cocos nucifera

The coconut shown here arrived in New Zealand through the mail! A Wellington address is written on the outside, and the remains of two stamps indicate that it was posted from Rarotonga around 1920. The addressee, Mrs Paterson, donated a collection of Cook Islands material to the Dominion Museum in 1954. She and her husband visited the islands between 1922 and 1924, and she probably posted this coconut during one of those visits.

In earlier times, visitors to Stewart Island often posted large, dried leaves of muttonbird scrub (*Brachyglottis rotundifolia*) as a memento of their visit; the Post Office prohibited this practice in 1915. Today, our biosecurity laws would also prevent plant material like coconuts or leaves being received in New Zealand from overseas.

But coconuts can still get here naturally. They regularly appear on northern New Zealand beaches, and Te Papa's herbarium contains examples from the Kāpiti Coast. They are very buoyant in salt water and can be carried large distances by ocean currents from their native tropical Pacific islands. However, prolonged immersion in sea water kills them, and they almost never germinate. Even if they did, New Zealand's climate is not warm enough now to sustain them.

Once, coconuts did grow here naturally. Small fossilised coconuts about the size of a walnut are regularly washed up on Coopers Beach in Northland, eroded from reefs lying offshore. They are also found occasionally from Gisborne to South Canterbury, and are typically black from carbonisation and flattened by burial in sediment. The presence of these fossilised coconuts in rocks dating back twenty million years is evidence that New Zealand's climate was once much warmer than it is today. PB

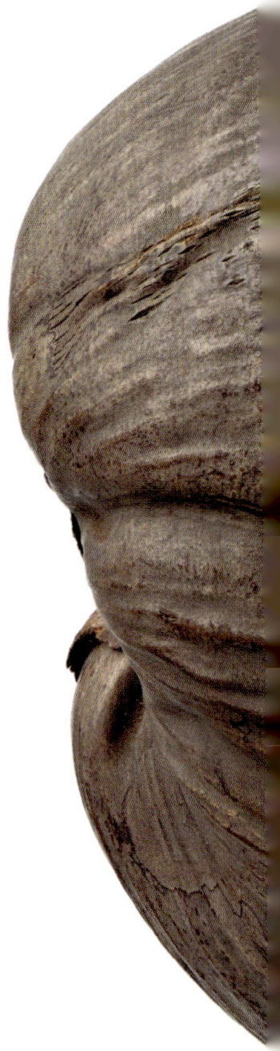

Coconut (*Cocos nucifera*), FE013158, height 215 mm, width 290 mm, depth 200 mm, posted to Mrs J Paterson, 374 Evans Bay Road, Wellington, from Rarotonga, c. 1920.

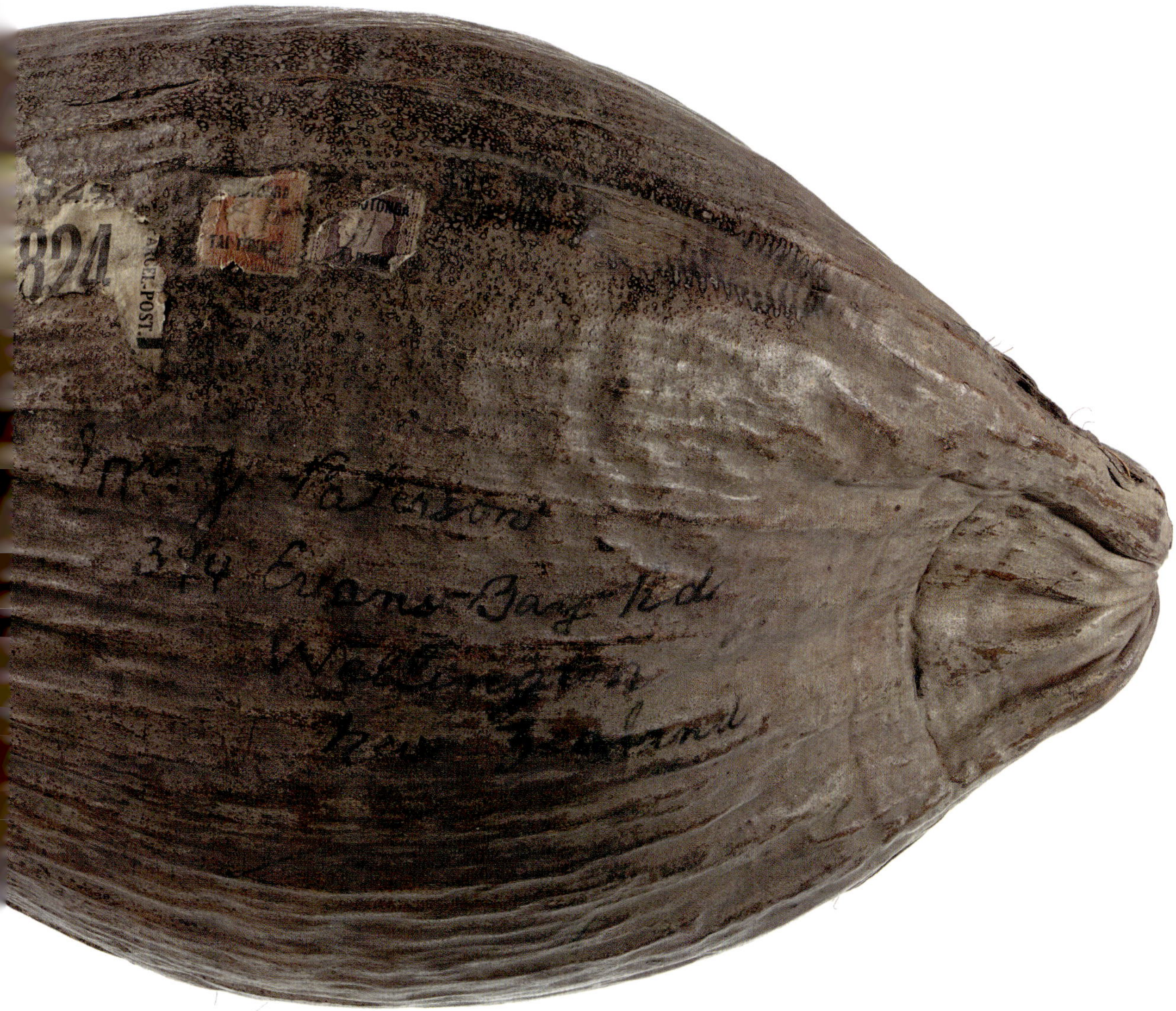
824
Mrs J Paterson
344 Evans Bay Rd
Wellington

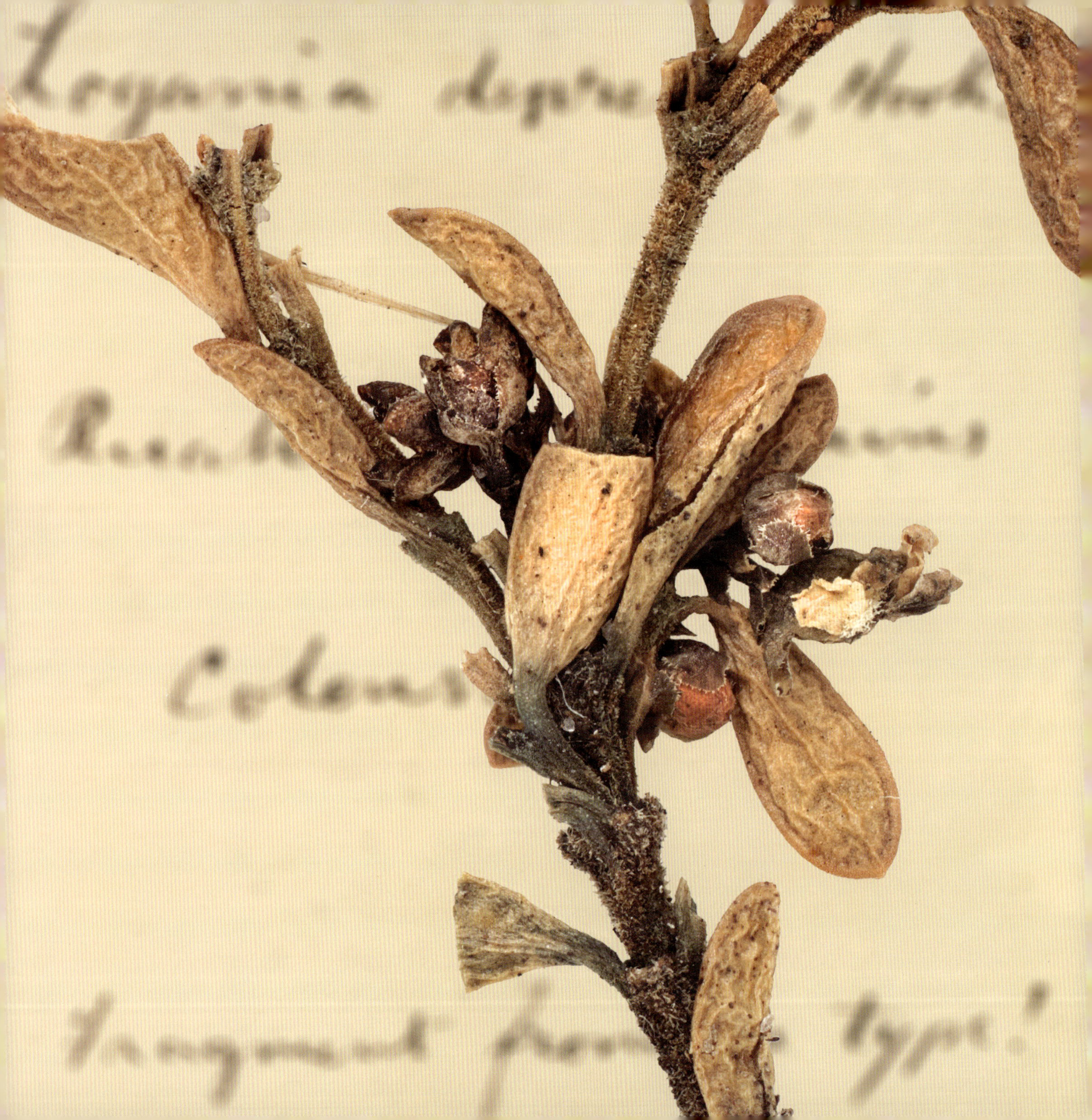

Colenso's extinct shrub

Logania depressa

New Zealand has lost a large number of species through recent extinctions, including such familiar species as moa (order Dinornithiformes), Haast's eagle (*Aquila moorei*) and huia (*Heteralocha acutirostris*). Our plants have fared relatively well compared with our birds, and only a handful or so of recent extinctions have been recorded.

One extinct New Zealand plant that has long fascinated me is *Logania depressa*, a species that nearly escaped being discovered at all! *Logania depressa* is a small, sprawling shrub and its closest New Zealand relative is hangehange *(Geniostoma ligustrifolium)*. The botanist and explorer William Colenso stumbled across the only known plant of *Logania depressa*, growing in a hollow within tussock grassland in the Central Plateau, on 22 February 1847 – and no one has seen it since, despite extensive searches. Colenso collected a specimen, which he sent to the herbarium at Kew, in the United Kingdom. Te Papa is fortunate to have in its collections a small fragment of this original specimen.

If Colenso had been walking even a few metres to the left or right, he might have missed *Logania depressa* entirely and its extinction would be unrecorded. How many other species have we lost that we never knew existed? Many of New Zealand's species have naturally restricted distributions, and it is not difficult to imagine that many unrecorded endemic species went extinct during early land clearance. LS

***Logania depressa*, SP004635, collected by W Colenso, Central Plateau, North Island, 22 February 1847.**

Colossal squid

Mesonychoteuthis hamiltoni

Cephalopods are outstanding among invertebrates due to their intelligence, which rivals that of mammals and birds. When looking into a cephalopod's outlandish eyes, you can almost glimpse that spark of intellect; it can be a very unsettling experience. Now imagine that the creature in question was larger than you. How much more disturbing would that be?

019 The two largest invertebrates known are both cephalopods:
the giant squid (*Architeuthis dux*) and the colossal squid
100 (*Mesonychoteuthis hamiltoni*). It is not surprising that encounters with the former struck fear into the hearts of Nordic seafarers, spawning the legend of the Kraken. *Architeuthis* is the longest invertebrate, reaching up to 20 metres with its tentacles extended; it is long and slender, with thin arms and a delicate silhouette. *Mesonychoteuthis hamiltoni* (shown here) is the bulkiest, weighing up to 500 kilograms; it is massive, with a menacing appearance. Te Papa's colossal squid has been the museum's most popular exhibit since its debut in 2008, and the staff's videos and blog posts have attracted nearly one million viewers.

Both the giant and the colossal squid live in the deep oceans, from 1000 to 2000 metres below the surface; thus, they are rarely seen or captured. As a result, very little is known about them, especially *M.hamiltoni* because of its Antarctic habitat. Despite receiving its official scientific name in 1925, researchers have only recently been able to start more in-depth studies about this animal. This is where Te Papa's specimens play a major role, by providing scientists with a wealth of data. RS

Colossal squid (*Mesonychoteuthis hamiltoni*), M.160614, total length 5.4 m, weight 300 kg, collected by R Guild on the FV *San Aotea*, Ross Sea, Antarctica, 2003.

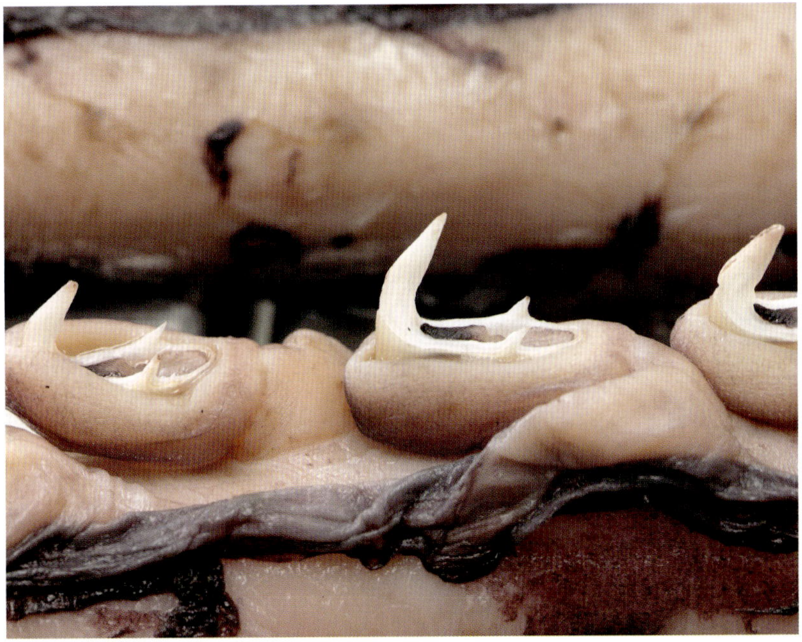

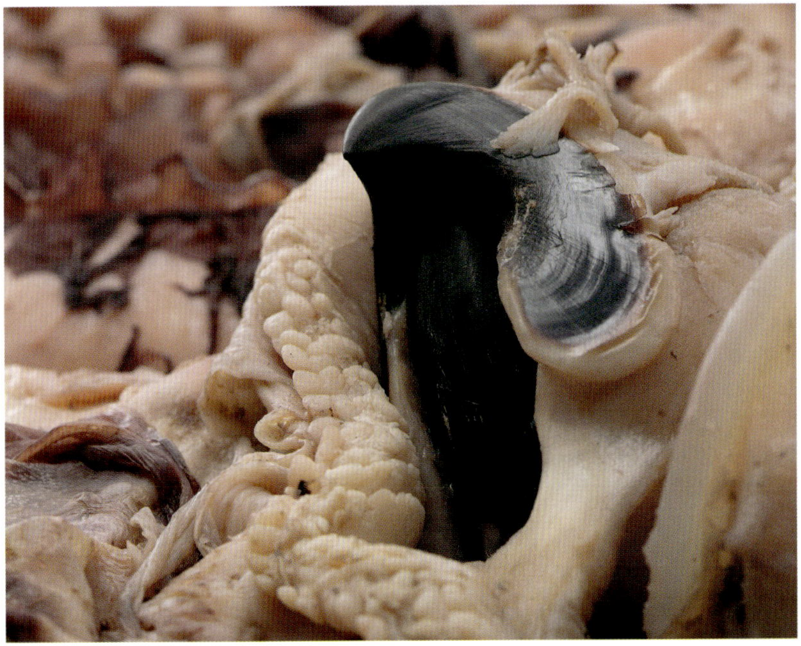

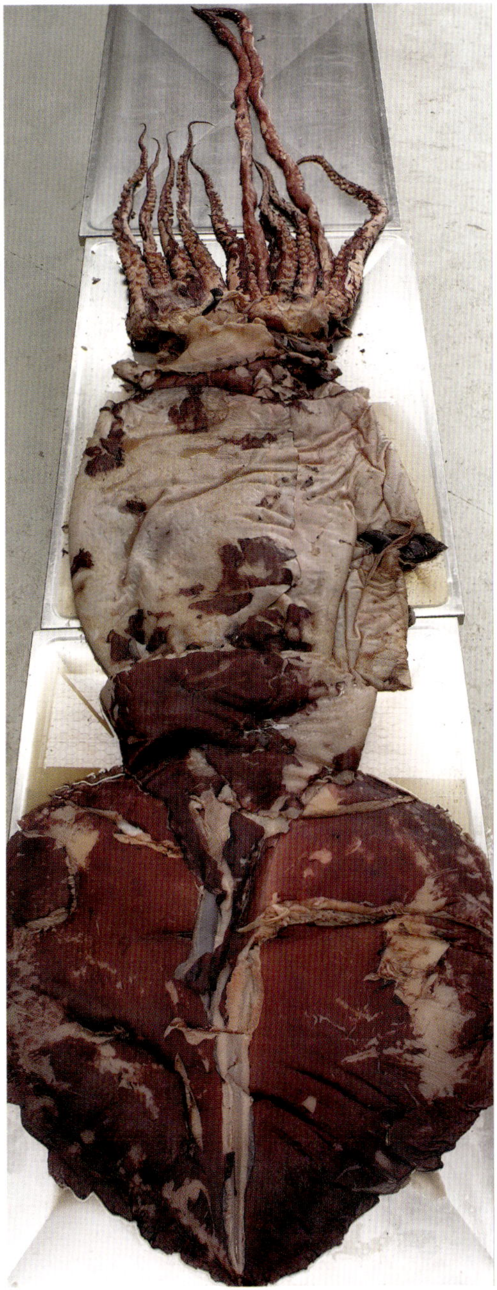

020

100

Crabeater seal skull

Lobodon carcinophaga

The crabeater seal (*Lobodon carcinophaga*) has some of the most remarkable teeth of any mammal. Their intricate shape has evolved from the basic three-cusp pattern found in other related seals; when the crabeater seal's jaws close, the two rows of teeth form a sort of sieve that gives a clue as to the seal's specialised diet.

'Crabeater' is a misnomer, though, for it is another crustacean that is on the menu. Crabeater seals feed almost entirely on Antarctic krill, sucking in mouthfuls of krill-laden seawater, then using their muscular tongues to pump it back out through their unusual teeth – sieving out the krill in the process. Their perforated pearlers perform a similar function to the baleen of the great whales.

Crabeater seals rarely venture north of the Antarctic pack ice, with only eight animals recorded in New Zealand between 1885 and 2015. Bizarrely, five of these animals either swam up the Hutt River or were found on nearby Petone Beach in Wellington Harbour. The teeth shown here came from the first of these, which came ashore in April 1916.

Is this unusual pattern of records an artefact of search effort (are other crabeater seals coming ashore elsewhere but not being spotted?), or is it an actual phenomenon? One possibility is that the shape of New Zealand's coastline guides any animal swimming north towards the Wellington coastline. Once within Wellington Harbour, if a wayward seal keeps swimming north it will bump into Petone Beach or be funnelled up the Hutt River. But this still doesn't answer what compels some crabeater seals to swim north away from the swarms of Antarctic krill that they are dependent on. CM

Crabeater seal (*Lobodon carcinophaga*), MM000387, skull length 245 mm, unknown collector, Petone, Wellington, 1 April 1916.

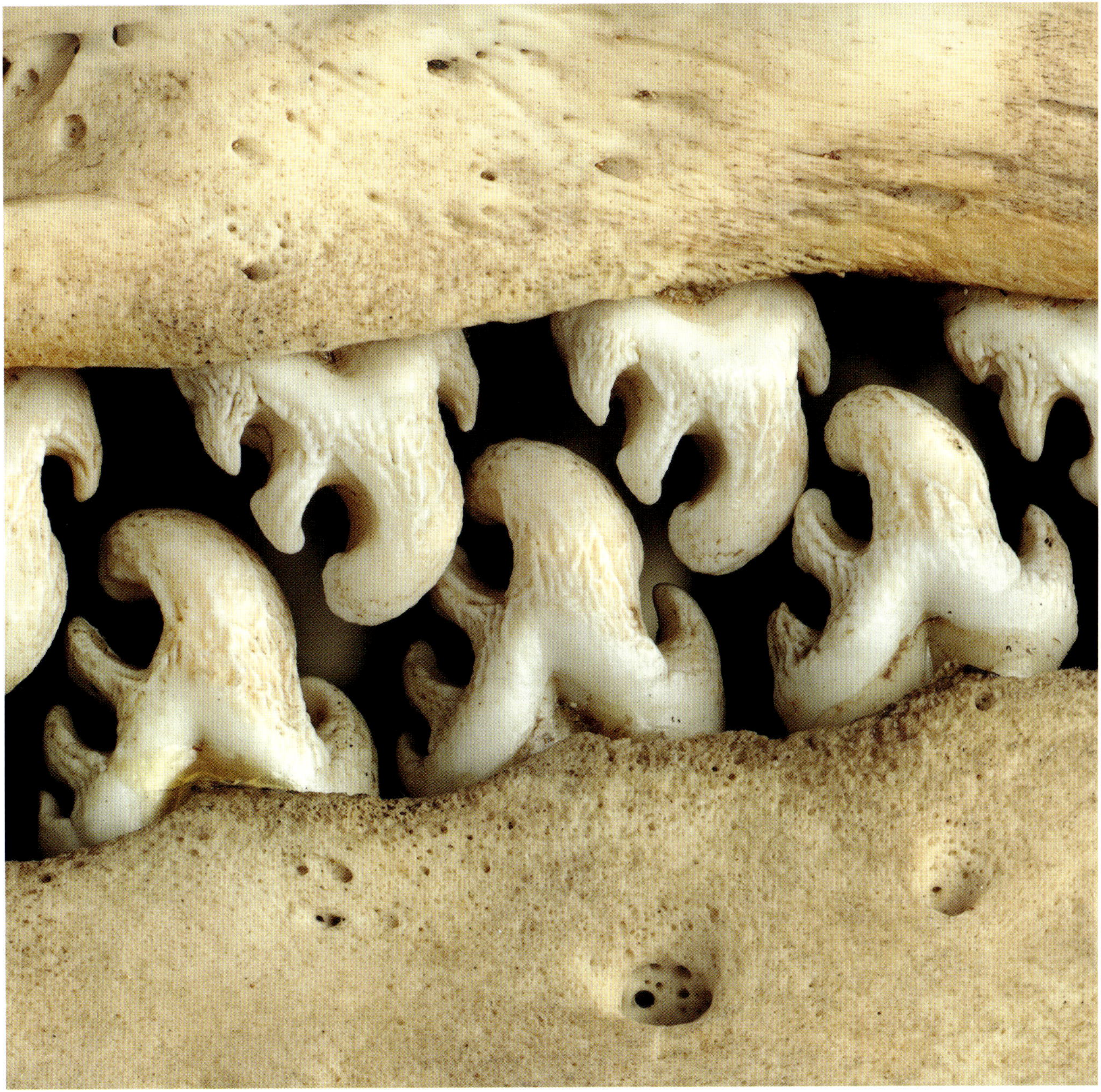

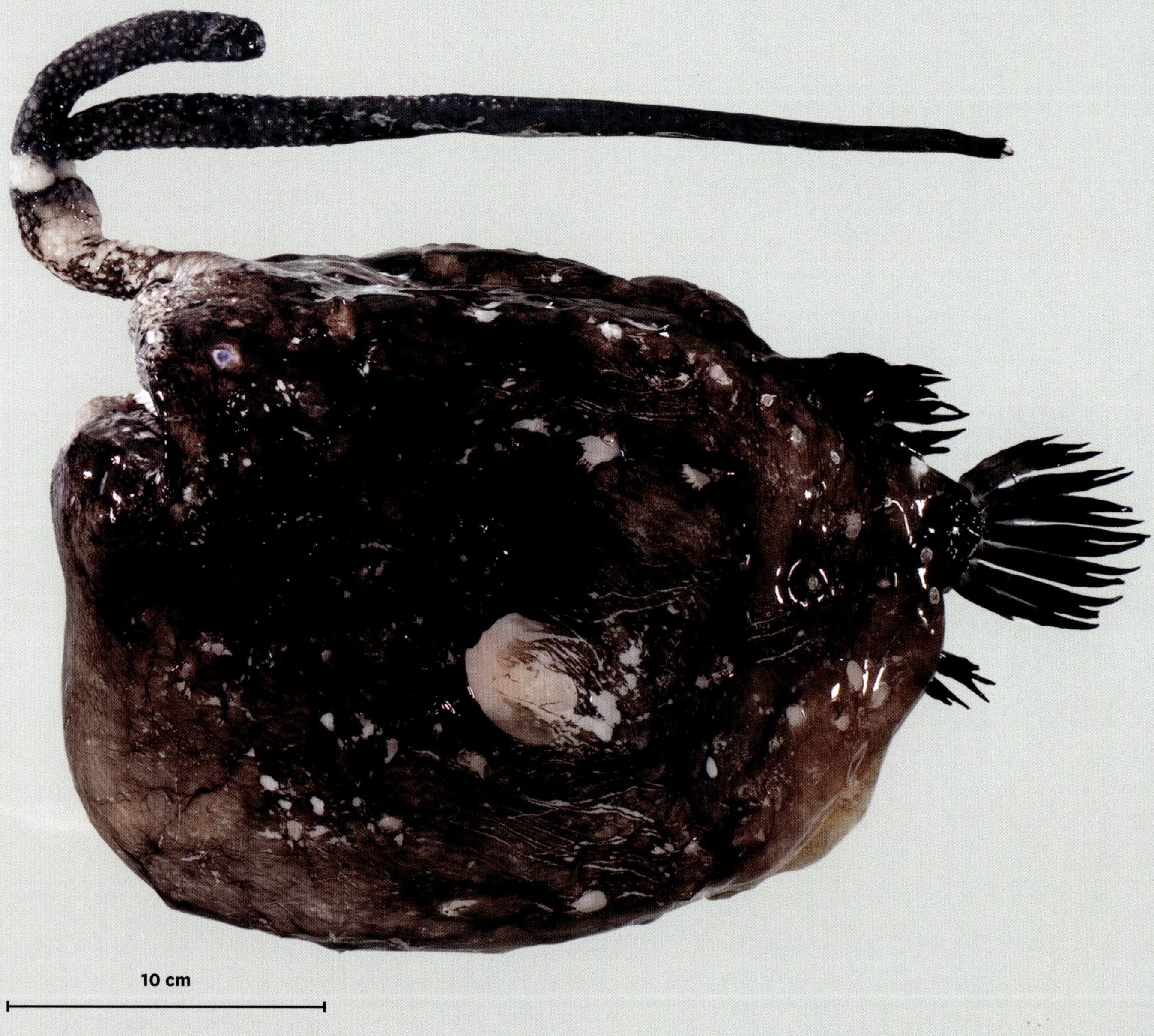
10 cm

Deep-sea anglerfish

Lophiiformes order

Deep-sea anglerfishes are classic examples of the biological and morphological (body shape and structure) extremes found in deep oceanic fish life. These anglerfishes are usually found in the mesopelagic zone, at depths of 1000–4000 metres, extending from the Arctic to the Antarctic. Females can be over fifty times larger than males. They have rows of enormous fangs in their large jaws, along with the characteristic, highly modified first dorsal fin ray that has an ornate light-emitting lure at its tip. The number, sizes and relative positions of filaments and appendages on the lure are used to accurately identify most of the 165 species known. Densities of females in the water mass are very low, and they are only caught occasionally.

The first comprehensive taxonomic revision of deep-sea anglerfishes of the New Zealand Exclusive Economic Zone, an oceanic area of more than 4,000,000 square kilometres, was published twenty years ago by Stewart and Pietsch, in 1998. Since then, an additional ten species (one new to science), four genera and one family have been found in our waters – the latest in 2016 – bringing our total to twenty-nine known species. This list is by no means complete, and more discoveries can be expected.

The plain prickly anglerfish (*Himantolophus litoceras*) shown here was caught in a commercial trawl off the north-east coast of the North Island in 2005. Although found in quite shallow water for deep-sea anglerfishes (654 metres) and relatively large at 276 millimetres long, only one specimen has been caught so far. The all-important lure was damaged, but enough remained to determine that this fish was new to science, and the species was named and described in 2010 by Stewart and Pietsch. CR + AS

Plain prickly anglerfish (*Himantolophus litoceras*), P.042004, holotype, standard length 276 mm, bottom-trawled by T Turton, FV *Seamount Explorer,* at depth of 654 m, South Cavalli Seamount, east of North Cape, 7 November 2005.

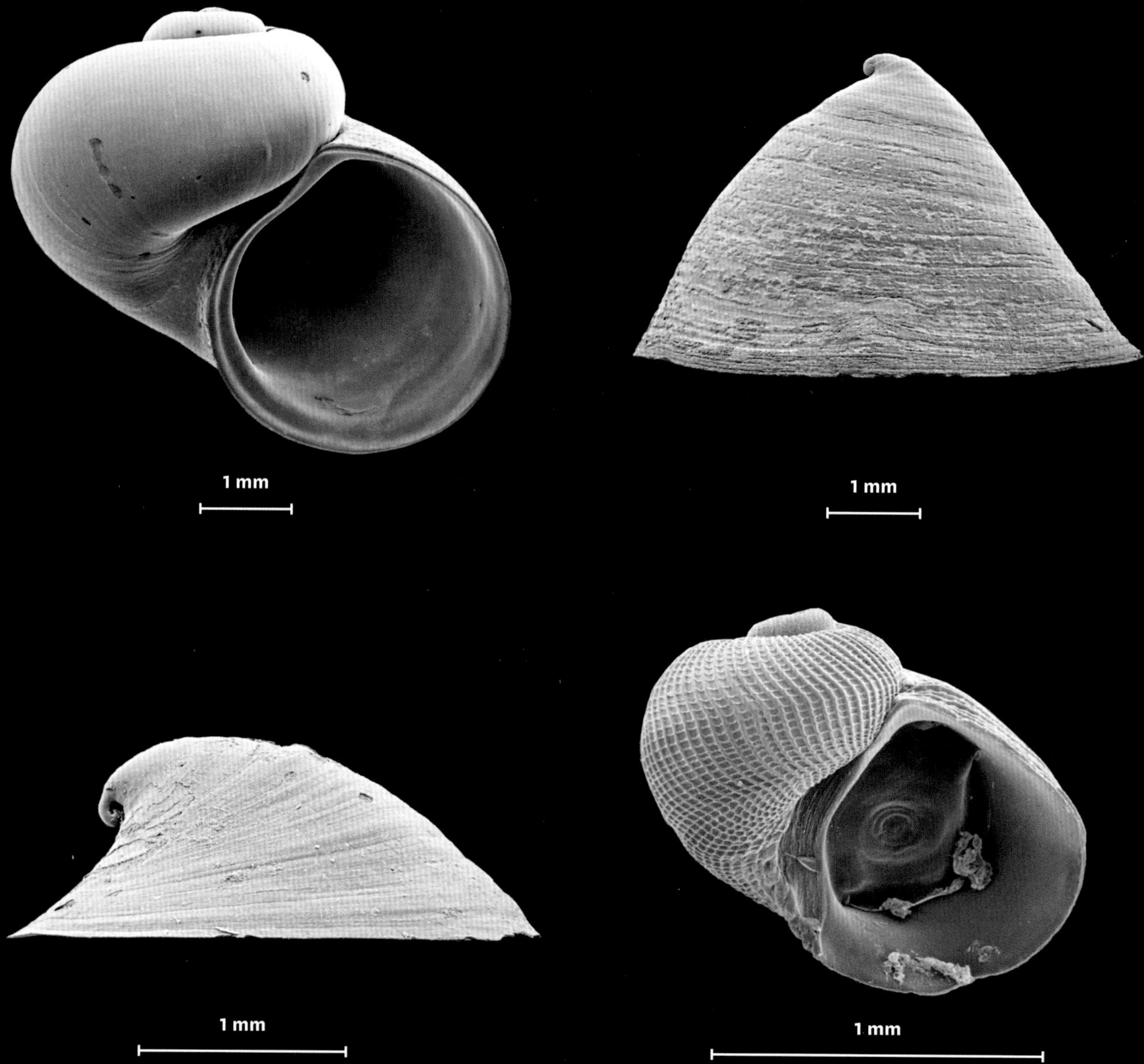
1 mm
1 mm
1 mm
1 mm

Deep-sea molluscs

Lepetelloidea and Cocculinoidea superfamilies

In 1976, while sorting a sample dredged at a depth of 13 metres in Whangaroa Harbour's main channel, I spotted what appeared to be a match-stick covered with tiny limpets in shallow pits. This proved to be a fish bone carrying a new species of *Cocculinella*. The only other *Cocculinella* species taken alive was from the Bay of Bengal, and it too was on a fish bone. This was the first record of a limpet living attached to fish bones, which they eat with highly modified teeth.

Later, a large decaying log riddled by boring bivalves and covered with hundreds of small molluscs was trawled from a depth of 1075–1100 metres off White Island. I received the log frozen, and after thawing it and breaking it up was astonished to find thousands of minute limpets and snails representing twenty-two species, six new genera and a new subfamily – all new to science.

All of these molluscs are specialised eaters of sunken wood, which, like the bone limpets, they eat with modified teeth, supported by symbiotic bacteria within their guts. These discoveries led to further searches for deep-sea molluscs. Sunken algal holdfasts yielded a new species of snail and a limpet. Decaying whale bones supported two new snails and two new species of limpet, representing a new genus in a new family. Spent mermaid's purses (shark and skate egg cases) supported a snail and two new species of limpet. Two more species of limpet live on the carapaces of living stone crabs. Old squid beaks, which litter our deep-sea floor, support a remarkable fauna of tiny limpets, all to be formally described. The search continues for further new species. BM

Top left: *Dillwynella ingens*, M.076097, holotype, collected by RV *James Cook* from sunken wood off Hokitika, 9 August 1983. **Top right:** *Osteopelta praeceps*, M.116970, holotype, collected by RV *Tangaroa* from decaying whale skull at depth of 372–379 m off Mernoo Bank, Chatham Rise, 29 January 1992. **Lower left:** *Tentaoculus balanteophaga*, M.118318, holotype, collected by RV *Tangaroa* from spent skate egg case at depth of 1335 m off Castlepoint, 2 April 1993. **Lower right:** *Bathyxylophila excelsa*, M.075126, holotype, collected by FV *Kaltan* from decaying wood at depth of 800–810 m off Mernoo Bank, Chatham Rise, 28 September 1982.

023

100

Elephant bird bones

Mullerornis agilis and *Aepyornis maximus*

It is uncertain how these bones, which belong to the large, extinct flightless elephant bird (*Mullerornis agilis* and *Aepyornis maximus*) from Madagascar, came to Te Papa, but it is likely that they arrived via trade with European museums, perhaps in the 1930s. Te Papa has nine elephant bird bones, and they just happen to have the best-preserved DNA of any elephant bird bones tested worldwide. As Te Papa's dedicated geneticist I regularly take samples from specimens in our collections for DNA analysis. I am always amazed to see how scientists used to take samples when the study of ancient DNA was in its infancy. The bones shown here provide examples of large 'windows' where bone was (cut out) and then ground up to obtain DNA. Modern techniques mean that DNA samples can now be taken with minimal damage to our precious specimens.

DNA from these particular bones was critical in determining the family tree of the ratites, a group of largely flightless birds that includes moa, kiwi, emu and ostrich. The relationships of this group have fascinated scientists for years. How did the different species of flightless ratite come to inhabit the far-flung land masses of the southern hemisphere? Are they relics of the break-up of the supercontinent of Gondwana, or did they somehow arrive later across the oceans?

The genetic family tree of the ratites showed that elephant birds did not have a close relationship to neighbouring African ostriches, but instead were most closely related to the New Zealand kiwi. This suggests that the ancestors of ratites reached their present distribution by flying and only later became flightless. LS

Elephant birds: (Left two bones) *Mullerornis agilis*, **S.038300, unknown collector, Beloha, Madagascar, 1933. (Right two bones)** *Aepyornis maximus*, **S.038302, unknown collector, Betioky, Madagascar, 1933.**

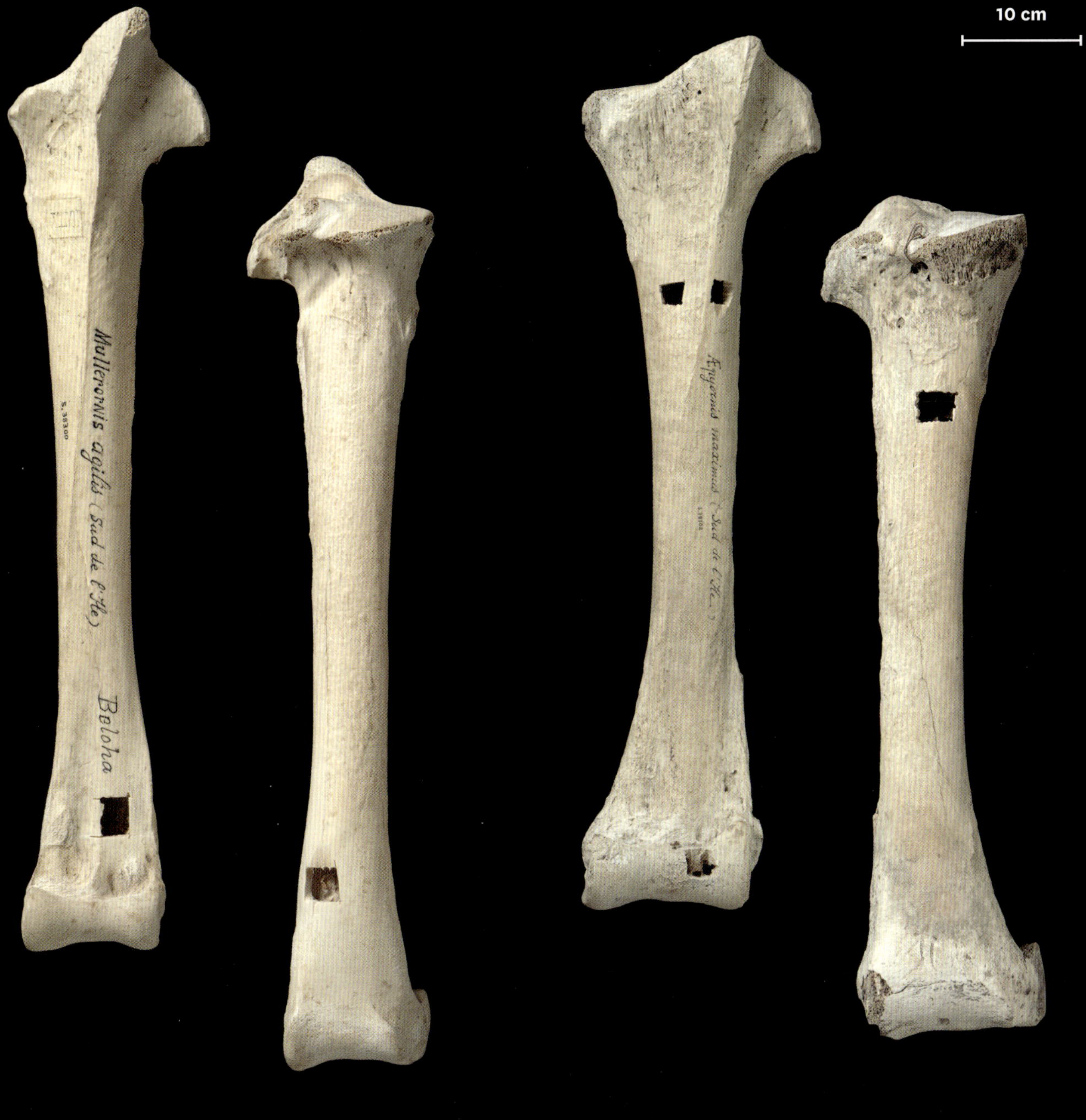
10 cm
Mullerornis agilis (Sud de l'Ile)
Beloha
Æpyornis maximus (Sud de l'Ile)

Elephant seal louse

Lepidophthirus macrorhini

The rather odd-looking creature shown here is the elephant seal louse, *Lepidophthirus macrorhini*. Like many other lice, it is intimately associated with one host species (the elephant seal, *Mirounga leonina*) and can't survive away from it for long. Compared with bird lice living in nice warm plumage, this species has to cope with some pretty tough conditions. The elephant seals of Australia's Macquarie Island, lying halfway between New Zealand and Antarctica, only come ashore for a few weeks twice a year, so their lice have to survive in icy waters the rest of the time. The seals also dive to depths in excess of 1500 metres, subjecting the lice to enormous pressure.

These lice have several adaptations that help them endure such extremes. First, they prefer living inside the hair follicles of the hind flippers. These are areas of high blood flow when the seal surfaces, ensuring that the lice have an ample food supply. Second, by living in the hair follicles, the lice are difficult to dislodge. The lice are also covered with a coating of thick scales and spines. The scales can trap air, but these insects are also able to extract enough oxygen from sea water to help them survive submergence, especially as their metabolic rate drops substantially in cooler temperatures. The spines trap sebaceous fluid from the hair follicles, which might help the lice resist the cold.

These lice only breed when the seals go ashore. The hind flippers are still the best place to be, as they are the warmest part of an elephant seal's body. This helps provide the temperatures needed for the lice to increase their numbers. **PS**

Elephant seal louse (*Lepidophthirus macrorhini*), AI.019436, collected by DS Horning from an immature elephant seal (*Mirounga leonina*), south end of isthmus, Macquarie Island, 24 December 1977.

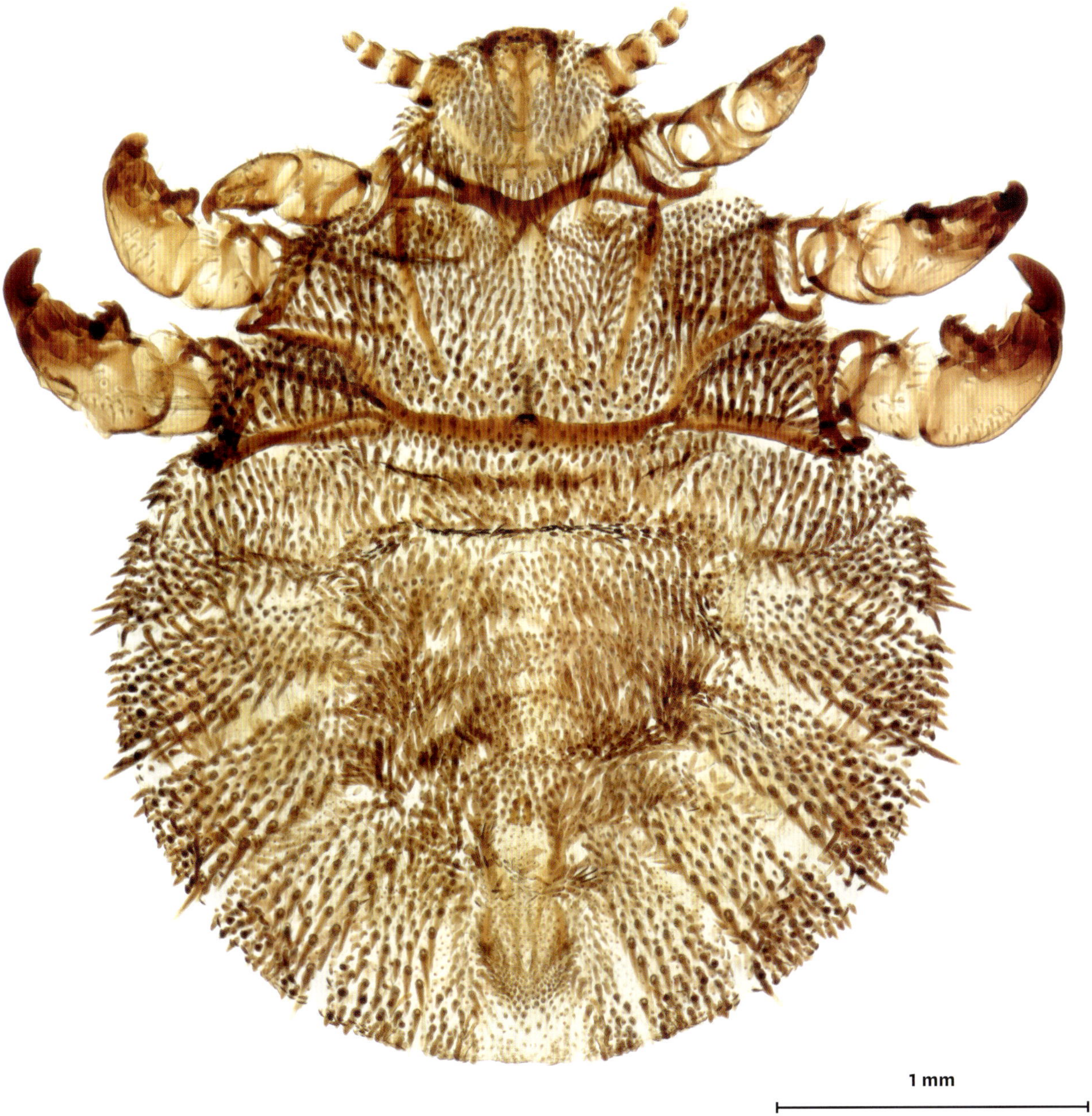
1 mm

Endemic yellow-petioled hebe

Veronica flavida

Hebes are among New Zealand's most well-known and well-loved plants, with over eighty different species. In addition there are the lesser-known but equally fascinating snow hebes, speedwell hebes, sun hebes, semi-whipcord hebes and heath hebes, ranging in size from alpine cushions to small trees. Several Te Papa botanists have collected and studied hebes, and as a result their evolutionary history is well understood. All are now placed taxonomically in the genus *Veronica*, our largest plant genus with 122 native species, though we continue to use 'hebe' as the common name. Most hebes are found naturally only in New Zealand, with a few occurring elsewhere in the Pacific; others are popular overseas as cultivated plants.

Te Papa holds over 7000 specimens of *Veronica* in its herbarium. The specimen shown here is a favourite because it is a complete specimen, comprising leaves, woody branches, and flowering and fruiting shoots at various stages of development. The collectors even put some fruits and flowers in ethanol to preserve their shape. The dried collection shows thoughtful mounting, with two leaves in one corner showing the upper and lower surfaces, enabling straightforward observation of important taxonomic characters. The labels also reveal the evolution of the plant's name, starting with the informal *Hebe* 'Waima', then its first scientific name *Hebe flavida* and finally its current name *Veronica flavida*.

The inclusion of this specimen in this book is a tribute to former Te Papa botanists Mike Bayly and Alison Kellow, who wrote the definitive book about New Zealand hebes after many years of collection and research, based on field work, morphology, genetics, chromosome counts and flavonoid biochemistry. Together with colleague Peter de Lange, they described and illustrated this species in their book. HM

Endemic yellow-petioled hebe (*Veronica flavida*), SP081886, collected by MJ Bayly and A Kellow, Waima Range, Northland, 27 May 2000.

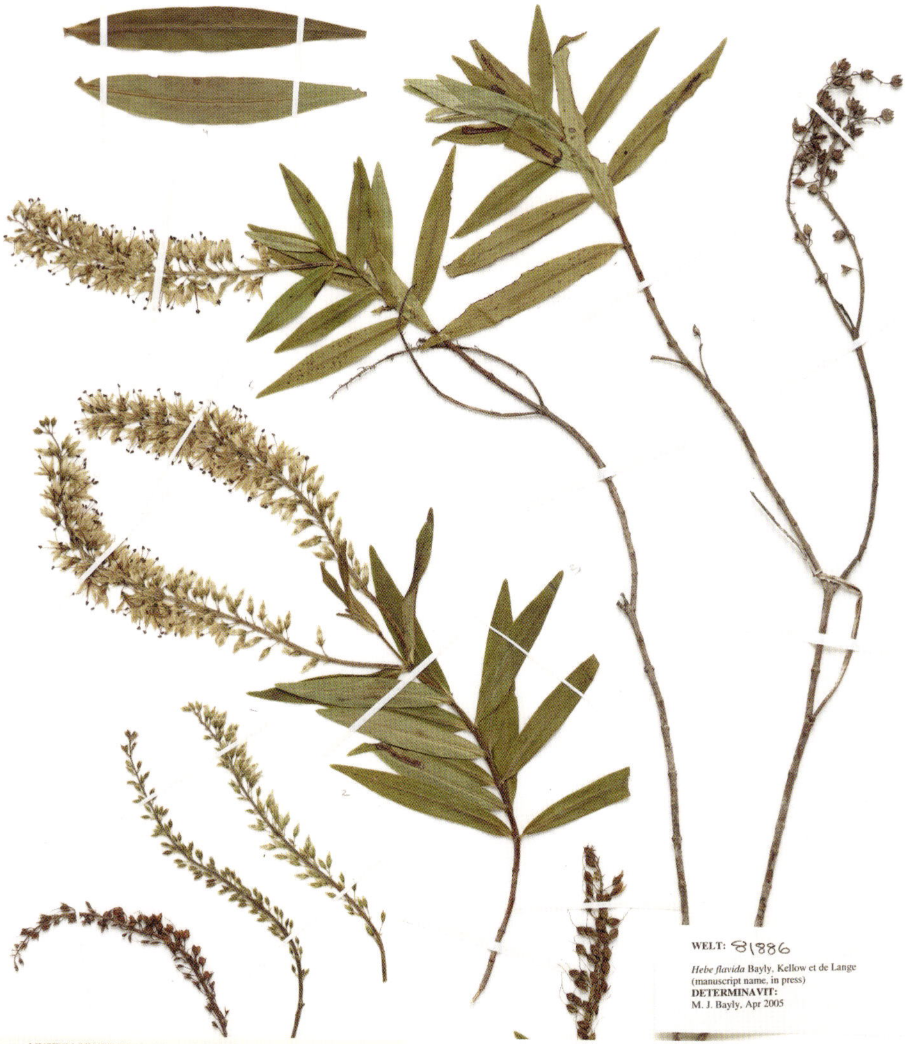

WELT: 81886

Hebe flavida Bayly, Kellow et de Lange
(manuscript name, in press)
DETERMINAVIT:
M. J. Bayly, Apr 2005

MUSEUM OF NEW ZEALAND TE PAPA TONGAREWA, HERBARIUM (WELT)
WELT SP081886
Veronica flavida **(Bayly et al.) Garn.-Jones**

Seen for eFlora: PJ Garnock-Jones 2017

HERBARIUM, MUSEUM OF NEW ZEALAND TE PAPA TONGAREWA
WELT SP081886 **WELLINGTON, N.Z.**

Hebe "Waima"

Loc. New Zealand, North Island, North Auckland, Waima Range, beside Six Foot Track.

Map NZMS 260 O06 51- 33- **Lat./Long.** 35°30.00' S 173°27.00' E
Hab.

Alt.(m) 280
Coll. Bayly, Michael; Kellow, Alison [MJB 1419] **Date** 27.05.2000
Det. Bayly, Michael **Date** 27.05.2000

Collection Notes
Shrubs to c. 3 m tall. Frequent. Corolla lobes pink to mauve, fading white. Anthers purple. Filaments and styles white. Flowers faintly sweet scented.

Herbarium History

Specimen Notes
Undescribed species, tag names "aff. acutiflora" and Waima Yellow". Also wet collection. Voucher specimen for material sent to IRL for flavonoid analysis.

WELT SP081886

Objects

F—G

Kaitiaki Taonga Collection Manager for Molluscs Bruce Marshall, far left, Curator of Invertebrates Rick Webber and Postdoctoral Researcher Rodrigo Salvador in the mollusc storage area of Te Papa's Tory Street facility.

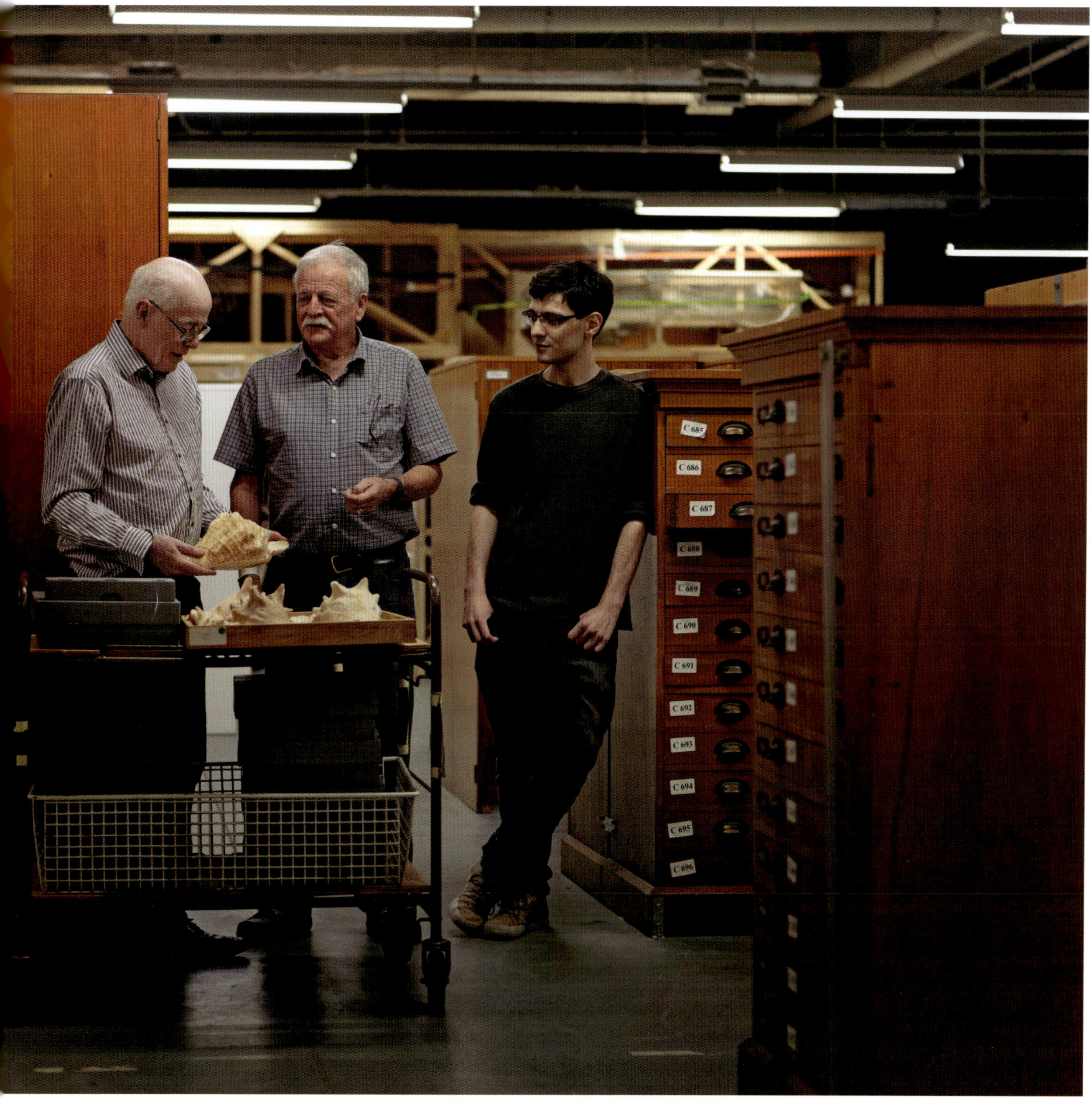
C 685
C 686
C 687
C 688
C 689
C 690
C 691
C 692
C 693
C 694
C 695
C 696

19 020 021
25 026 027
31 032 033
37 038 039
43 044 045
49 050 051

22 023 024

028 029 030

34 035 036

040 041 042

046 047 048

052 053 054

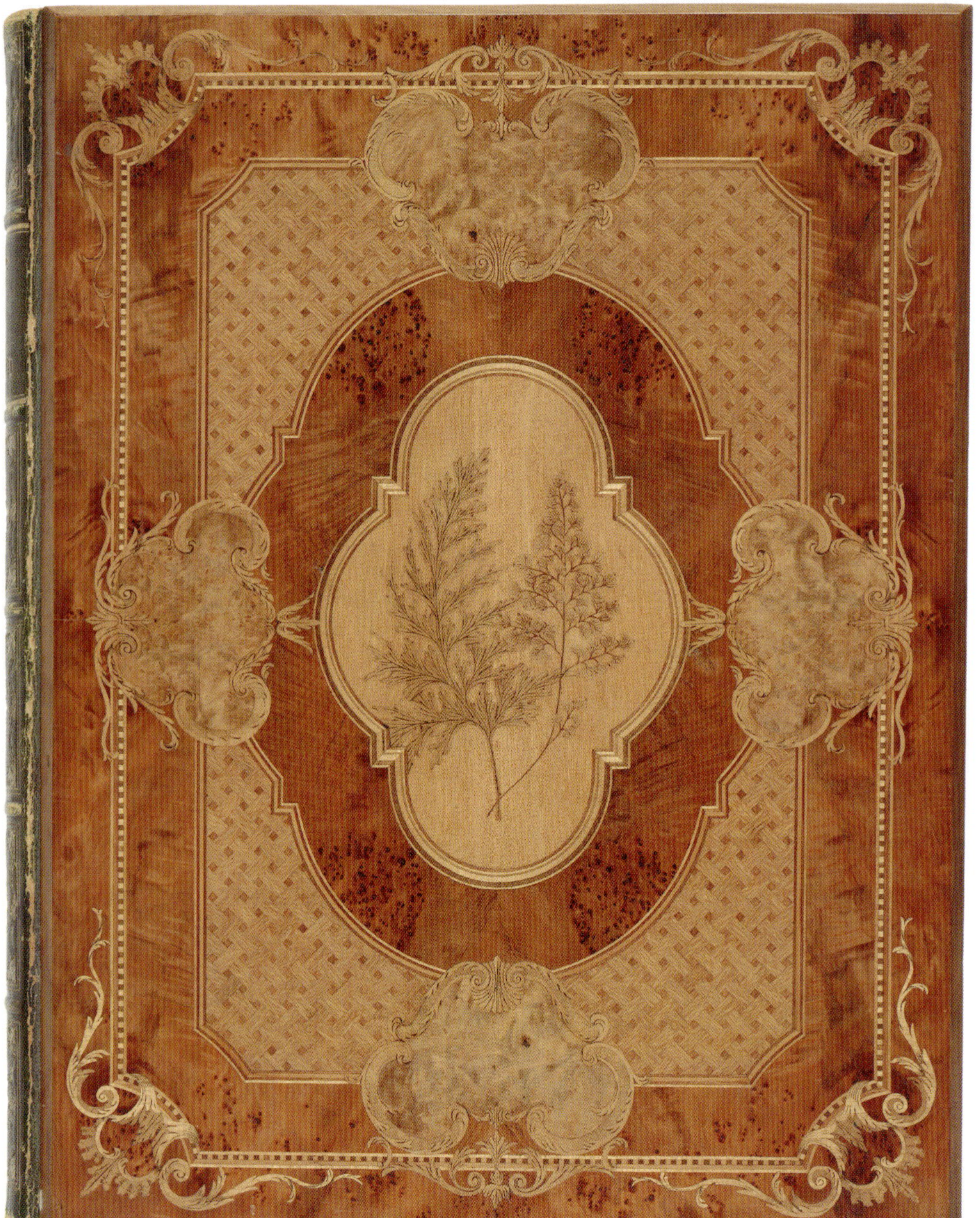

Fern album

Anton Seuffert

This exquisite example of inlaid marquetry work was produced by Bohemian-born craftsman Anton Seuffert, who worked as a cabinetmaker in Auckland from about 1860. He is responsible for some of the finest surviving pieces of early New Zealand furniture, using native timbers and often incorporating incredibly accurate images of ferns.

In the late nineteenth century, albums of pressed ferns became very popular with European colonists as presents to send 'home'. The ferns were evocative of the settlers' new homeland, and the covers presented an opportunity for artistic expression. Local dealers in the main centres offered a variety of albums in different sizes and styles, and contemporary advertisements show the range of products for sale. Budget albums could be assembled at home using pages, dried ferns, labels and cardboard covers supplied individually by the dealer. At the top end of the market, there were ready-made albums with carved kauri covers, and boxes containing small sheets of dried ferns.

Thomas Cranwell was an Auckland dealer who ran a business in Parnell. In addition to the usual range of albums, he produced a few deluxe items for a select clientele. The covers of these albums were crafted by Anton Seuffert, incorporating inlaid marquetry fern designs so accurate that they can be easily identified (as the native species *Hymenophyllum dilatatum* and *Adiantum aethiopicum* in this case). Inside the album were beautifully presented pressed fronds of different species, each apparently growing from a base made of mosses and pieces of fern, and discretely identified by the name on a small label. The lucky recipient of one of these albums could hardly have failed to appreciate the fern's cultural and ecological significance in New Zealand. PB

Fern album made by Anton Seuffert, c. 1875, GH003578, wooden cover with inlaid marquetry using native timbers, 480 mm x 375 mm.

Fiordland black coral

Antipathella fiordensis

The fiords of Fiordland were formed about 20,000 years ago by glaciers gouging out rock, forming U-shaped valleys and leaving dam-like deposits submerged at the mouths of the fiords. These stifle the exchange of fiord water with ocean water and, consequently, water circulation within the fiords.

Fiordland is famous for both its forests and its incessant rainfall. Rain water flows down through forest litter, becoming brown, and forms a layer on top of the sea water. Because mixing is minimal it remains there, its brownness acting like sunglasses to reduce the light penetrating the fiord water. Along with the shadows cast by the steep sides of the fiords, a dark zone forms close to the surface that imitates the darker, deeper waters of the open sea. The result is a remarkably stable aquatic environment in terms of light, temperature and salinity. Many unexpected or new invertebrate and fish species have been found in this unique and precious ecosystem.

This is the home of Fiordland black corals. Black corals are named for their black skeletons but, when alive, they are not black at all but are covered by a white tissue connecting a colony of thousands of tiny, six-tentacled polyps that build and live on the skeleton. Around 1980, two similar species (one of them *Antipathella fiordensis*) were found to be abundant on fiord walls in Fiordland at depths of 10-35 metres. This was very surprising because most black coral species live in much deeper water. A survey estimated over seven million colonies throughout Fiordland. There was speculation about the potential market value of this easily accessible, semi-precious resource, but fortunately, exploitation hasn't occurred and the black corals are now protected by law. RW

Fiordland black coral (*Antipathella fiordensis*), CO.007131, height 1.5 m, width 1.0 m, collected by a Department of Conservation diver, Doubtful Sound, Fiordland, 1990s.

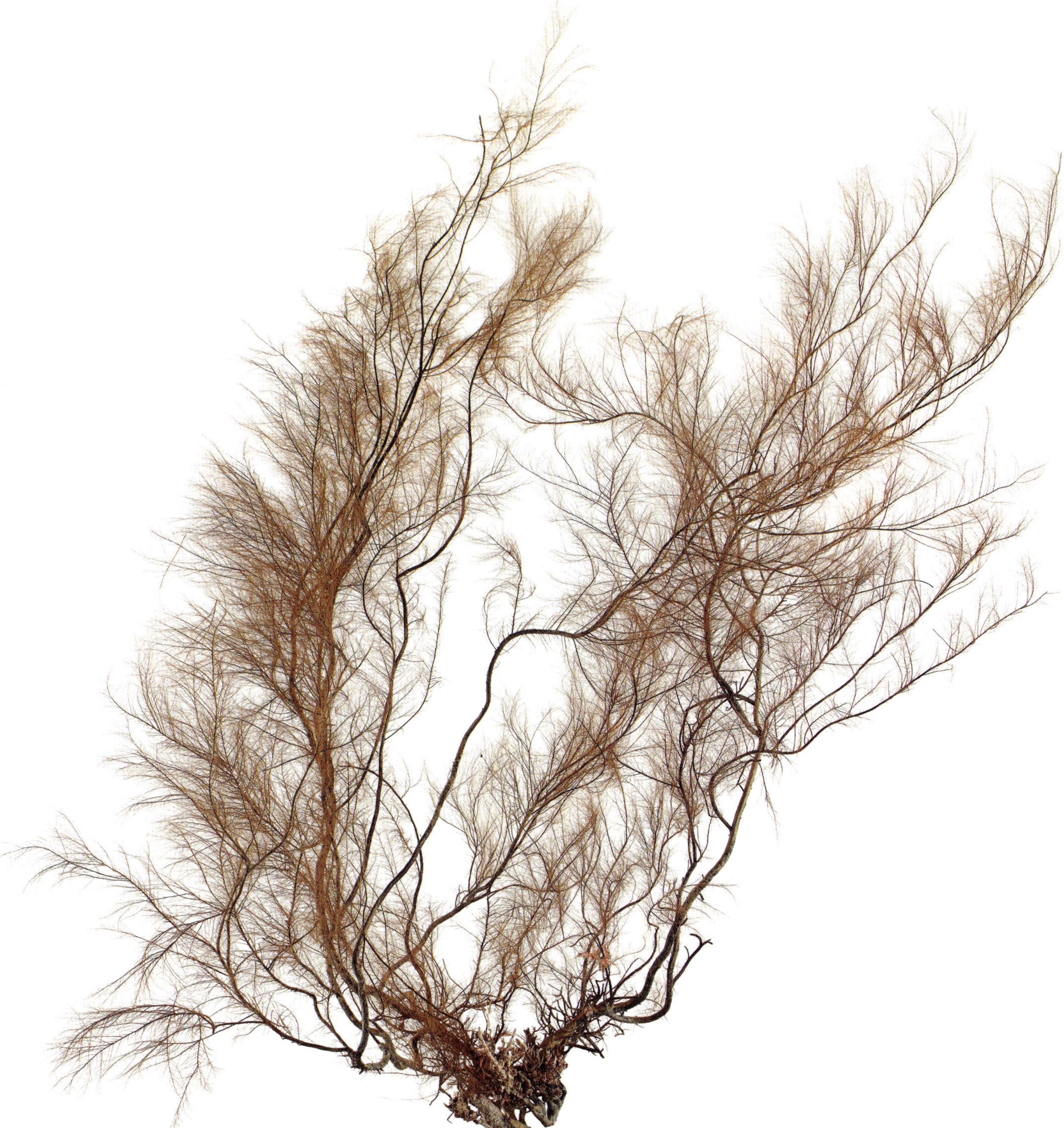

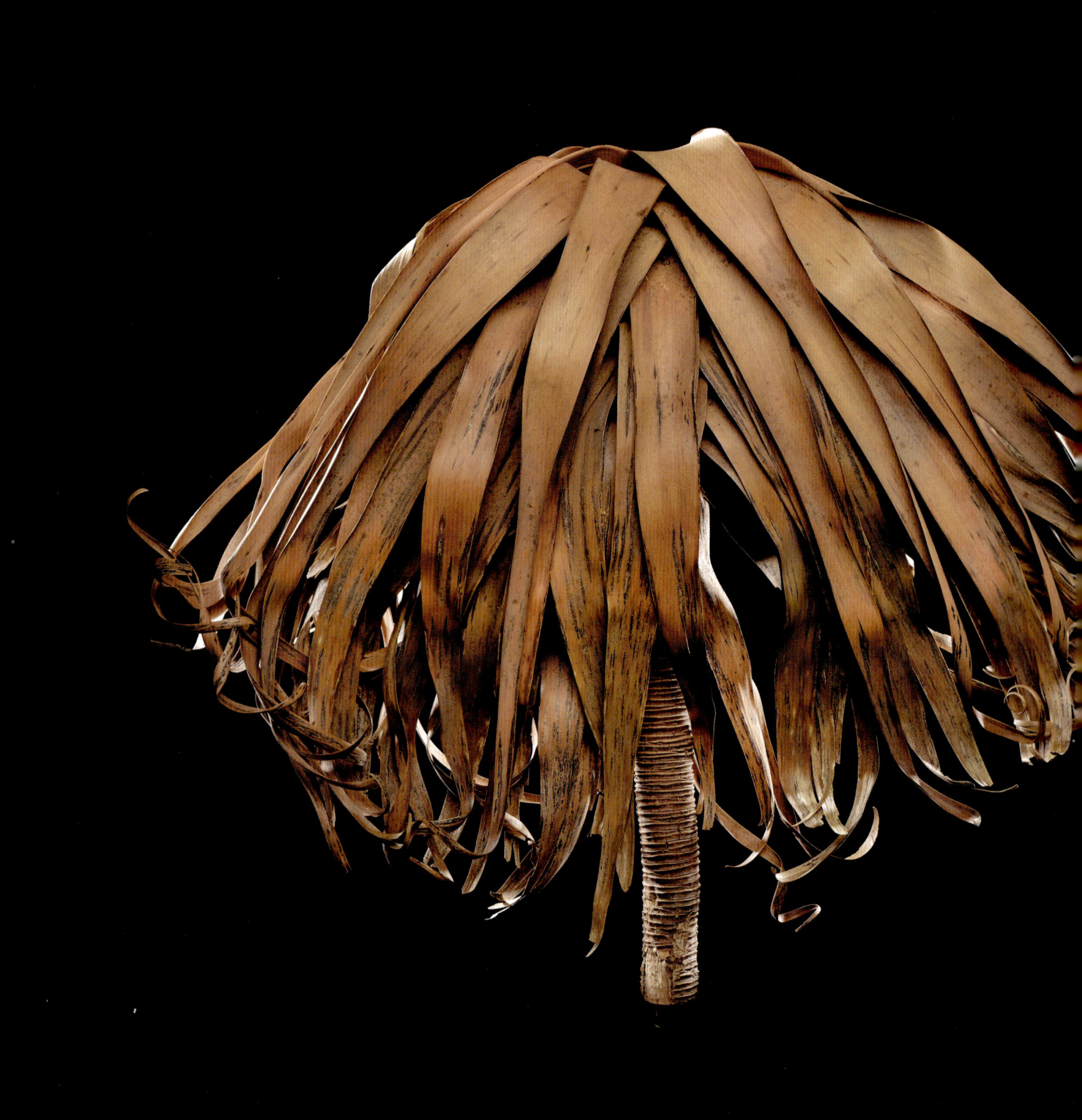

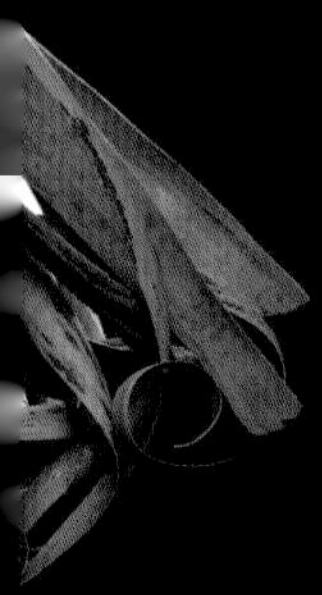

Fiordland grass tree

Dracophyllum fiordense

028/100

The distinctive appearance of the *Dracophyllum* tree shown here has led to it sometimes being called a pineapple tree or Dr Seuss tree. Specimens like this leafy head add a quirky character to a collection, but they're also a bit of a pain because they need to be treated differently to the 99 percent of specimens that conform to the collection standard. These 'regular' specimens fit easily on a piece of card just slightly narrower than A3 (or divided across several such cards) or are small plants that can be enclosed in an envelope made from A4 paper.

Standardised storage makes it efficient. Specimens of the same size fit into the same-sized folders, which fit into the same-sized boxes, which fit onto the same-sized shelves. Our sister institutions do the same, so we can easily exchange specimens, whether as loans or gifts. To fully document plant diversity, however, the storage rules sometimes have to be broken. To preserve the three-dimensionality of this leafy head, it has been stored in a special large box on a specially modified shelf.

This particular specimen of *Dracophyllum fiordense* was collected by WRB Oliver in 1927 from Fiordland's Wilmot Pass. Perhaps uniquely, there's a photo of a dapperly dressed Oliver returning with this specimen slung over his shoulder (see page 244). Today, we can use cameras to record features that do not fit onto a flat sheet, but that option wasn't easily available to Oliver. Even so, the challenge of collecting and storing large specimens often leaves us with knowledge gaps. For example, to get measurements of fronds for our recent revision of New Zealand's tree ferns, we had to make new collections. LP

Fiordland grass tree (*Dracophyllum fiordense*), SP055115, collected by WRB Oliver, Wilmot Pass, Fiordland, March 1927.

029 / 100

Flaming flagfin

Hime pyrhistion

One of New Zealand's most elegant and strikingly coloured fish species is the flaming flagfin, *Hime pyrhistion* (family Aulopidae). However, because it is both uncommon and lives at depths of 80–380 metres, few New Zealanders encounter it.

Flagfins get their common name from their large dorsal fin. Currently, there are about nine species of *Hime* known in the Pacific and eastern Indian Ocean, four of which have been scientifically described since 2013. The New Zealand species *Hime pyrhistion* was named after the Greek words *pyr* (fire) and *histion* (sail), in recognition of the colourful red-and-yellow-patterned dorsal fin.

While known for over thirty years, the flaming flagfin has only recently been formally described. First captured in 1982 by a recreational fisherman near the Cavalli Islands in Northland, it was originally believed to be *Aulopus japonicus* – a widespread Indo-Pacific species. In 1990, however, Japanese scientist Takeshi Shimizu compared it with specimens collected from the Norfolk Ridge by scientists on board the RV *Shinkai-maru*, and placed it in the genus *Hime* as a new species. Unfortunately, he was unable to publish his work. In 2000, the project was taken on by American researcher Bruce Thompson, who sadly died in 2007 without having published either.

In 2013, with renewed impetus to publish the flaming flagfin's scientific name in our guidebook *The Fishes of New Zealand*, I was fortunate to finally name and describe this wonderful and intriguing species together with my close colleagues Martin Gomon from the Museum of Victoria and Andrew Stewart from Te Papa. CS

Flaming flagfin (*Hime pyrhistion*), P.054047, holotype, standard length 236 mm, collected by E Daysh at depth of 80 m, west of Poor Knights Islands, Northland, 19 June 2012.

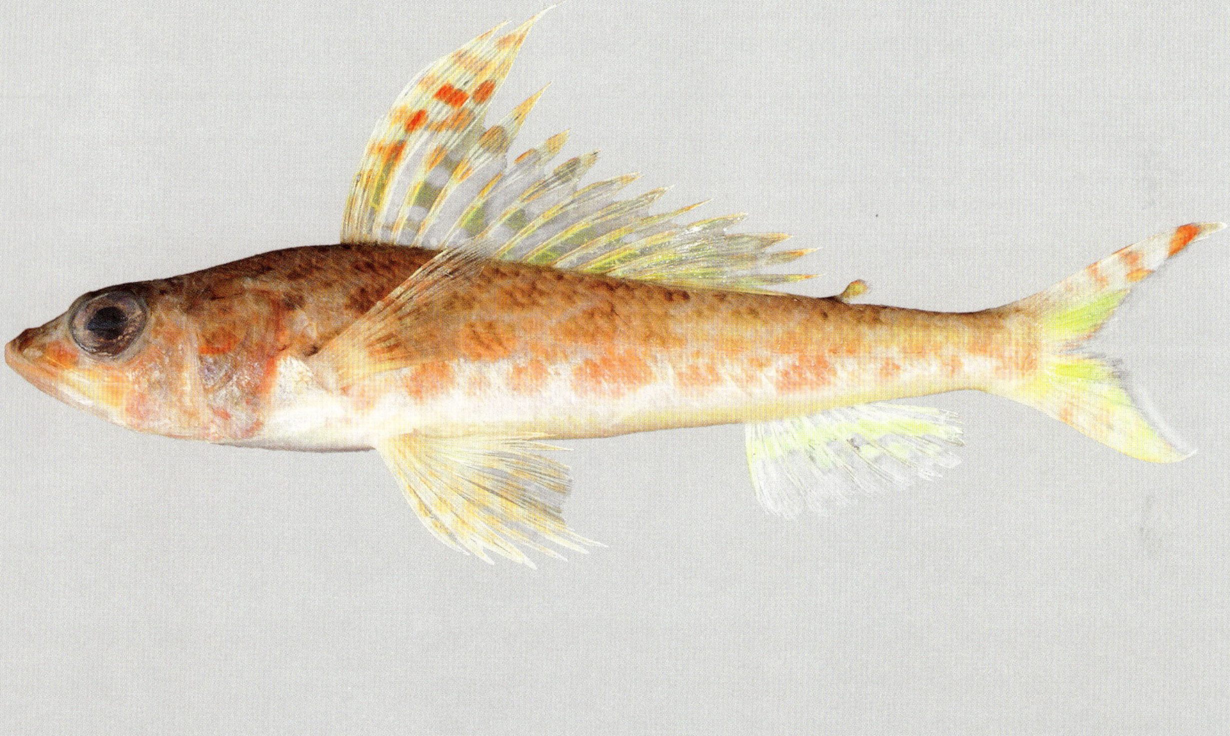
5 cm

FLIT fly sprayer

Waiter brand manual fly spray dispenser

Inventions to repel parasitic insects date back to times when humans still lived in caves. But the science of insecticides really flourished from the time European settlers arrived in North America, bringing a variety of parasites with them to join the new ones already awaiting them. Pests thrived in overcrowded environments and many of the new settlers were soon overrun by head lice, bedbugs and fleas. Later, in the early 1900s, there was a huge outbreak of yellow fever (transmitted by mosquitoes) during the construction of the Panama Canal. And during World War One soldiers faced malaria (carried by mosquitos), typhoid (carried by blowflies) and trench fever (carried by body lice).

030/100

Over the years, many different formulas of insecticide have been invented to rid ourselves of such unwanted insects. FLIT, launched in 1923, is probably the most famous brand. It came in a handy pump-action sprayer, was based on mineral oils, and was made by the Standard Oil Company of New Jersey – later known as ESSO and then ExxonMobil. A FLIT advertising campaign that ran for many years used cartoons drawn by Theodor Seuss Geisel – better known as Dr Seuss. The style of these ads, and many others Geisel drew for other products, are instantly recognisable by anyone familiar with his children's books.

In the 1930s, fighting crop-eating bugs became a colourful affair. 'Paris Green' got its colour from the copper–arsenic compound it contained. 'London Purple' contained coloured calcium arsenite left over from the production of aniline dyes in London. Countless other brands have also been developed. Interestingly, one dispenser labelled Waiter Brand, made in Hong Kong in the 1930 and 1940s, closely resembled the early FLIT design.

Later on, chemists added lead to the arsenic compounds, and for the first half of the twentieth century these were the most popular insecticides – with minimal effectiveness against foe but maximal capacity to poison friend. JK

Vintage Waiter brand manual fly spray dispenser, AEO.000001; round red tin (diameter 80 mm x height 80 mm) soldered by the lid to a hand pump with wooden handle; made in Hong Kong in the Tat Yuen Factory, c. 1930–1940s.

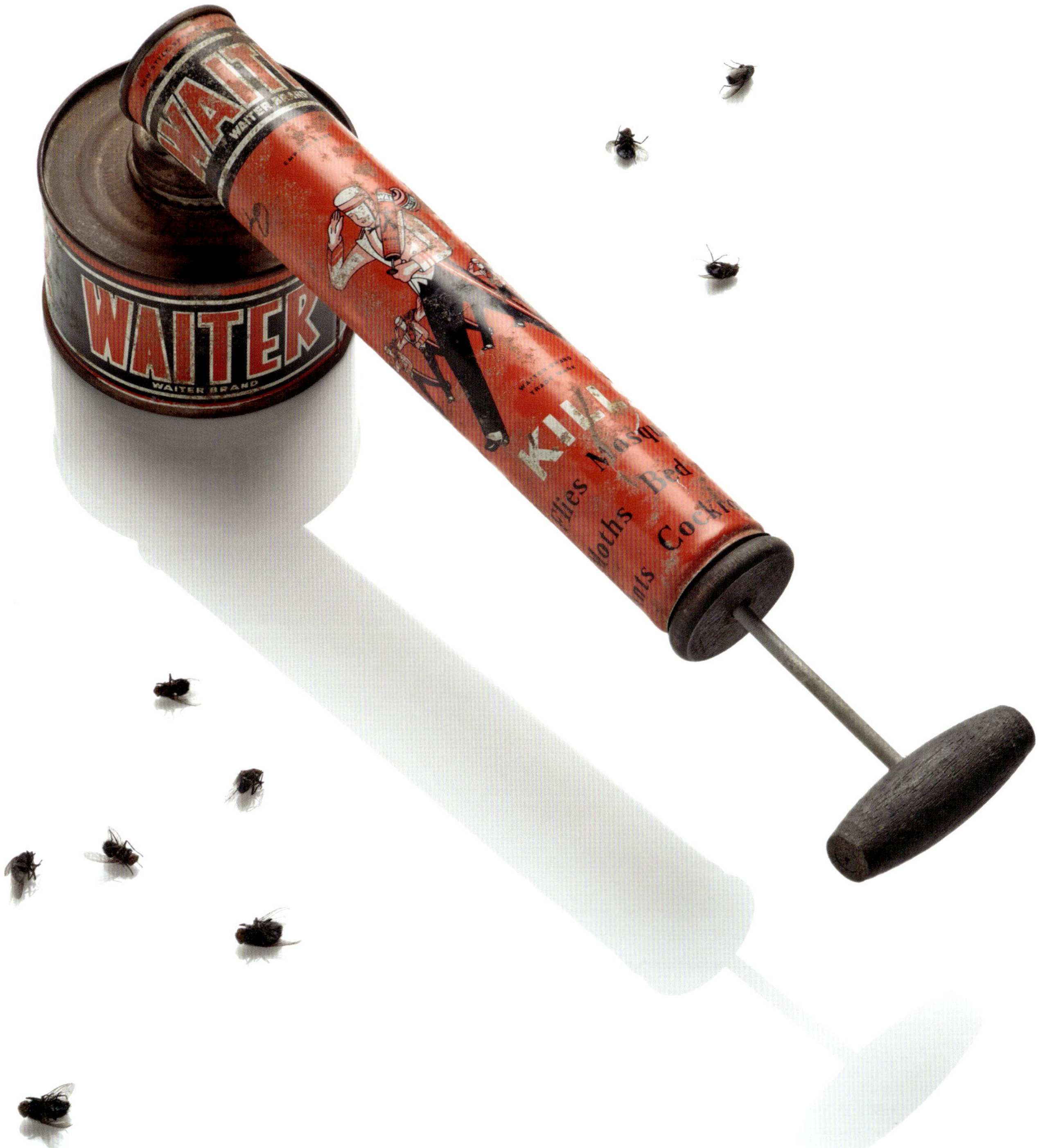
WAITER
WAITER BRAND
KILL
Moths
Bed

031
100

The Forest Flora of New Zealand

Thomas Kirk

Thomas Kirk knew plants well. His interest had been fostered at an early age, and after migrating to New Zealand in 1863 he set about collecting and learning the flora almost from the moment he arrived. His passion for botany never faded; he was still actively writing *The Students' Flora of New Zealand* when he died in 1898. Kirk published some 150 papers and described around 200 new species, accumulating a huge herbarium, the main part of which is now at Te Papa. He exchanged specimens with overseas botanists, and these are found in major herbaria all over the world.

Working with his herbarium today, I admire the energy that went into its collection, preservation and identification. Kirk's plants are beautifully presented, and labelled in a neat and tidy hand that deteriorated little with age. Above all, his identifications are reliable and consistent, reflecting his broad understanding of the New Zealand flora.

Kirk's single most important work was *The Forest Flora of New Zealand*, published in 1889. As a former timber merchant and later Chief Conservator of Forests, he was superbly well-qualified for the task. This is not a dry and technical Flora; there are clear descriptions, discussion about names, a flowing text full of historical detail, information about cultivation, properties and uses, and over 150 plates illustrating the species at life-size or larger. The toothed lancewood (*Pseudopanax ferox*) was described and illustrated here for the first time, based on a specimen in Kirk's herbarium. *The Forest Flora* is a work that reflects Kirk's passion for New Zealand forests, and his legacy of 800,000 acres of reserves established during his time as Chief Conservator. PB

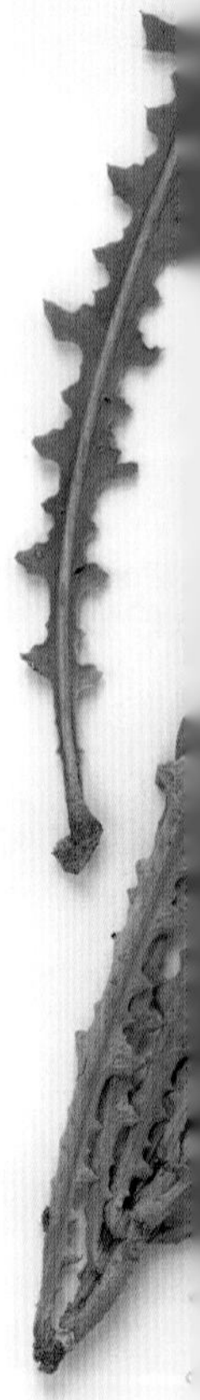

Toothed lancewood (*Pseudopanax ferox*), SP025008, syntype, collected by T Kirk Sr, Wairoa, Nelson, no date. Toothed lancewood (*Pseudopanax ferox*) illustrated as Plate 24 in T Kirk, *The Forest Flora of New Zealand* (Government Printer, Wellington, 1889), drawn by the Survey Department, 1886, to accompany the original description.

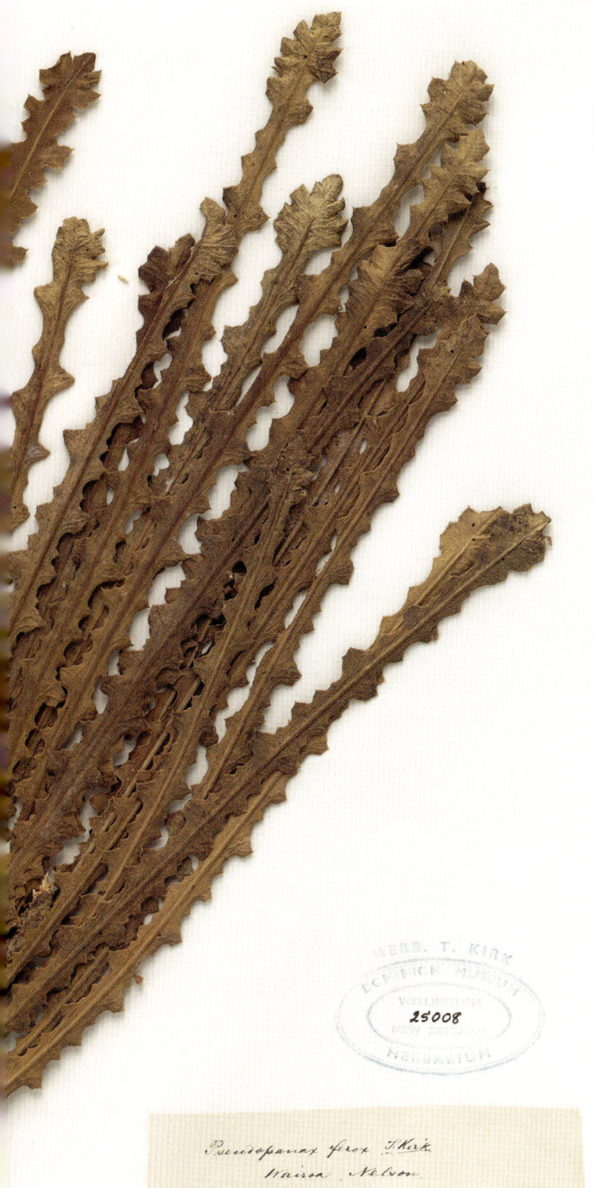

HERB. T. KIRK
DOMINION MUSEUM
WELLINGTON
25008
NEW ZEALAND
HERBARIUM
Pseudopanax ferox T. Kirk
Wairoa Nelson
T. Kirk
Forest Flora N.Z., t.xxiv

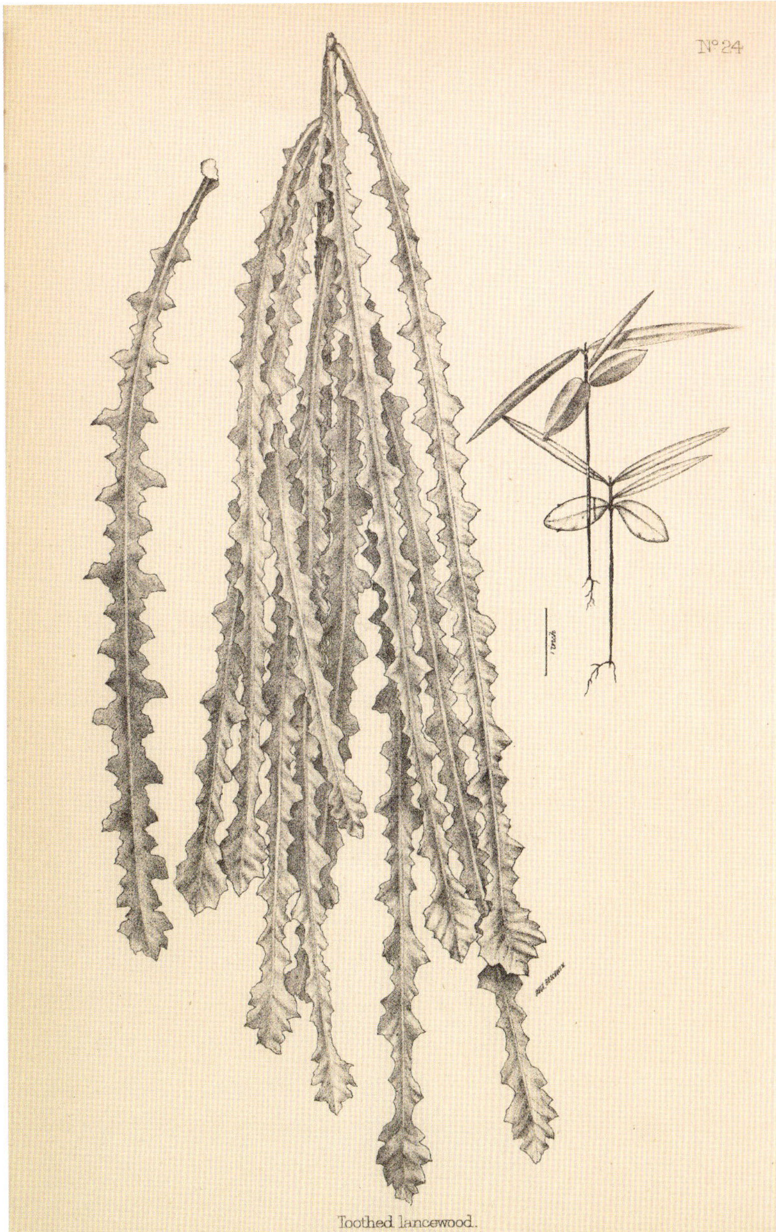

N°24
Toothed lancewood.
(PSEUDOPANAX FEROX)
Deflexed leaves.

032

100

Forget-me-not pollen

Myosotis brevis

There is something exquisite about this image. Perhaps it is the intricate texture and symmetry of the egg-shaped objects; perhaps the stark contrast of the orbs floating in a jet-black void, a cosmic image that a telescope might relay back to Earth from deep space. The image is not in fact a telescopic one; rather, a scanning electron microscope was used to photograph the surfaces of these three pollen grains.

The grains come from a Te Papa specimen of the world's smallest forget-me-not, *Myosotis brevis*. The image was one of hundreds taken as part of a study of the pollen of southern hemisphere *Myosotis* species. The plant produced these and other pollen grains to pollinate the ovules in the flower, eventually producing seeds and thus enabling the plant to reproduce. (Unfortunately for this particular plant, pollination and reproduction were interrupted in the name of science!)

Myosotis pollen grains are tiny – each of the ones shown here is just over 10 microns (0.01 millimetres) long, making them only slightly larger than human red blood cells. These minute parts of a plant specimen that is itself very small contain vital clues to its identification. Although they are microscopic, pollen grains have many taxonomic characters that are useful for classifying different families and genera of plants, and sometimes for distinguishing plants at species level. This *Myosotis brevis* pollen has a unique combination of features that allows us to identify the plant using pollen alone. Pollen characteristics may even be a better way to categorise species within the genus *Myosotis* than the larger, more obvious characteristics that have been used previously. HM

Pollen grains of the forget-me-not *Myosotis brevis*, SP102761, scanning electron microscope images, specimen collected by JM Prebble, Bannockburn, Central Otago, 8 October 2013.

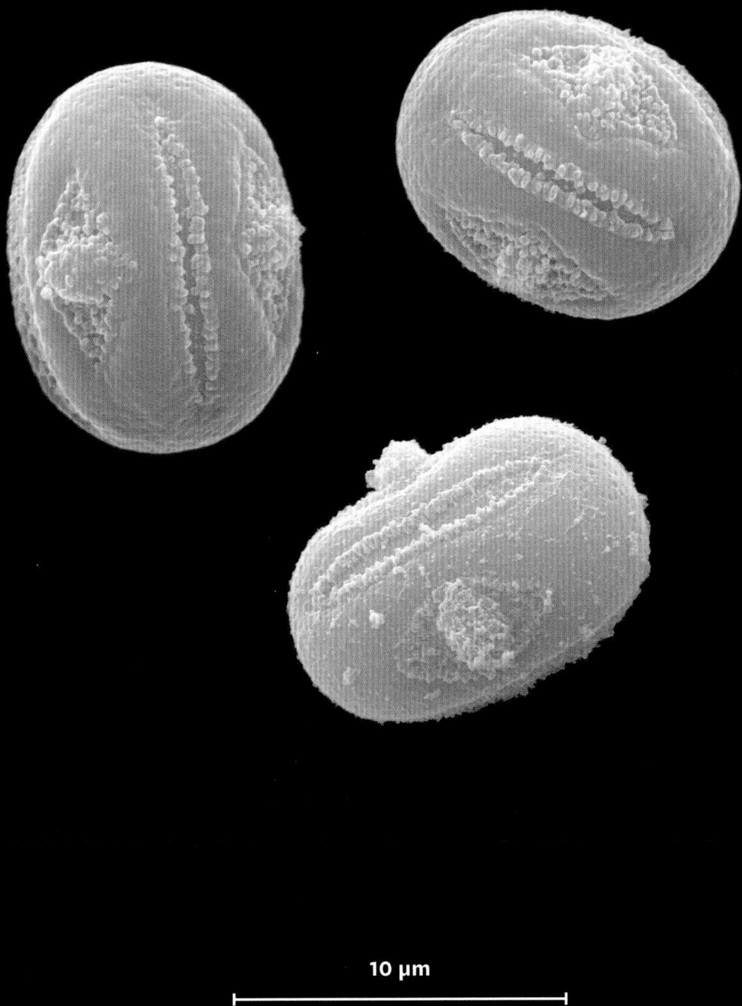
10 µm

Freshwater limpet

Latia manuherikia

033 / 100

Fossils found at the site of an ancient lake, Lake Manuherikia, near St Bathans in Central Otago, include vast numbers of mineralised fish bones, as well as the bones of birds, bats and a crocodile-like reptile – all now extinct. The enormous lake existed during the late Miocene, when temperatures there were considerably warmer than at present, probably as warm as modern-day Queensland.

Among these fossils are some from the phylum Mollusca: abundant freshwater snails, scattered bands of freshwater mussels (ancestral kākahi), and rare specimens of several species of small land snail that had been washed into the lake. Most exciting – and one of the rarest finds – was the discovery of a species of endemic freshwater limpet, named *Latia manuherikia* after the prehistoric lake from which it came. The genus *Latia* belongs to our only endemic family of molluscs, the Latiidae, which is most closely related to the Chilinidae of southern South America. Since neither *Latia* nor Chilinidae have a marine larval stage – which would allow them to spread via the oceans – it seems clear that their common ancestry dates from the late Mesozoic era, when New Zealand and South America were connected via Western Antarctica as part of the ancient super-continent Gondwana.

Of all the molluscs worldwide, *Latia* species are unique in secreting a cloud of luminous mucus when disturbed, presumably to startle or otherwise discourage predators. The two *Latia* species living today, a northern and a southern one, occur commonly on rocks and stones in streams and lakes throughout much of the North Island. They have evidently become extinct in what is now the South Island. BM

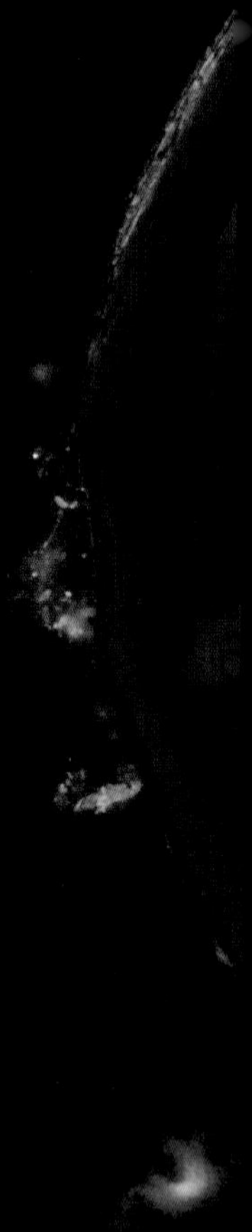

***Latia neritoides*, M.325933, underside of a living specimen emitting luminous slime, collected by J-C Stahl and B Marshall, Wainuiomata River, Rimutaka Ranges, 6 March 2018.**

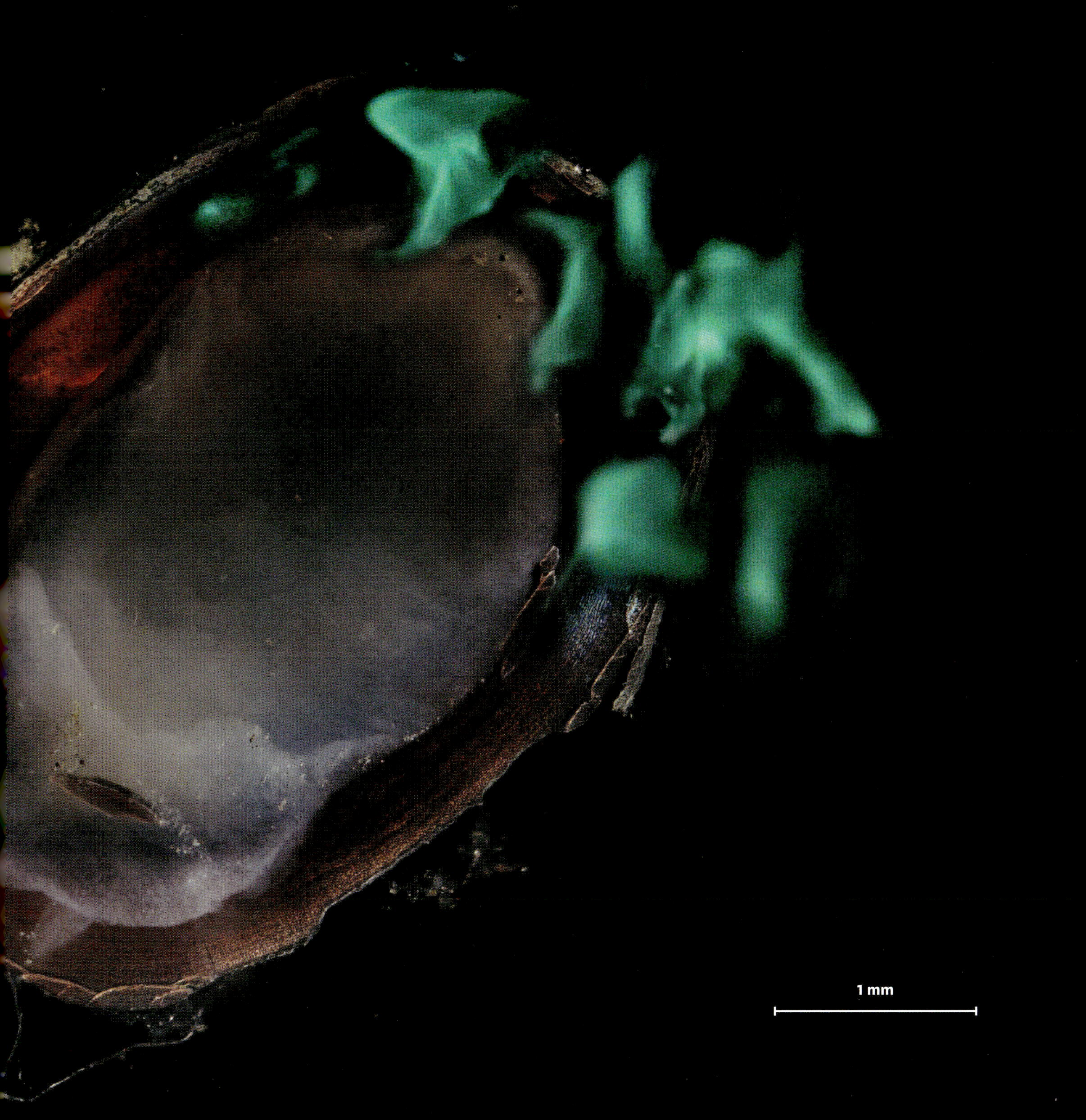
1 mm

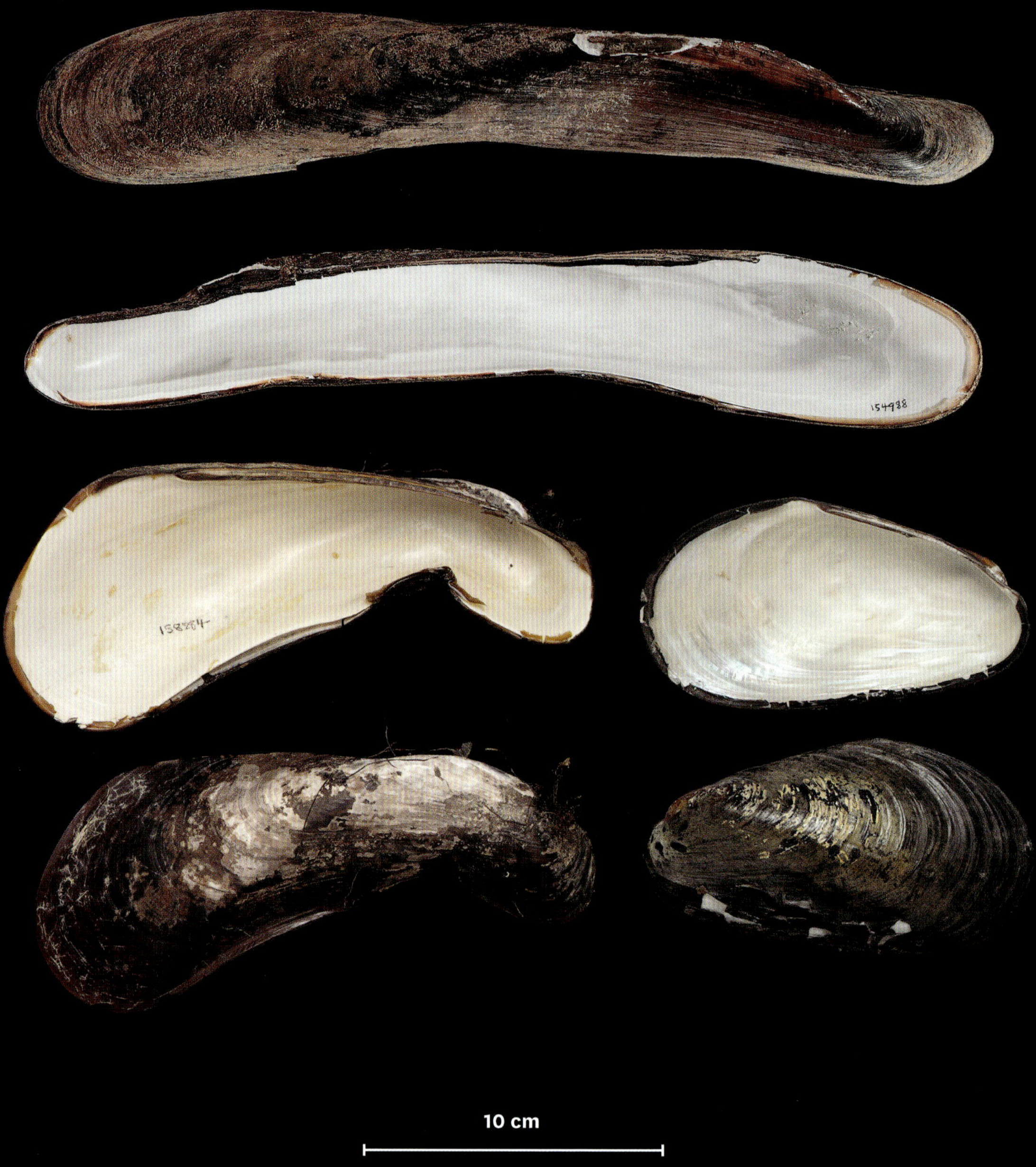
154988
10 cm

Giant mussels

Gigantidas gladius, Vulcanidas insolatus and *Gigantidas tangaroa*

In the 1960s, when I was a boy living in Whakatane, I heard local fishermen tell of giant mussels reeking of sulphur that were found entangled in grouper longlines set in deep water on seamounts off White Island. Regrettably I never obtained one of these; had I done so, I could have been first in the world to acquire a giant deep-sea mussel related to those taken from the Galapagos Rift Zone by the submersible *Alvin* in 1977.

In 2001, I joined other researchers on the RV *Tangaroa* to search for deep-sea animals associated with venting on submarine volcanoes on the Kermadec Ridge, to the east of White Island. Underwater cameras revealed a variety of fishes, crabs and molluscs, including large, scattered but dense beds of giant black mussels. Our attempts to obtain mussels using a huge dredge was hit-and-miss, as it rapidly filled up with scoria, but we eventually obtained a sample. The mussel proved to be a new species belonging to a new genus, and the second-largest true mussel in the world. We named it *Gigantidas gladius*, and introduced it together with a second, smaller species from a deep-sea cold seep off Cape Kidnappers.

Subsequently, a second giant mussel – although somewhat smaller – was taken elsewhere on the Kermadec Ridge. We had found not only another new genus and species, but also the most primitive and shallowest-living vent mussel. We named it *Vulcanidas insolatus*, meaning 'volcano mussel in sunlight'.

Two of these mussels occur in areas of volcanic venting with elemental sulphur, and obtain their nutrients using sulphur-oxidising symbiotic bacteria living within their tissues. One or both of them was presumably the source of the reports of my youth. BM

Top two: *Gigantidas gladius*, M.154988, paratype, collected by RV *Tangaroa* at depth of 360–755 m, Rumble V volcano, Kermadec Ridge, 24 May 2001. **Lower left two:** *Gigantidas tangaroa*, M.158284, holotype, collected by S Donker on the FV *Tasman Viking* at depth of 920–1205 m, cold seep off Cape Kidnappers, 26 May 2001. **Lower right two:** *Vulcanidas insolatus*, M.296656, holotype, collected by RV *Tangaroa* at depth of 140–200 m, Macauley Caldera, Kermadec Ridge.

035

100

Giant penguin fossil

Kumimanu biceae

Fossils often take years to extract from the rocks that conceal them, and this was the case with what might be the world's largest ever penguin.

In 2004, while searching for bird fossils on a beach in North Otago, some colleagues and I found a couple of bone scraps encased in rocks, neither of which seemed of earth-shattering significance. One fossil appeared to be small fragments of turtle bones, appearing too large and solid for any bird. In 2015, the 'turtle' rock was finally sent to Al Mannering of Christchurch for extraction; he is one of the few people who can do this laborious, specialised work. As Al chipped away tiny rock flakes from the 'turtle' bones, he realised that he was looking at the coracoid (a shoulder bone) of an enormous penguin.

Eventually Al revealed several bones of the flipper, body and leg. After studying the remains in detail, several colleagues and I concluded that this penguin would have been 1.77 metres long and weighed 101 kilograms – a good size even for a human! We named it *Kumimanu biceae* – *kumi* being a large monster in Māori mythology and *manu* meaning bird in Māori, with *biceae* honouring my mother Bice Tennyson, who fostered my interest in natural history. Enormous fossil penguins had already been found in the southern hemisphere, but *Kumimanu* was outstanding for its size (rivalled only by some fragments from Antarctica) and its incredible antiquity at fifty-five to sixty million years old – considerably older than any other really giant fossil penguin. We now know that enormous penguins evolved at the dawn of penguin evolution and that modern penguins are much smaller (the largest living penguin is the emperor, weighing in at 30 kilograms). AT

Bice's penguin (*Kumimanu biceae*) fossil bones still partly encased in rock, S.45877, holotype, collected by A Tennyson, P Scofield and P Duncan, near Oamaru, North Otago, 5 July 2004.

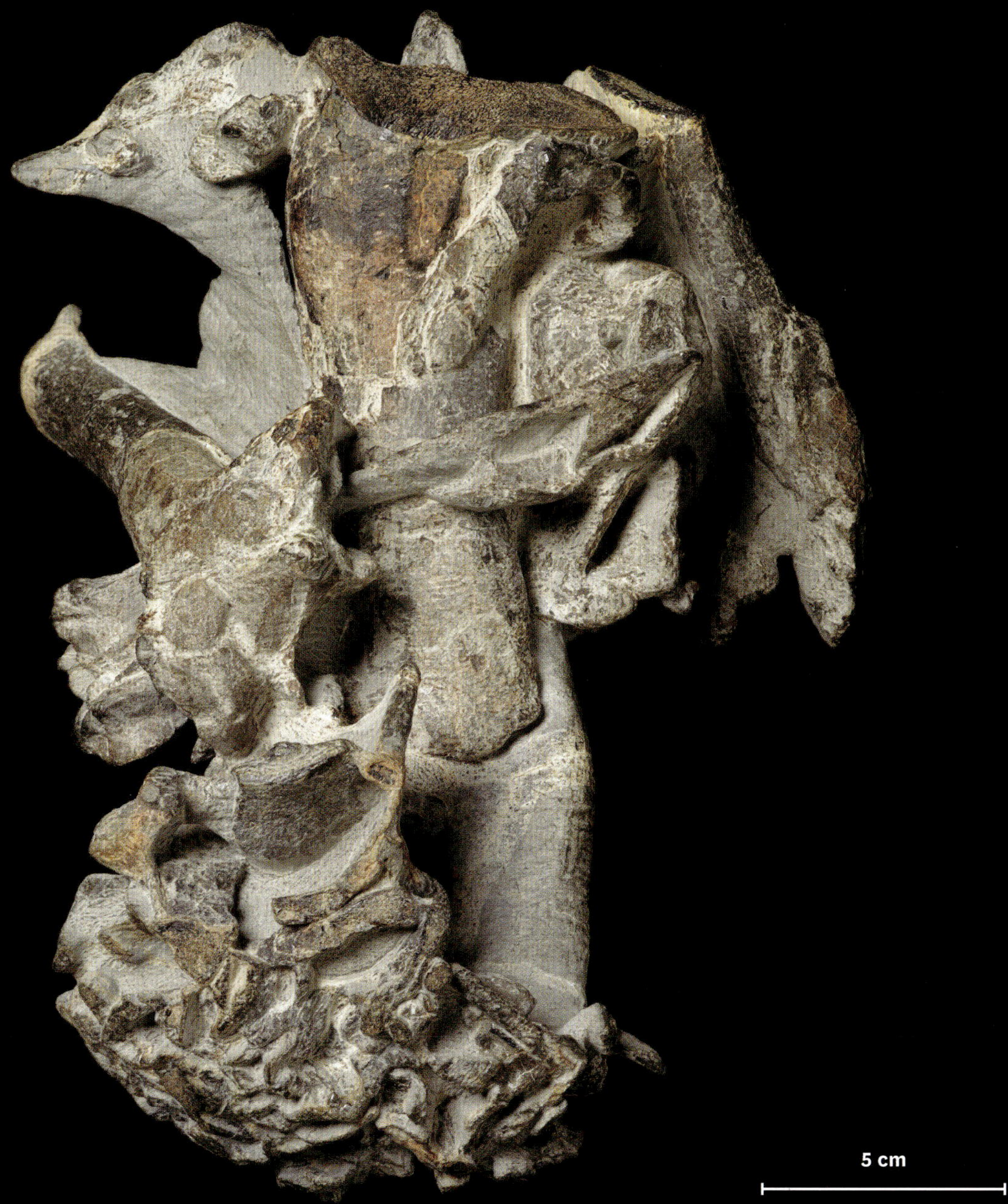
5 cm

Glow worm

Arachnocampa luminosa

Most New Zealanders – and many tourists – have experienced the awe inspired by our sparkling glow worms, though not everyone has seen the actual animal behind the light. Māori call them titiwai, which refers to lights reflected in water. The scientific name, *Arachnocampa luminosa*, is almost as lyrical: *Arachnocampa* means 'spider-worm', after the way the carnivorous larvae hang threads of sticky silk to catch their food: small flying invertebrates.

036
100

It's now well known that *A. luminosa* is not a worm, but an insect that produces bioluminescence to attract its prey. Their true nature was discovered by George Vernon Hudson in the 1880s, after raising glow worm larvae in a tank. The larvae went through a pupa stage and then emerged as a fungus gnat, a type of fly. The tiny brownish flies only live long enough to reproduce – just a few days.

Seen in daylight, the white larvae, which can grow up to 3 centimetres long, are not very spectacular. In the dark, however, the glow worm's tail-light shines brightly. The glow results from a chemical reaction between molecules of luciferin and oxygen, taking place in an organ equivalent to the human kidney. An air bag surrounding the light-producing organ provides oxygen and also acts as a reflector that concentrates the light.

So, in short, they are maggots lying in spider-web-like hammocks, with a chemical reaction at their rear that turns a colony of them into a sky of stars – while the adults are barely noticeable. However, after millions of years evolving in a dynamic landscape, this extraordinary creature is now threatened by human interference through habitat destruction. JK

Glow worm (*Arachnocampa luminosa*) adult, AI.034829, reared and collected by GV Hudson, Wilton Botanical Garden, Wellington, April 1889.

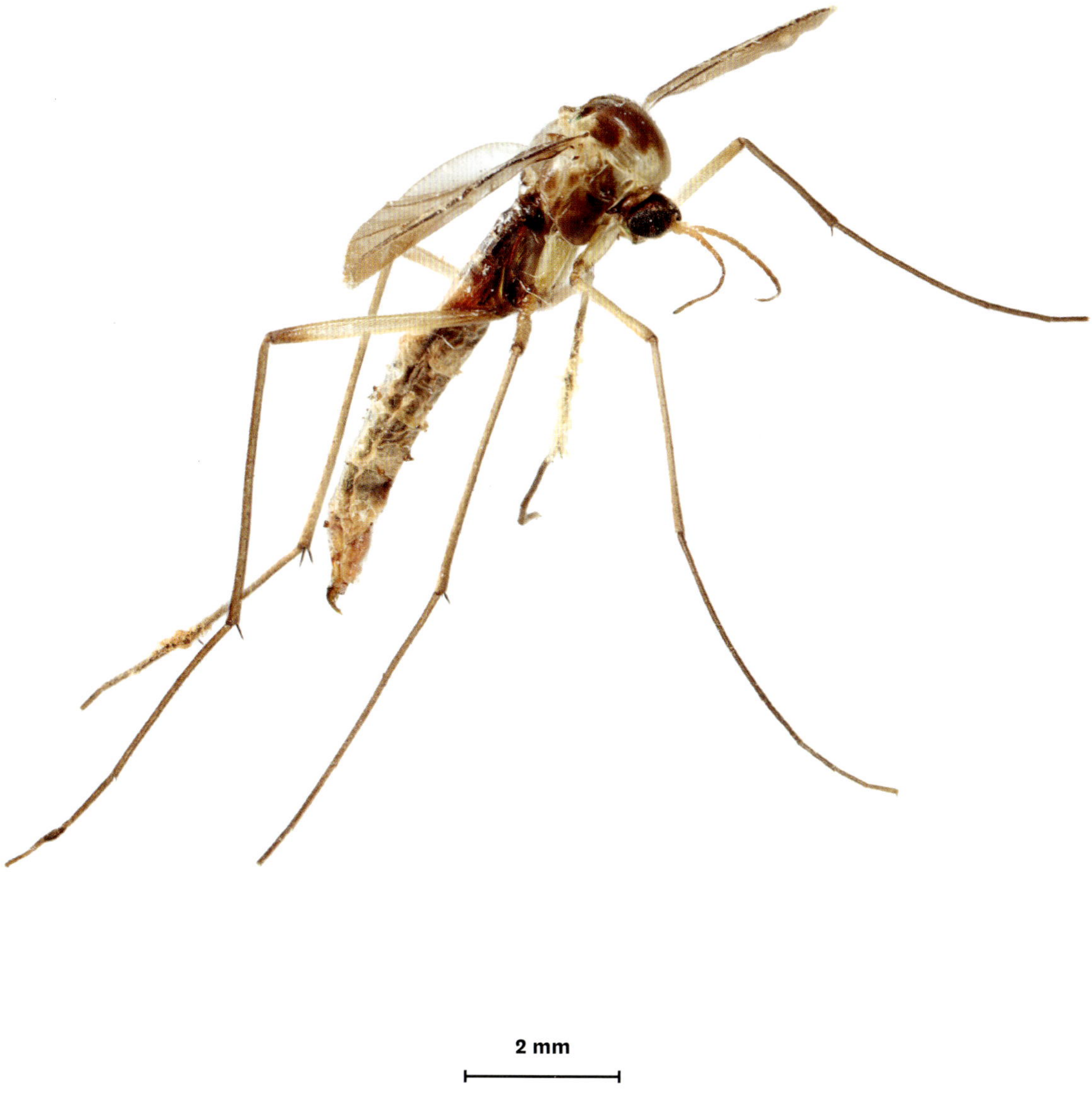
2 mm

Gordian worm

Parachordodes species

According to legend, the Gordian knot was impossible to untie, although the person who could unravel it would become ruler of Asia. Alexander the Great, skilled at thinking outside the box, simply sliced the knot open. Today, a Gordian knot is a tricky problem that can be solved by thinking creatively.

037/100

The legend provides an apt name for the Gordian worm. Not only does this aquatic nematomorph curl up into a knot but the worm also has a parasitic life cycle and requires two different hosts – one of which is not aquatic. This is quite a tricky problem! The adult worms, which may grow to between 20 and 30 centimetres long, live, mate and lay their eggs in freshwater. On hatching, the larvae either burrow into a juvenile aquatic insect host (for example caddisflies or mayflies) or are eaten by them. So far, so good.

When the host emerges as an adult it leaves its freshwater home, taking the juvenile worm with it, dormant as a cyst. Eventually the host gets eaten, or dies and then gets eaten, in New Zealand by wētā (*Deinacrida* and *Hemideina* species). The worm bursts free of the first aquatic host straight into the gut of the new terrestrial host. After feeding on the body of the wētā, the worm matures and needs to get back into water. To do so, it changes the wētā's behaviour and makes it drown itself; dead wētā can often be found in puddles or ponds.

Many questions remain. For example, tree wētā will consume more than one cyst, yet only one nematomorph will emerge. What is the winner's secret? How do the worms make the wētā crave water? It is up to us to untie those knots! JK

Gordian worm (*Parachordodes diblastus*), TMP031942, no collection data.

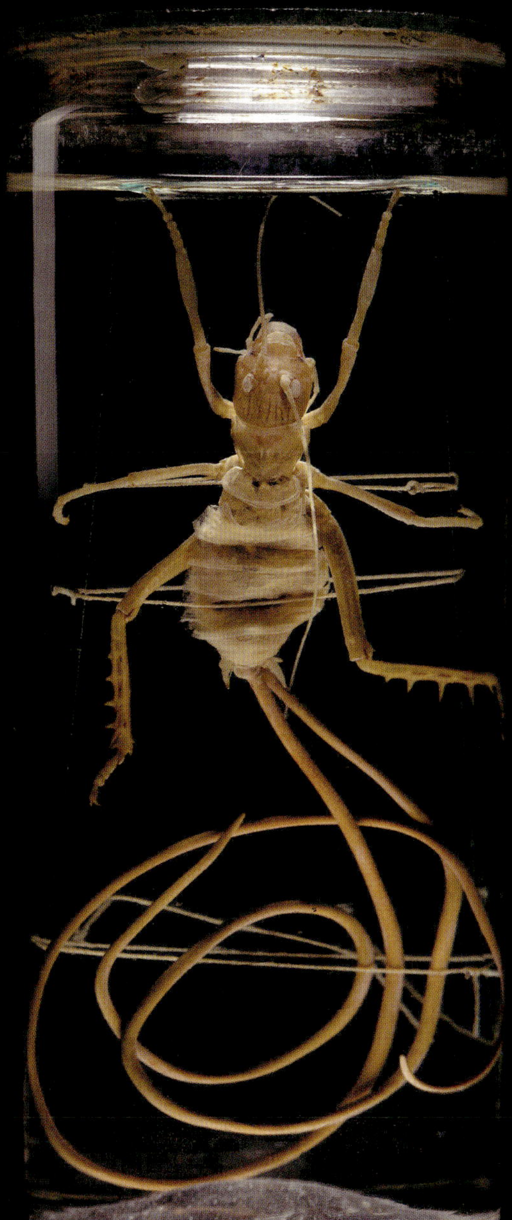

Gorse

Ulex europaeus

Te Papa herbarium staff collect native and adventive plant species from each district in which they grow. Taken together, these specimen records provide a full representation of what grows where in New Zealand, and show how the physical characteristics of the plants vary in time and space. New Zealand now has more adventive species than native plant species.

038 / 100

Collected by Te Papa Research Fellow Patrick Brownsey, this specimen of gorse (*Ulex europaeus*) is representative of a well-known adventive species. Gorse was introduced to New Zealand by early settlers in an effort to replicate the neat gorse hedgerows used instead of fences in England. Gorse grew vigorously in New Zealand's warmer climate, and is regarded as one of our earliest – perhaps our first – pest plant species, spreading quickly into open farmland with devastating economic impact.

Today, however, gorse also plays a positive ecological role in the regeneration of native New Zealand forest. If left undisturbed, gorse plants fix nitrogen in the soil, lay down a nurturing mulch layer, and shade regenerating native species, which eventually take over as the gorse ages. New Zealand botanist Hugh Wilson expertly demonstrated this process over decades of passive management at Hinewai Reserve on Banks Peninsula.

This gorse specimen is one of only seven currently in the Te Papa collection. We may still have some work to do if we are to provide a true representation of this species for current biodiversity studies and for the scientists and public of the future. AK

Gorse (*Ulex europaeus*), SP060703, collected by P Brownsey, South Karori Road, Karori, Wellington, 8 August 1978.

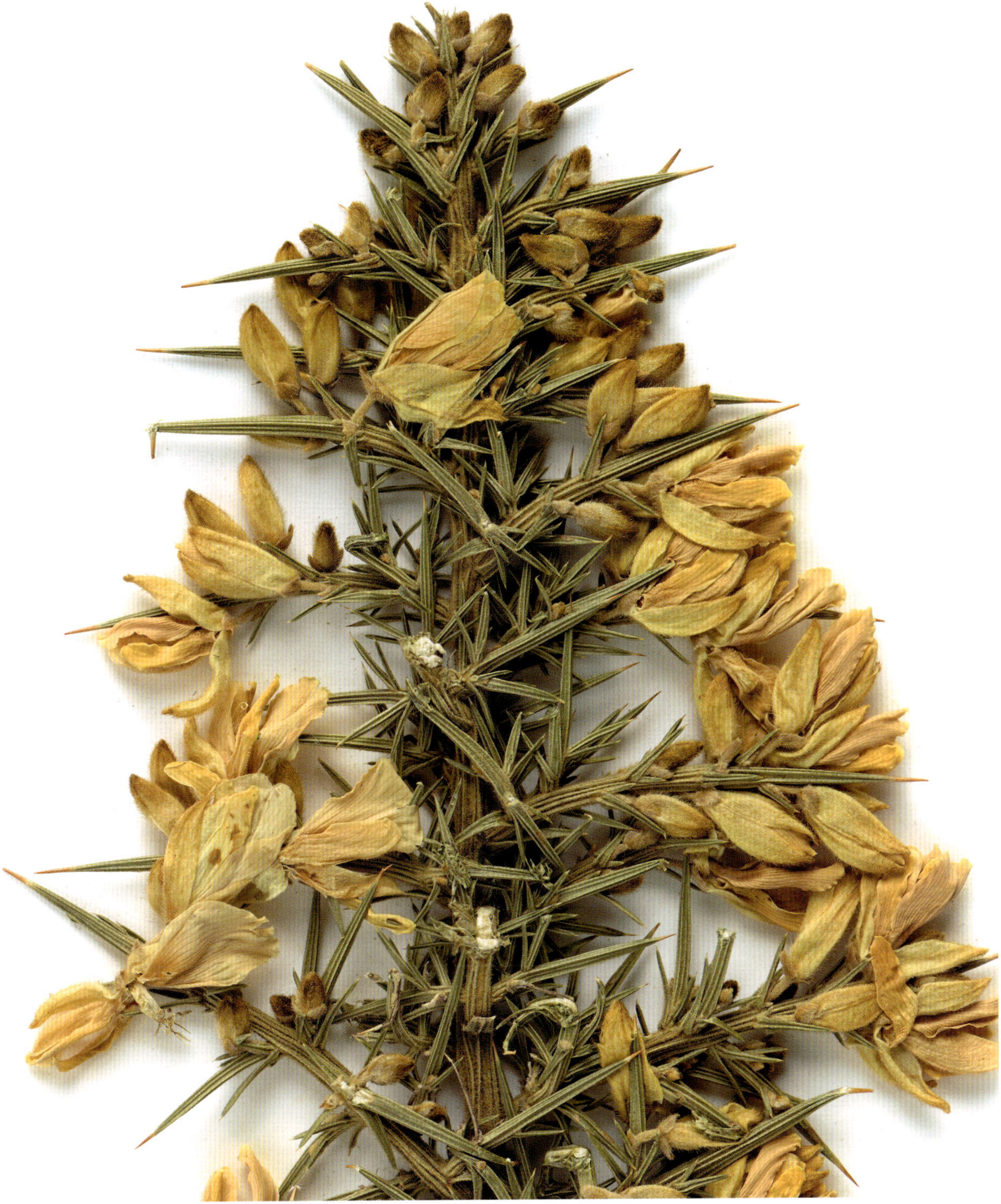

Great white shark jaw

Carcharodon carcharias

The great white shark (*Carcharodon carcharias*) – first described by Linnaeus in 1758 – is found in most temperate and tropical seas. In the New Zealand region it has been seen in shallow coastal waters, including large northern harbours, and offshore from the Kermadec Islands to Campbell Island. Depths range from the surface to more than 1200 metres. A highly migratory shark, juveniles and adults move from New Zealand to the Coral Sea and south-west Pacific during winter.

039

100

It is the largest predatory shark, with males growing to at least 5.5 metres long and females to 6.4 metres. Food includes ocean fishes, turtles, seabirds, squids and marine mammals. The great white shark has been protected in New Zealand waters under the Wildlife Act 1953 since 2007.

At Te Papa, we have developed the capability to preserve large species for our national fish collection. Temporary wooden fixing tanks lined with plastic sheeting, capable of holding over 2000 litres of fixative, along with permanent stainless-steel storage tanks holding up to 3300 litres of 50 percent alcohol, enable us to store scientifically valuable whole specimens for long periods of time.

In 2011 a shark was accidentally caught in a moki set net that had been left overnight because the sea was too rough to recover it safely. It was brought to Te Papa, and dissection of the fresh specimen (see page 255) was recorded on video and in photographs and is now used for education and public programmes. An accurate set of 91 key measurements were taken, along with samples of muscle tissue, liver and skin; external and internal parasites were collected; and stomach contents were identified – all to support a range of international research projects. CR + AS

Great white shark (*Carcharodon carcharias*) jaws, P.007384, collected by Mrs Kelly, Cook Strait region, July 1979.

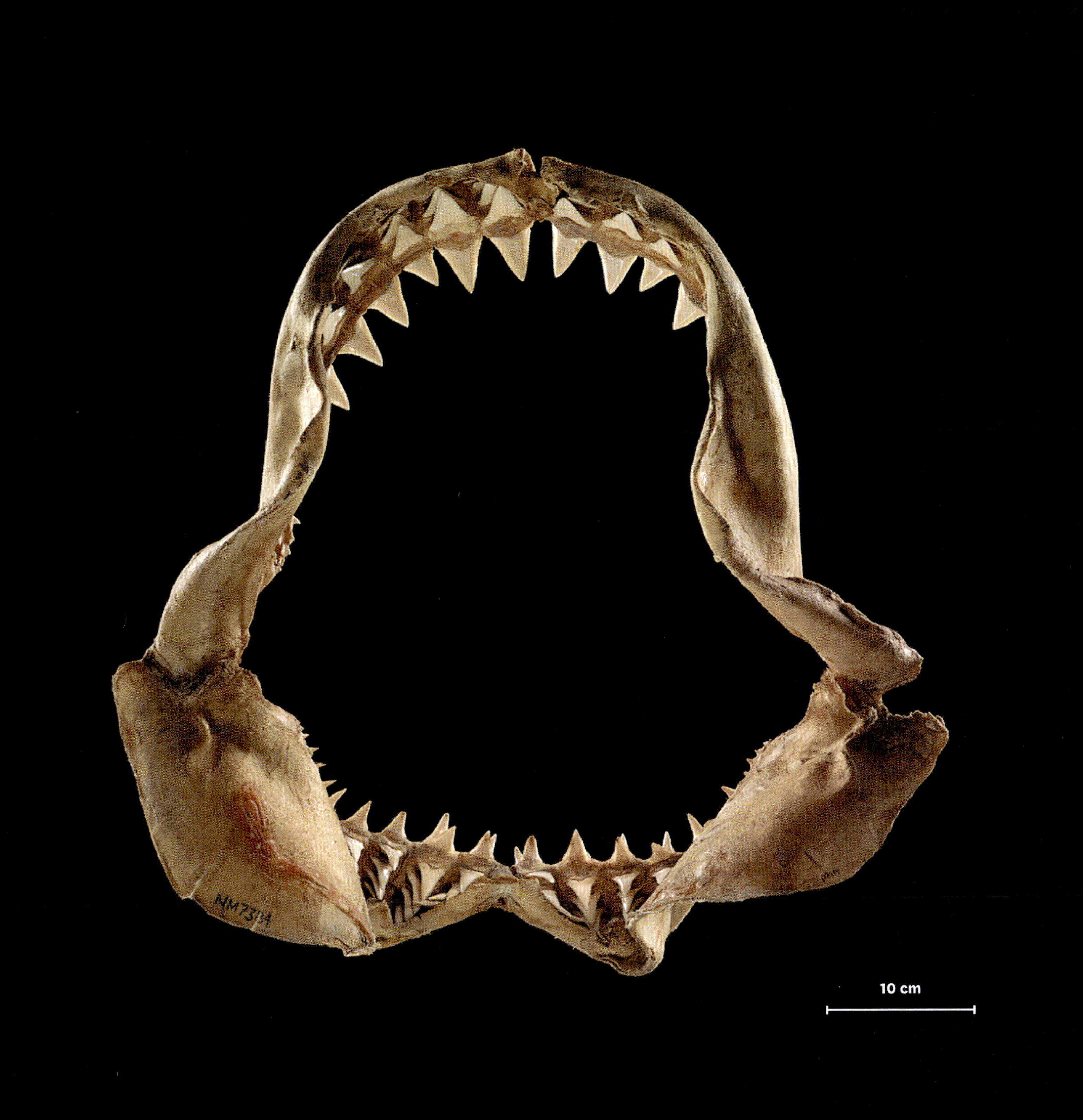
10 cm

Objects
H—M

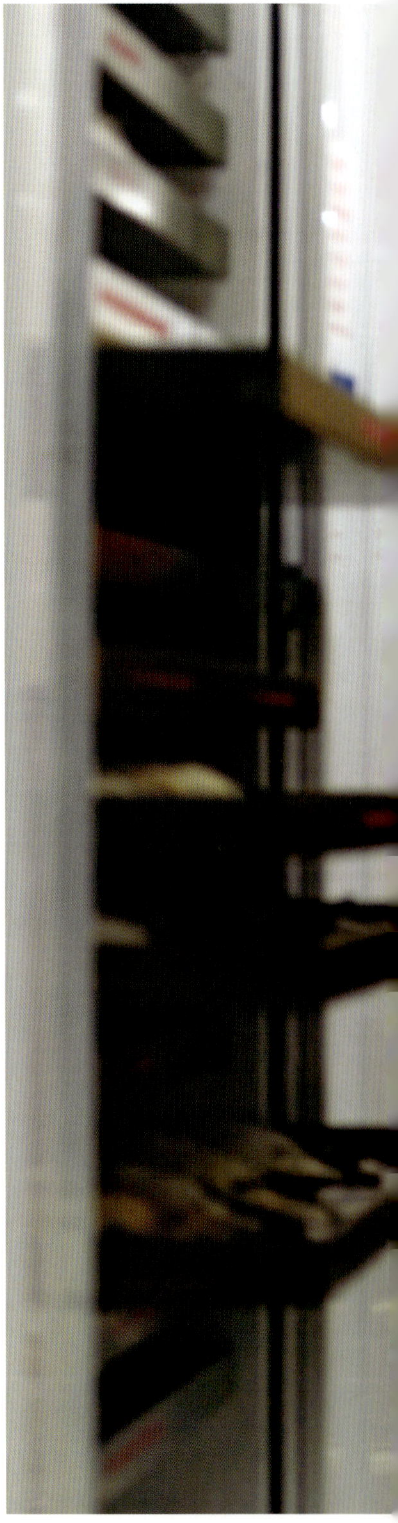

Curator Vertebrates Alan Tennyson, far left, Geneticist Lara Shepherd and Curator Vertebrates Colin Miskelly in the vertebrate storage area of Te Papa's Tory Street facility.

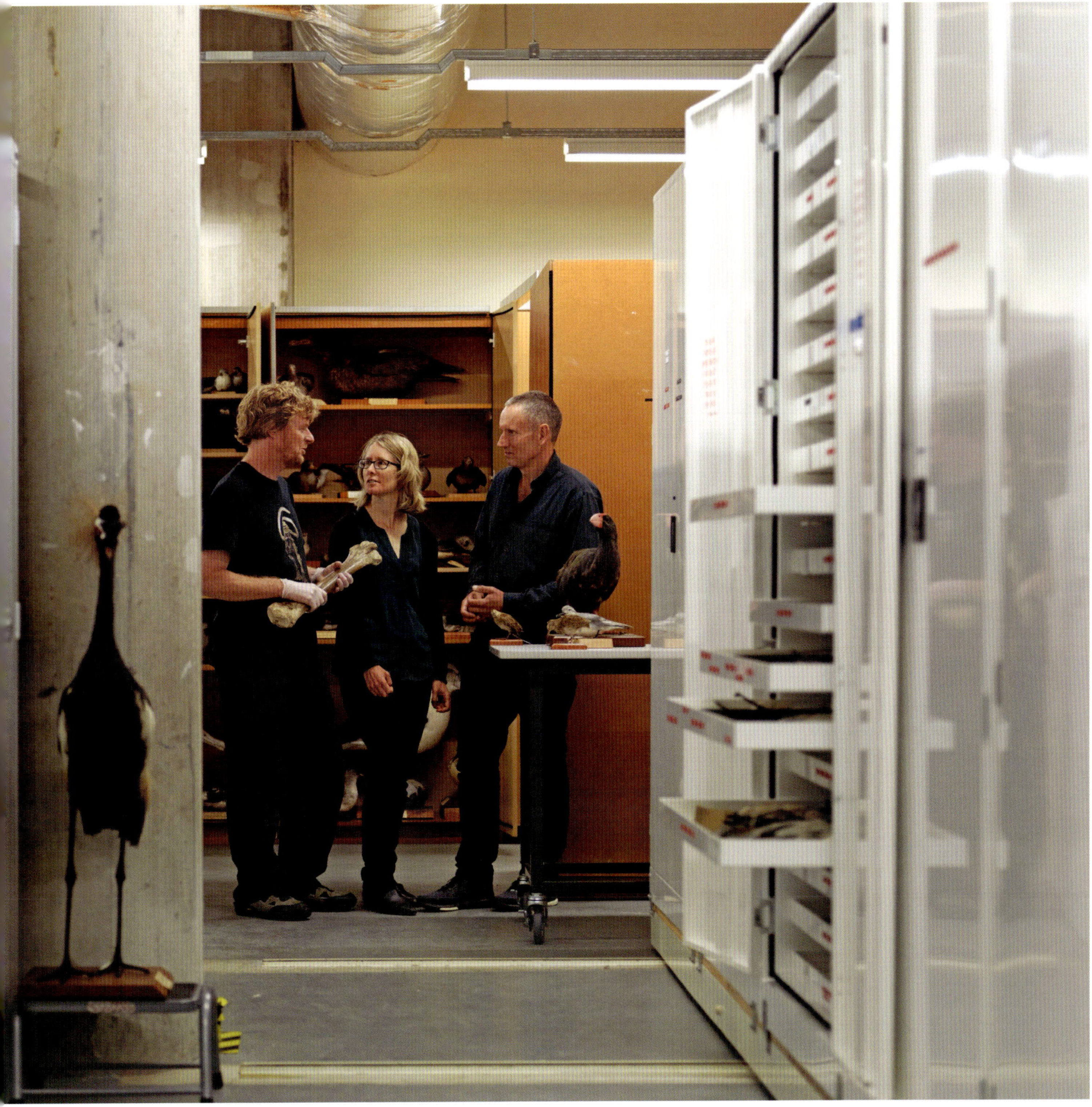

035 036 037

042 043 044

048 049 050

054 055 056

060 061 062

066 067 068

38 039 040 0

045 046 047

051 052 053

057 058 059

063 064 065

069 070 07

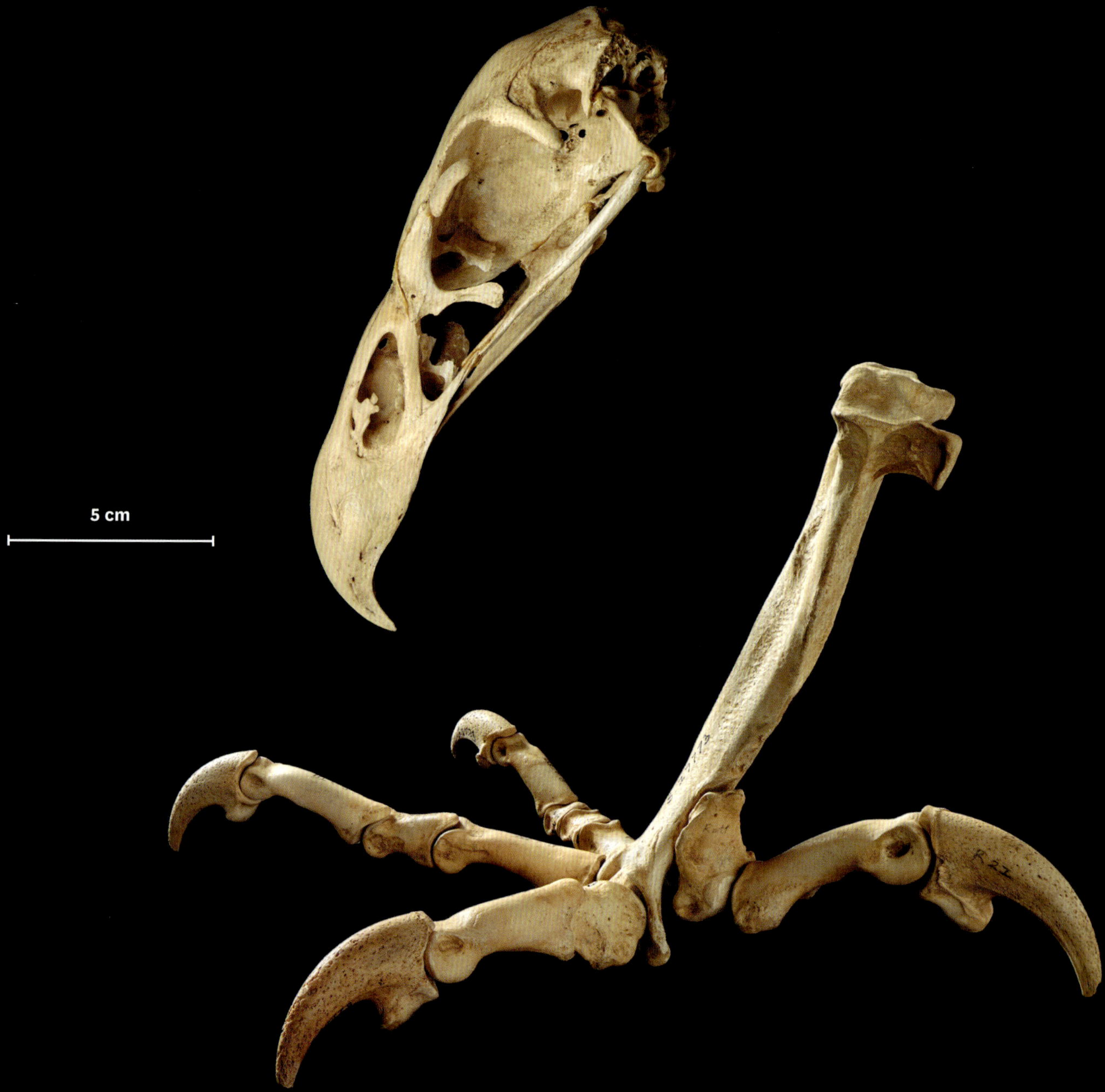
5 cm

Haast's eagle skull and foot

Aquila moorei

Eagles frequently symbolise courage and strength, and are often considered to be the kings of the skies. When Polynesians first arrived in Aotearoa New Zealand from East Polynesia, they would have been shocked and awed to find a giant aerial predator like nothing they had seen before.

No other eagle species in the world is or was as large as the species that inhabited this country. With a wingspan reaching 2.6 metres in the largest females, Haast's eagle was a giant that had evolved to tackle an even more enormous prey – moa. What we know about this eagle has mainly been gleaned from their rare fossil remains preserved in caves and swamps. We know that they attacked and fed on moa because holes made by the eagles' huge talons can still be seen in the preserved pelvic bones of moa. The eagle had also evolved an unusually long beak to reach into the carcasses of these giant prey.

Exactly when and why the eagle became extinct is unclear, but there is no solid evidence of it surviving into the European era. When moa and other large birds became extinct as a result of hunting pressure, the eagle became extinct too. Without its main prey the eagle was probably doomed, but its demise may have been hastened by more-direct human impacts, too. Māori legends of the Pouākai swooping down and carrying off adults and children may refer to attacks on humans by Haast's eagle – such terrifying behaviour would quite understandably have made it a target of human persecution. AT

Haast's eagle (*Aquila moorei*) skull and foot, S.27773, collected by P Millener, T Worthy and D Smith, Mt Owen, Kahurangi National Park, 14 January 1990.

Hagfish

Myxinidae family

'Yuck!' is a common response when first encountering a hagfish. But these primitive jawless marine animals are fascinating – and very successful. As 'proto-vertebrates', they share a common ancestor with all vertebrates. Hagfish fossils date back to the Cambrian, over 500 million years ago, and there are over eighty known species alive today.

Modern hagfishes were originally thought to be scavengers only. However, in 2011, Te Papa scientists and collaborators filmed, for the first time, hagfishes actively preying on live fishes. These videos also revealed how hagfishes fend off attacks by sharks and bony fishes by releasing a gill-clogging slime – a unique behaviour that had also not been observed before (to see these videos, search for 'hagfish' on Te Papa's YouTube page).

When attacked or stressed, specialised glands expel tiny pellets that swell massively on contact with sea water, instantly producing enormous amounts of slime. This is made even more effective by small balls of silk-like threads that unravel to become stiff strands more than 10 centimetres long, creating a persistent, choking mat.

All eight species are endemic, with some confined to surprisingly small areas. Recent Te Papa research revealed four new species, including the world's largest and heaviest hagfish, *Eptatretus goliath*, which grows to at least 1277 millimetres long and 6.2 kilograms in weight. It lives on the continental slope at depths of 811–1192 metres, only between North Cape and Bay of Plenty.

Hagfishes eat their prey using a long, muscular tongue that has two sets of horny, pointed teeth on it. By pushing the tongue back and forward, the teeth rasp a hole, allowing the hagfish to burrow into the flesh of its prey. CR + AS

Common hagfish (*Eptatretus cirrhatus*), head and everted jaws of freshly thawed specimen. Photo Carl Struthers, Te Papa.

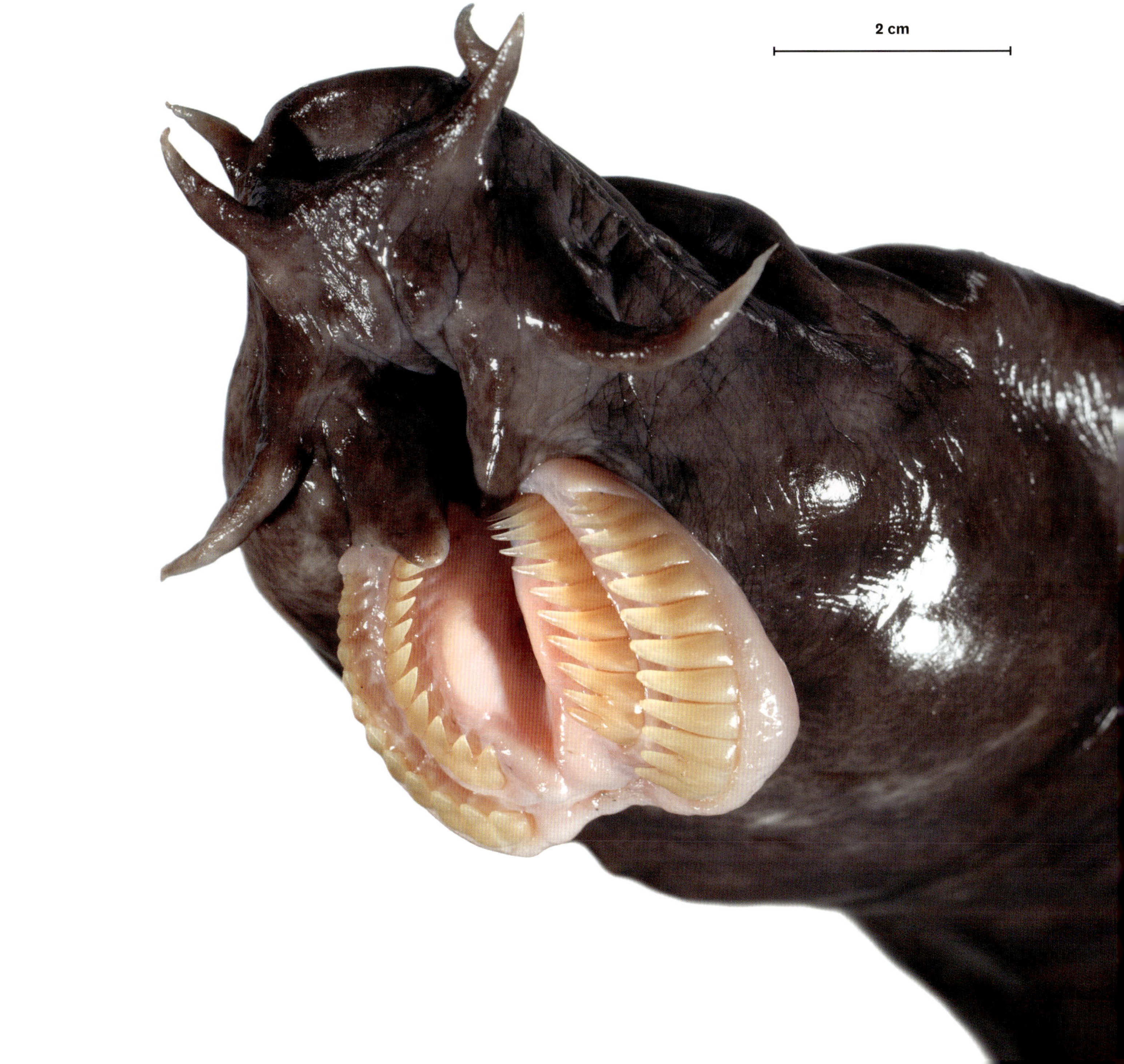
2 cm

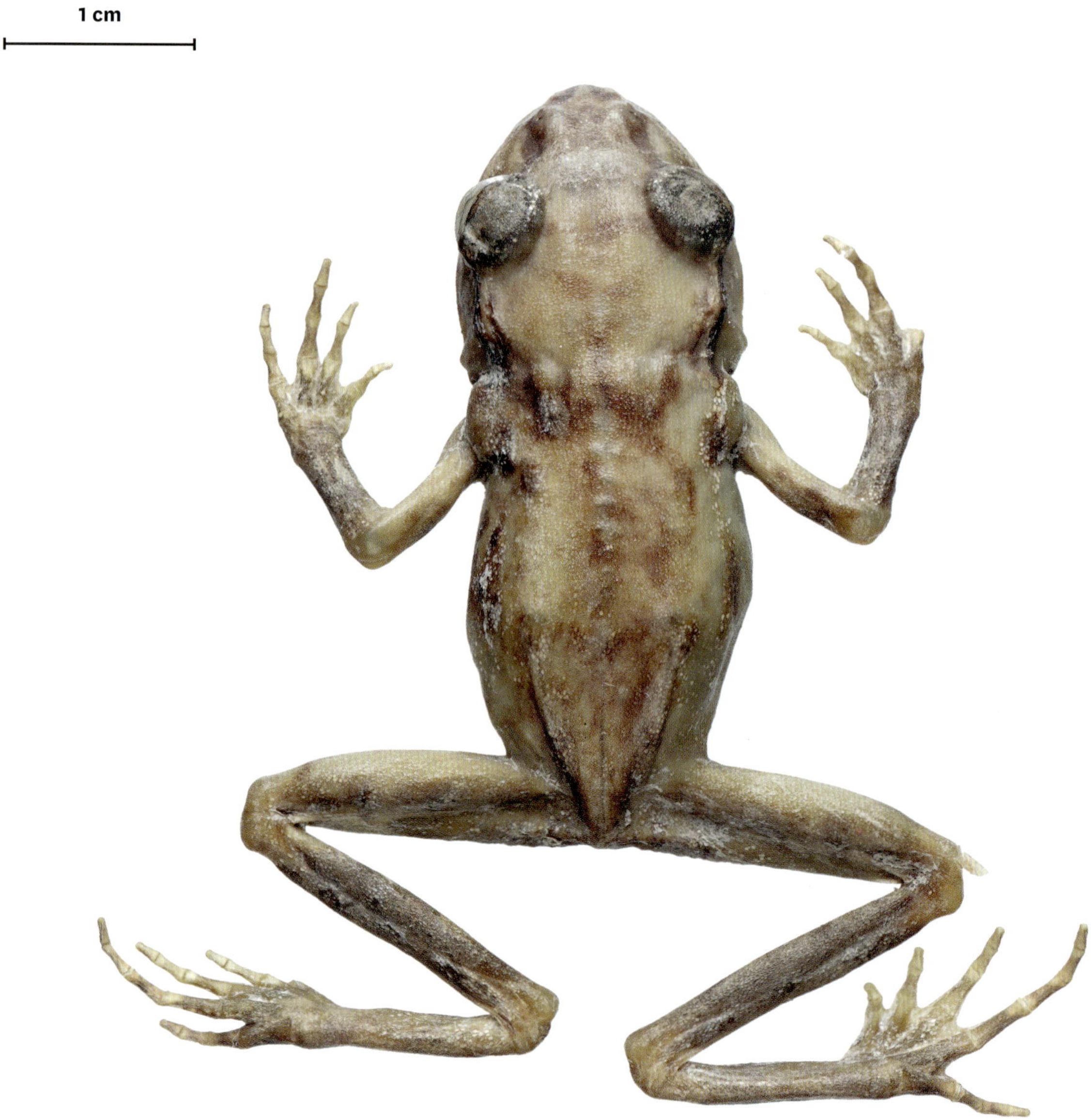
1 cm

Hamilton's frog

Leiopelma hamiltoni

Until 1915, New Zealand only had a single known species of native frog – Hochstetter's frog, named in 1861 and confined to forested areas in the northern North Island and Great Barrier Island. Then, in 1915, the Smith family went for a picnic on Takapourewa/Stephens Island, which marks the western entrance to Cook Strait. Tom Smith was an assistant lighthouse-keeper on the island, but it was one of his sons who made the momentous discovery. While moving rocks to make a seat, Richie Smith found a small frog. Realising that this was a notable find, the frog was sent to the Dominion Museum (now Te Papa).

Harold Hamilton, a zoologist on the staff of the museum, arranged to visit the island as soon as possible. Accompanied by amateur naturalist and collector Creagh O'Connor, he searched for frogs in the same small rock pile in September 1916. They found at least two frogs, but neither of them acted on their findings.

Further specimens were collected by Allan Thomson, Director of the Dominion Museum, in June 1918. Thomson forwarded two of the little frogs to Allan McCulloch at the Australian Museum, who named them as a new species. But somehow the history of their discovery became twisted, with McCulloch naming Harold Hamilton as the discoverer and honouring him in the species name (now *Leiopelma hamiltoni*).

The Smith family was offended that such a rare and notable species was not named after its rightful discoverer. Scientific names cannot be changed on these grounds, however, and it would be difficult to introduce the common name 'Smith's frog' after a century of this critically endangered species being known as Hamilton's frog. CM

Hamilton's frog (*Leiopelma hamiltoni*), AM.000012, length 39 mm, collected by H Hamilton, Takapourewa/Stephens Island, September 1916.

Hoodwinker sunfish

Mola tecta

Ocean sunfishes (family Molidae) are very large, iconic fishes found in oceans worldwide. For a long time, scientists believed that there were just two species – the oceanic sunfish (*Mola mola*) and the bump-head or short sunfish (*Mola alexandrini*). However, DNA and morphology analyses have uncovered a third, previously unknown species!

043/100

Sunfishes are rare in most museums because their size and habitat make them difficult to acquire, and they are even harder to preserve and store. Sunfishes are the heaviest of all bony fishes, with some weighing over 2 tonnes and growing to 3 metres in length. Through its networks, Te Papa has gathered one of the largest collections of *Mola* sunfish in the world – 15 whole specimens, ranging from 24 centimetres to 101 centimetres long. Most museum collections comprise only plaster casts and old mounted skins, or a very few small juveniles.

The new sunfish species evaded discovery for nearly three centuries, 'hiding' in a messy history of confused taxonomy. The extreme differences in appearance between juveniles and adults, coupled with a small number of specimens, make it difficult to accurately determine whether variation is occurring within a species rather than between two or more different species. Furthermore, the older casts and skins are poorly preserved and usually badly distorted, creating more confusion.

In 2017, the new species was named the hoodwinker sunfish, *Mola tecta* – from the Latin *tectus*, meaning 'disguised' or 'hidden'. The hoodwinker sunfish is the first addition to the genus *Mola* in 130 years. Despite its name, it is thought that this new species is similar in size and weight to the other two *Mola* species. CR + AS

Hoodwinker sunfish (*Mola tecta*), P.057679, holotype, total length 1012 mm, weight 52 kg, trawled by S Yeoman and FV *Ivan Golubets* from depth of 78 m, North Taranaki Bight, 25 December 2015.

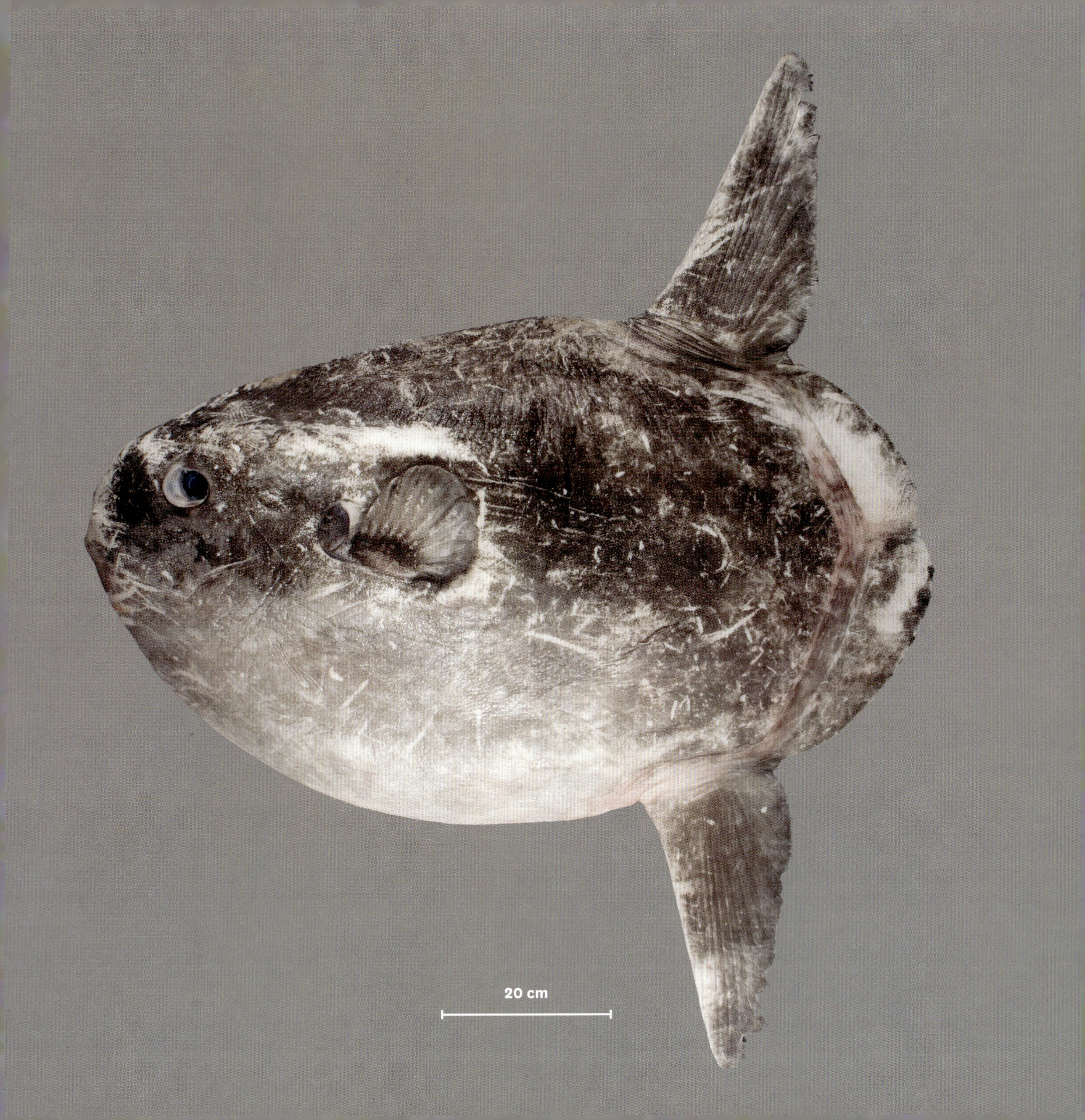
20 cm

Iguanodontid dinosaur tooth

Barilium dawsoni or *Iguanodon bernissartensis*

A truly remarkable specimen in Te Papa's collections is one of the first dinosaur remains ever recognised. In about 1820, Gideon Mantell and his wife Mary Ann obtained several large teeth in a quarry in Sussex, in southern England. Recognising that these teeth did not belong to any known animal, Gideon arranged for one to be examined by Baron Georges Cuvier, sometimes called the 'founding father of palaeontology'. Cuvier (incorrectly) concluded 'without hesitation' that the tooth was a rhinoceros incisor. However, by 1824 he had accepted that it was from a giant herbivorous reptile. A card still attached to the tooth bears annotations from both Gideon Mantell and influential geologist Charles Lyall that detail this important historical moment.

In 1825, Mantell concluded that the teeth were from a new kind of enormous reptile, which he called *Iguanodon* because the teeth bore some resemblance to those of living iguanas. It was not until 1842 that the famous anatomist Richard Owen coined the term 'dinosaur' for the group of huge, extinct reptiles that included *Iguanodon*. The identification of dinosaurs and other giant extinct animals firmly introduced the concept of extinction to the world and challenged the biblical view of the origins of life.

In 1859, the tooth was taken to New Zealand by Gideon and Mary Ann's son Walter. The specific identity of the tooth is a little uncertain, but in 2018 Cambridge-based iguanodontian expert David Norman suggested that it is most likely to belong to *Barilium dawsoni* or possibly *Iguanodon bernissartensis*. These are both large, robust, closely related herbivores that would have been between 8 and 13 metres long. AT

Iguanodontid fossil tooth, GH004839, acquired from W Mantell, 1930; collected Sussex, England, about 1820.

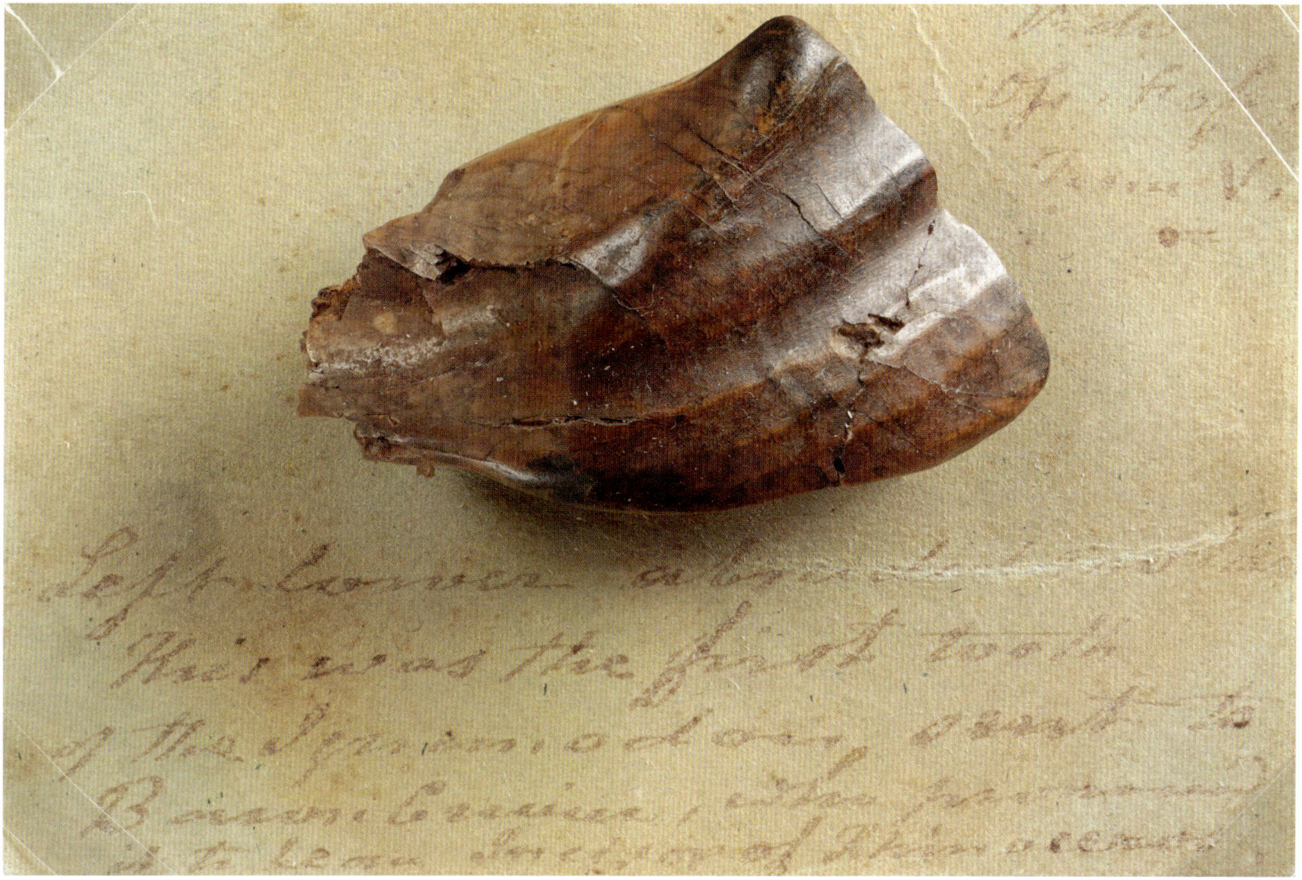
This was the first tooth

045 / 100

The Indigenous Grasses of New Zealand

John Buchanan

New Zealanders are often praised for their inventiveness and 'number eight wire' mentality. Such resourcefulness was demonstrated by John Buchanan, the first botanist and draughtsman of the Colonial Museum, when he was tasked with writing and illustrating a Flora of native New Zealand grasses.

Buchanan began his working life as a pattern designer in Scotland's calico textile industry, where he skilfully used patterns from plants to inspire new textile designs. Upon his arrival in New Zealand in 1852, he was thus perhaps naturally drawn into botany. Buchanan's combined draughting and botanical skills eventually led to his employment at the Colonial Museum in Wellington.

In 1876, the New Zealand House of Representatives approved the funding of *The Indigenous Grasses of New Zealand*. Nature printing was to be used to provide an accurate scientific record and practical guide for New Zealand farmers. Developed in Europe in 1853, this technique involved sandwiching each plant specimen between a steel plate and a lead plate and passing this between two rollers. A cast was then made from the imprint in the lead plate and used to print from.

Suitable materials were lacking in the colony, so Buchanan had to improvise. Once he had collected, pressed and dried his specimens, he imprinted these onto a lithographic stone, adding finer details by hand. The stone was printed onto paper through a press to produce a black image on a white background. Buchanan printed sixty-four plates in total.

The Indigenous Grasses of New Zealand became the most significant publication of nature printing in nineteenth-century New Zealand, winning Buchanan an award at the Melbourne Exhibition 1881 for excellence in lithographic technique. AK

Left: Kowhangatara/spinifex (*Spinifex sericeus*), female, SP059549. Part of an album of specimens used for the publication *The Indigenous Grasses of New Zealand*.
Right: Kowhangatara/spinifex (*Spinifex sericeus*), female. Plate IX, John Buchanan, *The Indigenous Grasses of New Zealand*, Government Printer, Wellington, 1880.

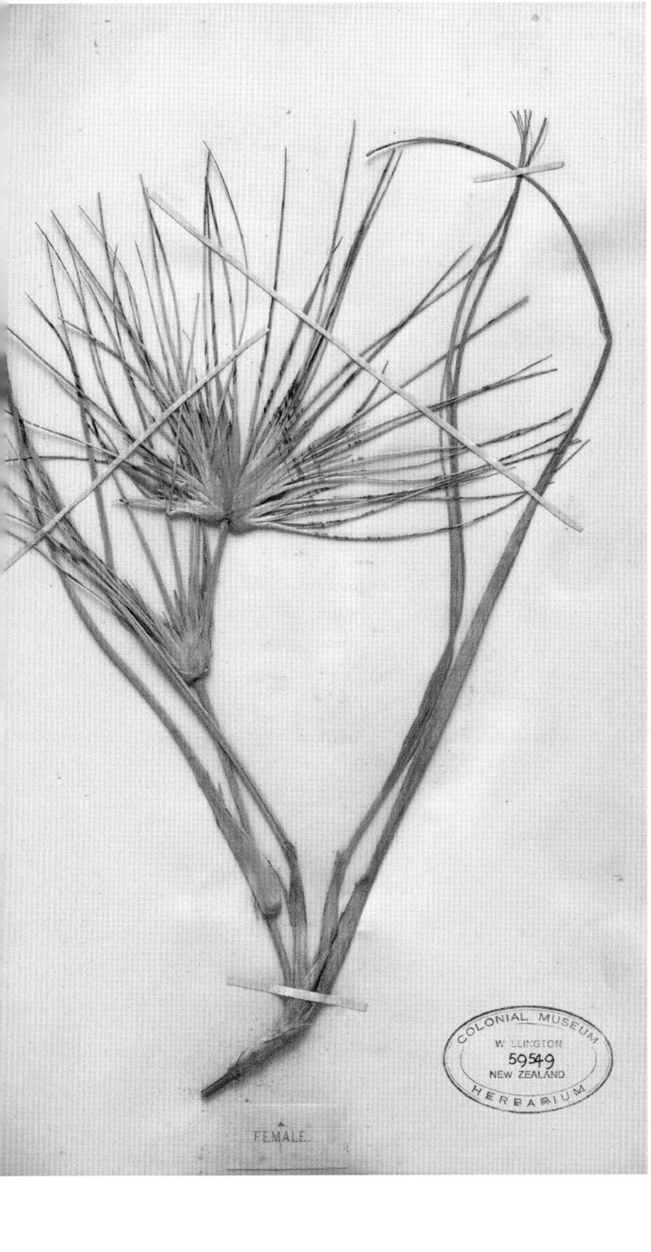

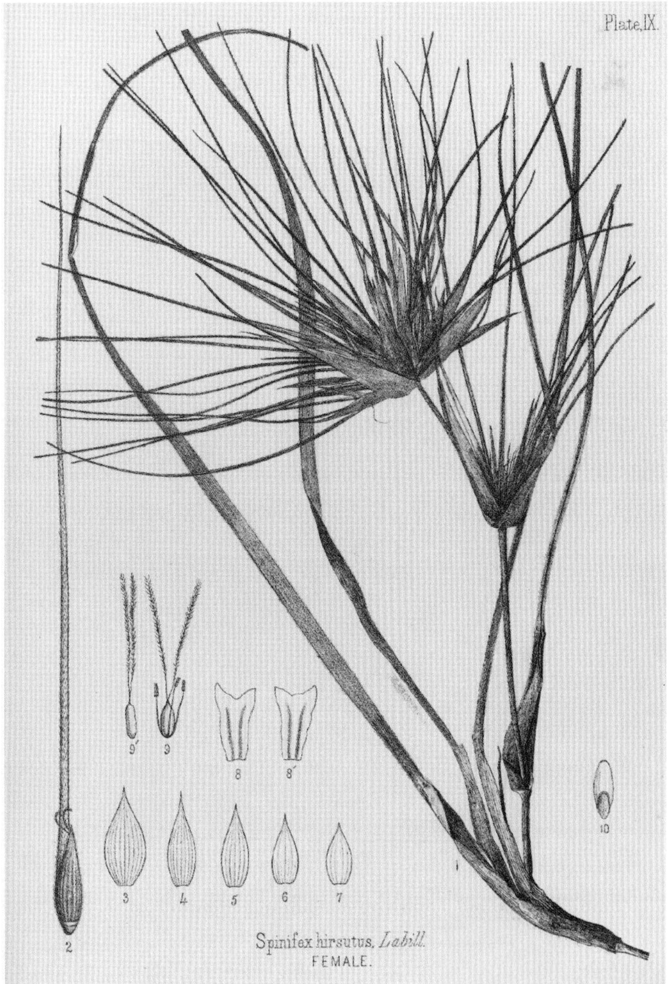

Spinifex hirsutus, *Labill.*
FEMALE.

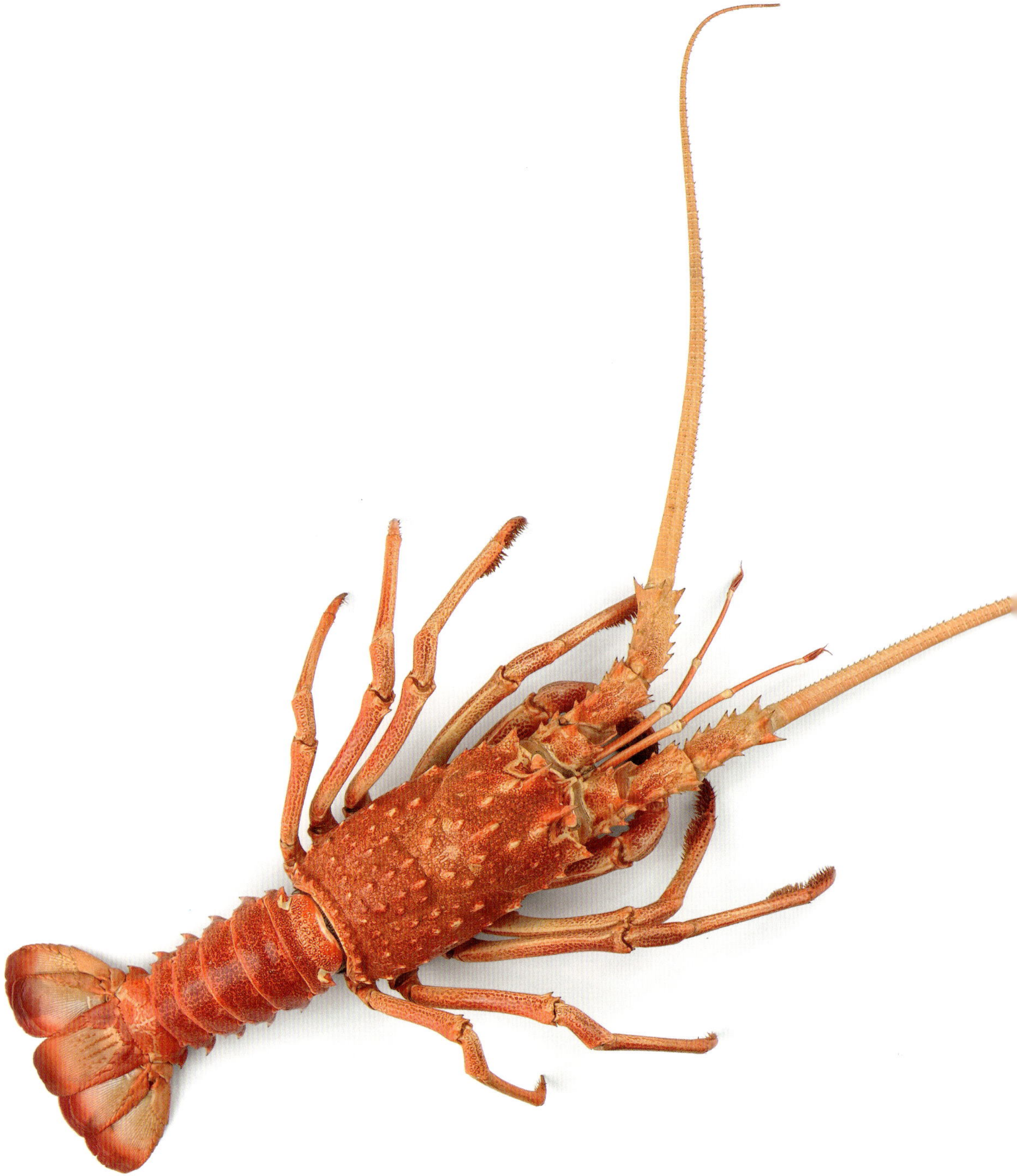

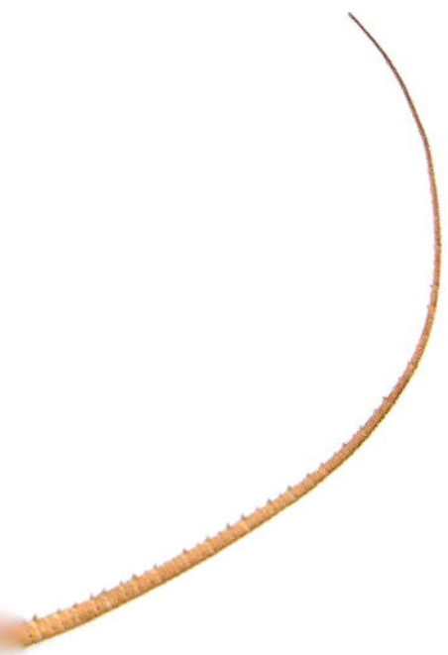

Joe Cave's lobster

Jasus caveorum

In 1995, Rakiura/Stewart Island lobster fisher Joe Cave sent his lobster-potting vessel on a two-month expedition to the east Pacific Ocean, to look for spiny lobsters at the underwater Foundation Seamounts. It paid off, and the vessel returned with several thousand frozen lobsters aboard. The best market for Joe's lobsters was America, but US Customs required an accurate taxonomic name for any imported marine species. When Joe sent us specimens for identification, it was clear that this was a new, unnamed species. John Booth of NIWA and I immediately set about publishing a name and description. 046 / 100

Named after Joe, Helen and Ernie Cave, *Jasus caveorum* is one of six species in the southern hemisphere genus *Jasus*, which includes our kōura (red rock lobster), *Jasus edwardsii*. Quite a lot is known about larval development in these other *Jasus* species. After hatching, the tiny, leaf-like larvae drift offshore to live and grow for one to two years in permanent oceanic eddies. These eddies eventually return the larvae close to shore, where they metamorphose into little lobster-like post-larvae, swim to the coast and moult into juvenile lobsters.

A distinct difference between Joe Cave's lobster and the other *Jasus* species is that it lives only on underwater seamounts, at depths of 140–180 metres and 2100 kilometres from the nearest land, Pitcairn Island. All the others live near land, which is important for the formation of the eddies used by the larvae. How, then, does *J. caveorum* maintain its location and not lose its larvae? Discovering the mechanism – perhaps currents, or modified larval duration or behaviour – will require a research trip to the seamounts to collect larvae, or lobsters from which to breed. RW

Jasus caveorum, CR.009966, body length 300 mm (from eyes to tip of tail), collected by M Wells in a lobster pot at depth of 100–200 m, Foundation Seamounts, south-eastern South Pacific, 1997.

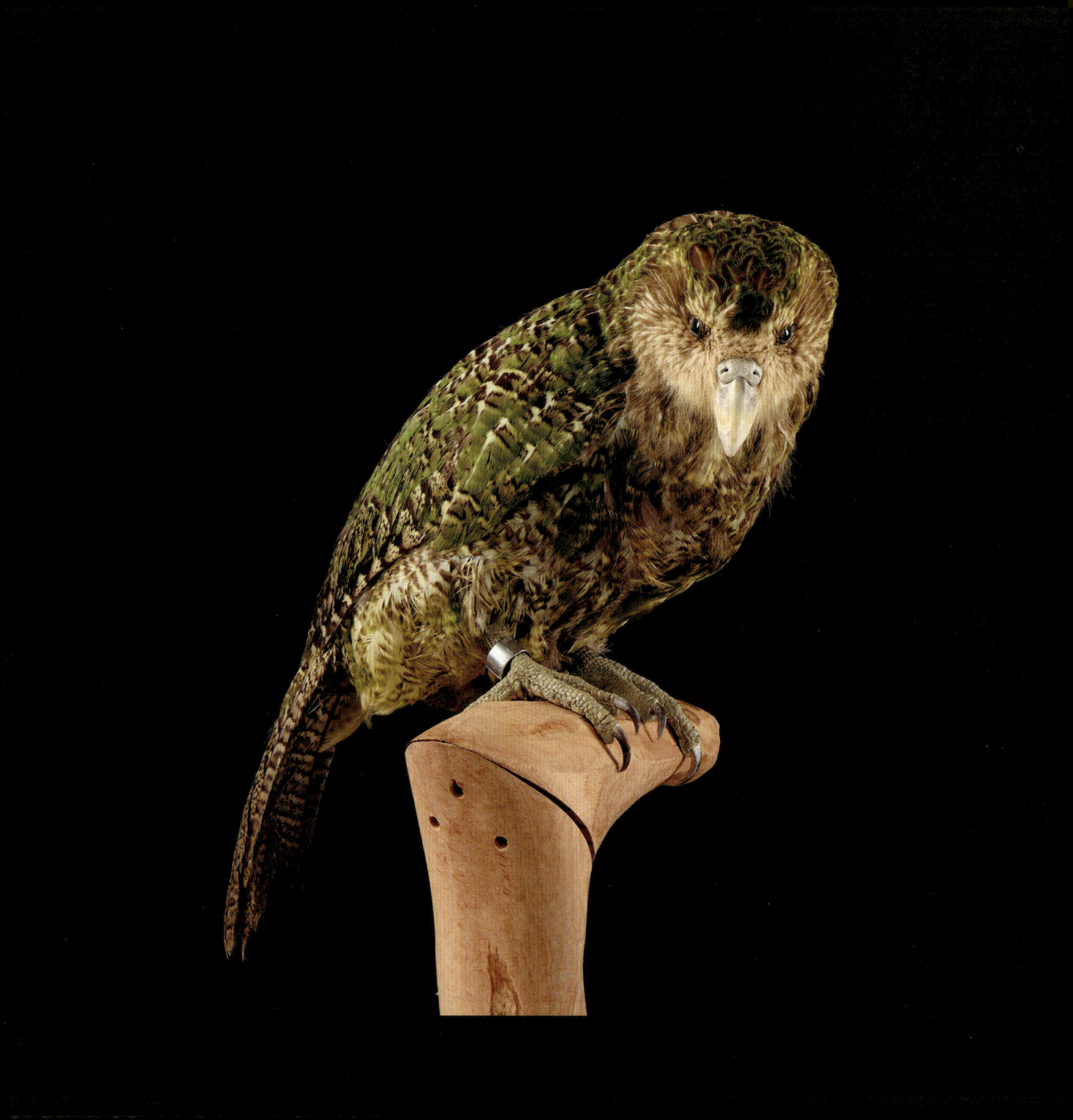

Kākāpō

Strigops habroptilus

In a land of weird, wacky and wonderful animals, the kākāpō is the undoubted rock star: a bird so rare and charismatic that each individual is named and births and deaths make national news. It is a bird of superlatives – the world's largest parrot, the only flightless parrot, the only nocturnal parrot – and it has taken a superlative effort to save it from extinction.

For much of the twentieth century, the kākāpō was considered so close to extinction to be possibly beyond reprieve. After mating with males at their display grounds on the flanks of some of Fiordland's highest peaks, female kākāpō undertook all nesting duties unaided, which exposed them to predatory stoats during the three long months of incubation and chick-rearing. Expeditions to the bird's last known stronghold in northern Fiordland between 1958 and 1976 found only a handful of males and no females.

Then kākāpō were thrown a life-line, with the discovery of birds in remote, untracked southern Stewart Island in 1977; within a few years, both females and nests were found. But all was not well: feral cats in this apparently pristine wilderness were slaughtering the kākāpō, and intensive efforts to control the cats failed. The only option was to catch every last bird and move them to islands free of predators.

Rhys was one of the last kākāpō rescued from Stewart Island, in 1992, and was released on Mana Island in May of that year. He thrived on his new island home for 14 months before apparently succumbing to organophosphate poisoning. Since then, only organically grown food has been used in the kākāpō supplementary feeding programme. CM

Kākāpō (*Strigops habroptilus*), OR.025019, length 460 mm, collected by D Merton, Mana Island, Wellington, July 1993.

The King Tāwhiao fern collection

Charles Jeffs

By 1888, the second Māori king, Tāwhiao, had become aware of the botanist Charles Jeffs, who was winning awards for his presentations of pressed plants. Tāwhiao commissioned Jeffs to collect twenty specimens of fern with important medicinal properties to be gifted to Dr JT Rennie, a Canadian physician working in New Zealand. On each sheet, Jeffs recorded the fern's usage in traditional Māori healing (rongoā Māori) alongside its name, and on a cover sheet he recorded 'in recognition of services rendered to the afflicted Maories [sic]'.

The 1880s were a low point in the history of the Tainui people. Two decades had passed since the Waikato wars, and Tainui were destitute, being mostly disassociated from their land – a grievance that was not addressed until 1995, when Tainui became the first iwi to reach a Treaty of Waitangi settlement with the Crown for the suffering, distress and deprivation caused by the wars and the land confiscations that followed. Diseases such as influenza and measles, formerly unknown in Aotearoa but introduced by European immigrants, were sweeping through Māori communities throughout the country. Rennie must have shown dedicated service in treating the people of Tainui to earn such appreciation from the King.

In the 1960s, a small tin box was discovered in an attic in Ontario, Canada. Inside were Jeffs' twenty fern specimens, well preserved. Today, the specimens, tin box and cover sheet are some of the most admired objects in Te Papa's botany collection. They are both a powerful link to a significant part of our history and a remarkable resource as a record of Tainui rongoā. AK

Mātātā/lace fern (*Paesia scaberula*), P020070/N, length 270 mm, width 170 mm, collected by C Jeffs, North Island, 1888.

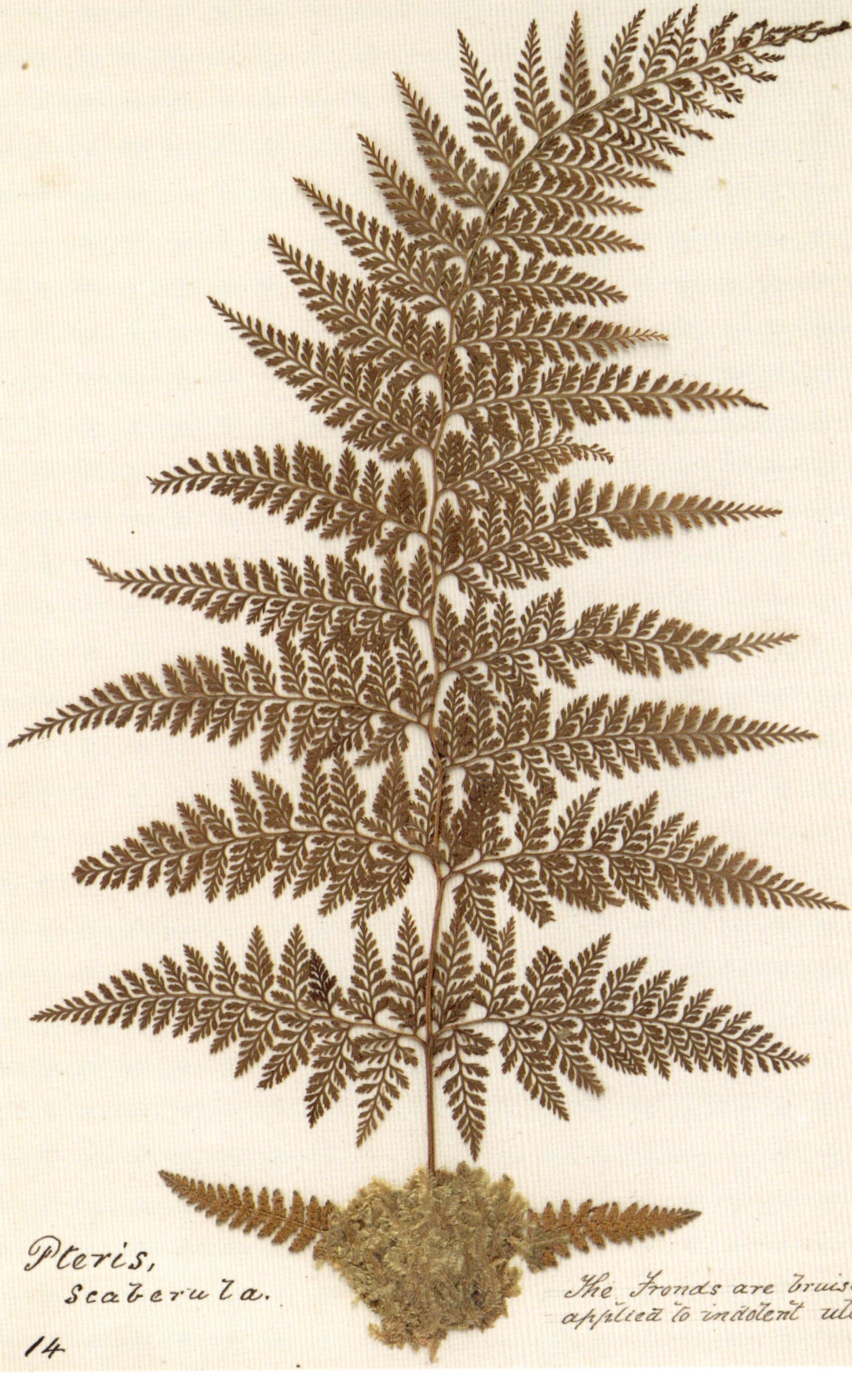

Pteris,
Scaberula.

The Fronds are bruised and
applied to indolent ulcers

14

049
100

Land hopper

Arcitalitrus dorrieni

This photo was a lucky shot – live land hoppers are difficult to photograph because when exposed to light they go into a frantic hop, dig and escape mode. Measuring about 9 millimetres on average, land hoppers are crustaceans and are closely related to the sand hoppers that we see on our beaches. Land hoppers occur mainly on continents and islands of the Indian and South Pacific oceans and are particularly common in New Zealand, Australia and New Guinea.

In New Zealand, land hoppers live, often in great numbers, among the fallen leaves of forests, alpine grasses, farms and gardens, eating the damp decaying leaves just above the soil. Here they also find shelter from light, and moisture to keep their gills functioning. Aotearoa has over forty species, of which just three are introduced, including *Arcitalitrus dorrieni*. The species are very similar to one another and difficult to distinguish. To make it even more difficult, land hoppers hatch from eggs in the mother's brood pouch as tiny adult replicas and grow to adulthood without obvious changes. Recent research using morphology and DNA analysis has revealed six species in one genus at the northern tip of Northland where there were thought to be three. There are likely to be many more undetected species.

Little biological research has been done on land hoppers in New Zealand, but it is obvious from local and overseas observations that they are major processors of leaf litter. One New Zealand species has been found at a density of 2670 hoppers per square metre. Land hoppers are also an important food for various native birds; for example, they were found in more than 90 percent of faecal samples collected from Snares Island snipe (*Coenocorypha huegeli*). RW

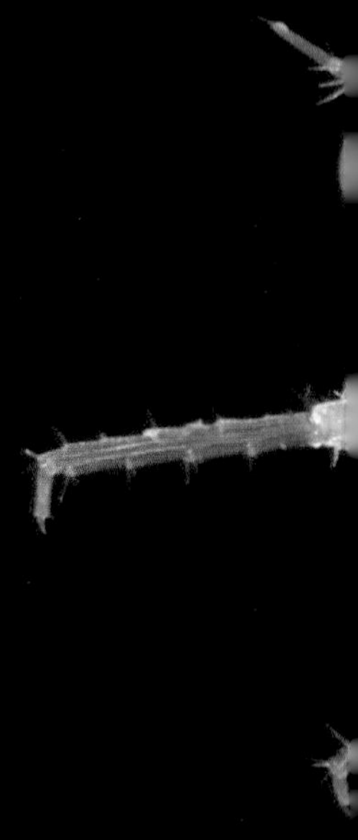

Land hopper (*Arcitalitrus dorrieni*), CR.025167, length 8.5 mm (front of body to tip of telson), collected by J-C Stahl from leaf litter, Central Park, Wellington, 16 February 2018.

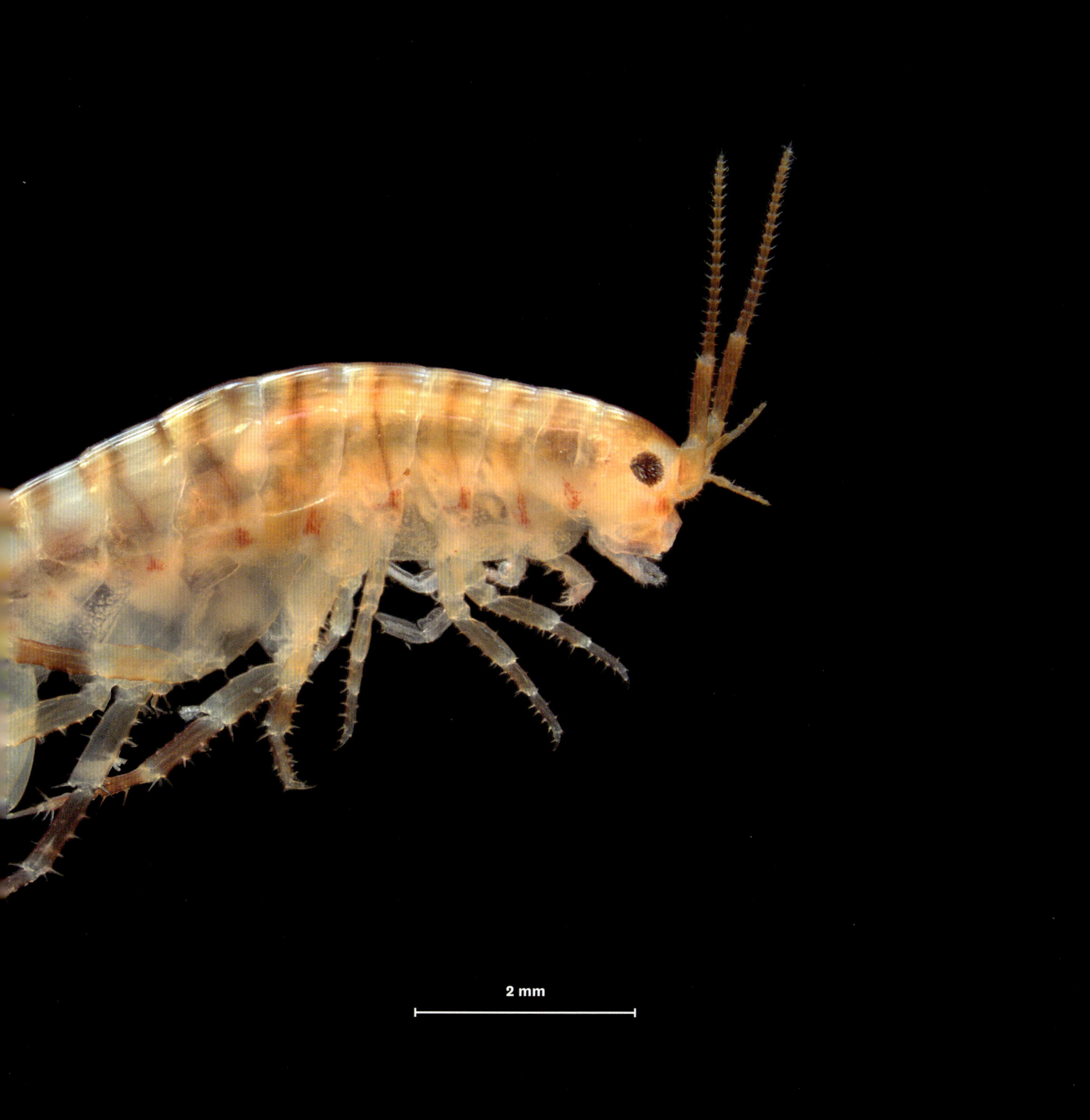
2 mm

Laughing owl

Ninox albifacies

In August 2015, I received a most unexpected phone call: Would Te Papa be interested in purchasing a specimen of a laughing owl? How could this be? This large owl became extinct in about 1914, and only 25 mounted specimens were known worldwide. Even more
050 surprisingly, the caller also offered an egg. Again, there were only
100 19 laughing owl eggs known to exist, and we thought we knew where every one of them was held.

As with most of New Zealand's extinct birds, there are more laughing owl specimens held overseas than locally. The last stronghold of the laughing owl was South Canterbury. As a result, Canterbury Museum holds the most specimens of any museum within New Zealand, with four mounted birds, six study skins and eleven eggs. Up until 2015, Te Papa held a single mount, one study skin and two eggs.

Had the specimens on offer been correctly identified? Was it a hoax? Had they been stolen? There was one person uniquely placed to answer the first two questions, and fortunately he lived within an easy drive of the caller. Noel Hyde does most of Te Papa's taxidermy, and had previously created two realistic laughing owl models for our *Blood Earth Fire* exhibition. Noel was able to confirm that the specimens were indeed genuine and in good condition. As all the other background checks came back positive, we were able to go ahead and buy them.

Unfortunately, neither bird nor egg has any provenance data. But now that they are in a public research collection, forensic techniques could be used to investigate where and when they might have been collected. CM

Laughing owl (*Ninox albifacies*), OR.030061, height 330 mm, no collection data.

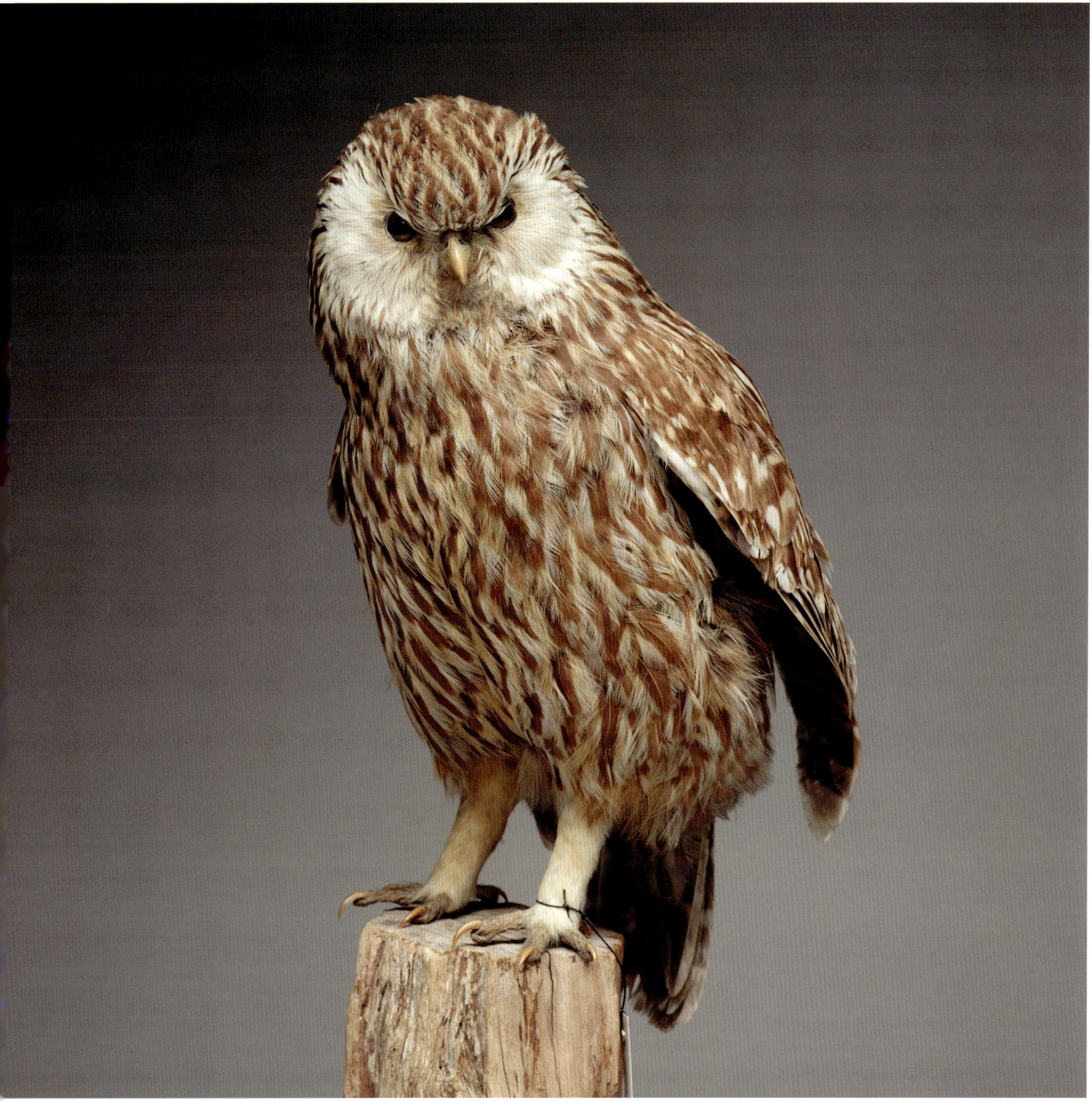

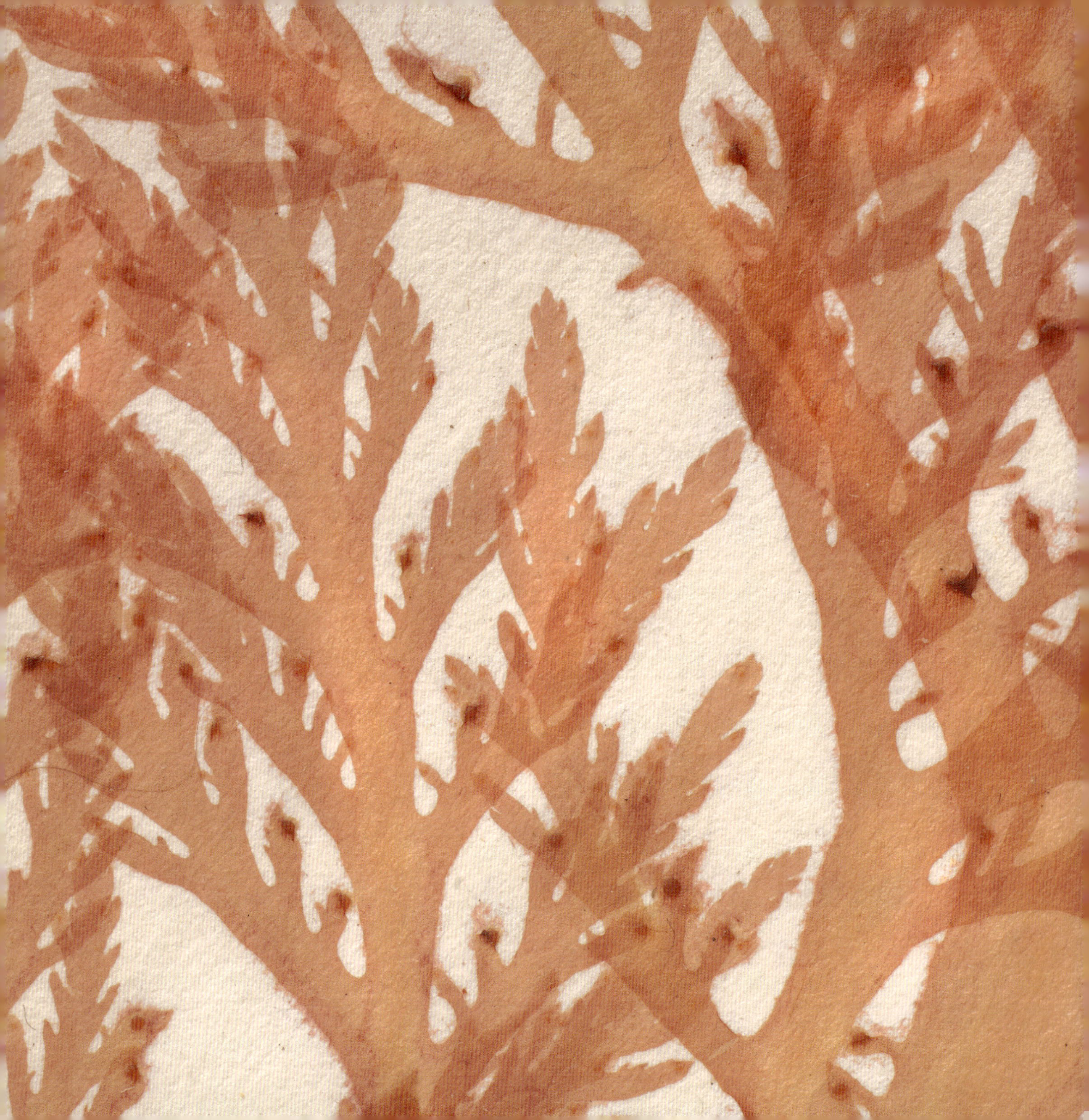

Lindauer's seaweeds

Ptilonia willania

From the time of Cook's first expedition, in 1769–70, specimens of New Zealand plants have been taken overseas for scientific study. Until the twentieth century, it was botanists in Europe who knew the most about the Linnaean classification of Aotearoa's plants. Even today, overseas expertise makes valuable ongoing contributions to the scientific naming and classification of New Zealand's biodiversity.

The specimen shown here is a red seaweed, *Ptilonia willania*, and number 218 in *The Algae Nova-Zelandicae Exsiccatae*, assembled between 1939 and 1953 by Victor Lindauer, son of famous painter Gottfried Lindauer. An exsiccatae is a set of dried plant specimens.

Victor Lindauer's exsiccatae of 350 algae – mostly seaweeds, plus some terrestrial and freshwater algae – comprised fourteen fascicles (instalments), with twenty-five sheets in each fascicle. About sixty copies of the exsiccatae were made – that's some 21,000 specimens, which is a lot of collecting, pressing, drying, mounting and labelling of algae! This massive enterprise apparently involved the whole family, and even intruded into the children's school lunchtimes. Most sets were seemingly sent overseas.

While the production of exsiccatae was fairly common overseas, few were prepared in New Zealand. The transfer of specimens overseas generally occurred on a smaller scale and was less systematic, such as a single set of specimens sent to a particular expert or institution. Lindauer's exsiccatae is acknowledged as a valuable reference set for subsequent study both abroad and in New Zealand. In recent years, I have overseen a couple of donations of partial sets of Lindauer's exsiccatae. Te Papa now has duplicates of most of the exsiccatae's specimens, increasing our flexibility in using them for both research and exhibition. LP

***Ptilonia willana*, A033372, collected by Eileen Willa, Stewart Island, 16 October 1945.**

Little spotted kiwi

Apteryx owenii

Sometimes, individual specimens can have a significant impact on our understanding of a species' history. As part of my PhD examining the past distribution of kiwi species, I identified three specimens that settled a scientific debate about when little spotted kiwi disappeared from the South Island.

Today, there are only around 1500 little spotted kiwi, and these are restricted to predator-free offshore islands and sanctuaries. However, deposits of bones, for example in caves, indicate that they were formerly widespread throughout New Zealand. By the time of European settlement, little spotted kiwi had largely disappeared from the North Island but were still common on the West Coast of the South Island. When exactly they went extinct in the South Island was unclear; confusion with their relative the great spotted kiwi added to the uncertainty. It was widely reported that South Island little spotted kiwi went extinct in the 1930s, but other researchers disagreed and believed that it survived on the West Coast for much longer.

052

100

The skeleton shown here is from a kiwi hit by a car in north-west Nelson in 1978. Using DNA, we were able to show that this specimen, as well as the two skins shown here that were collected in 1952 from central Westland and 1978 from south Westland, were indeed little spotted kiwi (as opposed to juvenile great spotted kiwi). This suggests that little spotted kiwi had indeed survived, and were widespread, in the South Island until much more recently than generally accepted. This result demonstrates how little we know about our native species, even the prominent ones like our (unofficial) national bird, the kiwi. LS

Little spotted kiwi *(Apteryx owenii)*: (left) OR.001174, collected by G Broughton, Rough River, Westland, 16 October 1952. **(middle)** OR.023043, collected by D Murray, Smyth River, South Westland, 3 August 1978. **(right)** OR.023036, collected by Mrs Wilson, Westhaven Inlet, Northwest Nelson, 25 July 1978.

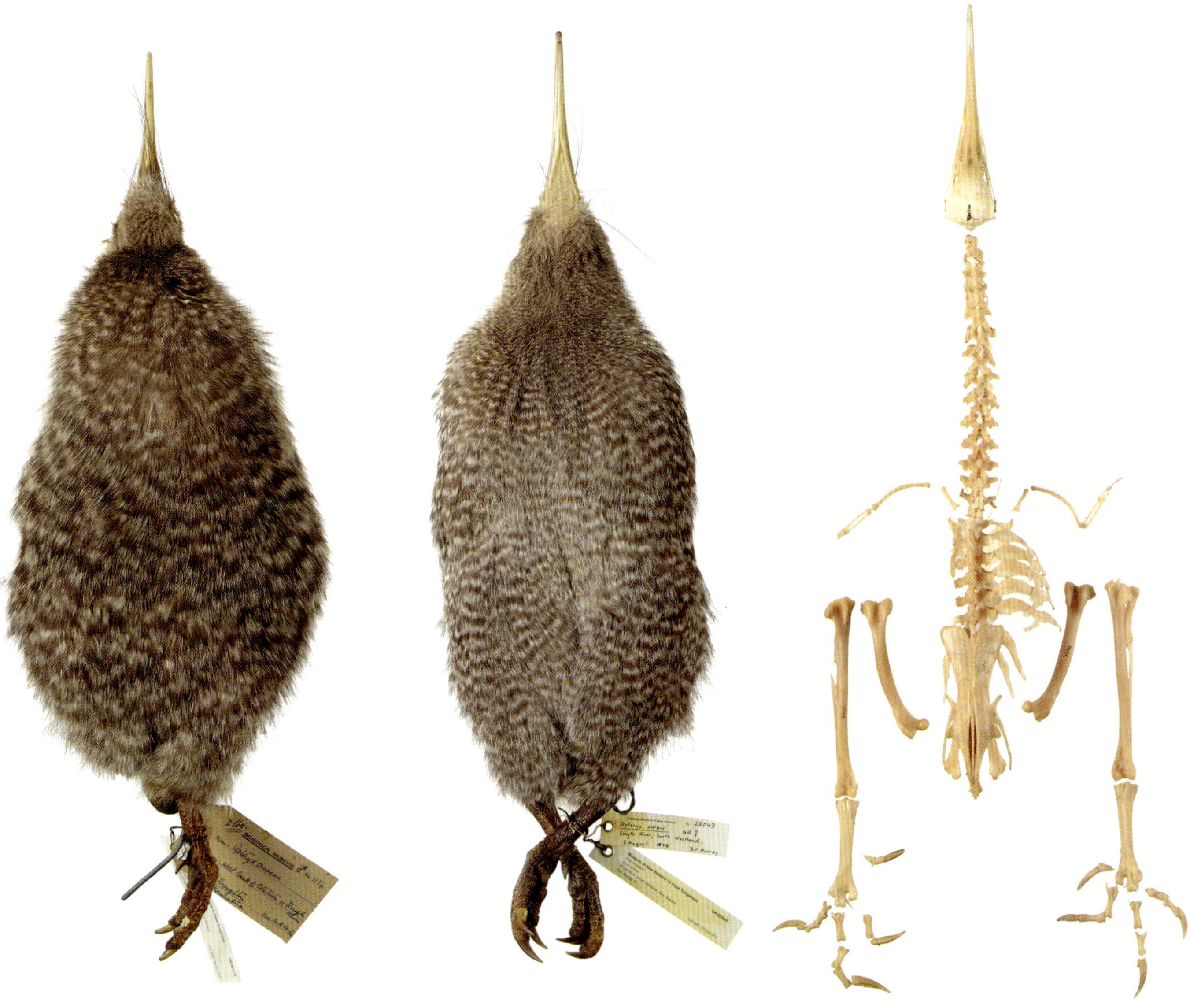
DOMINION MUSEUM
Apteryx oweni
28043
ad ♀
Smyth River, South Westland.
Museum of New Zealand Te Papa Tongarewa

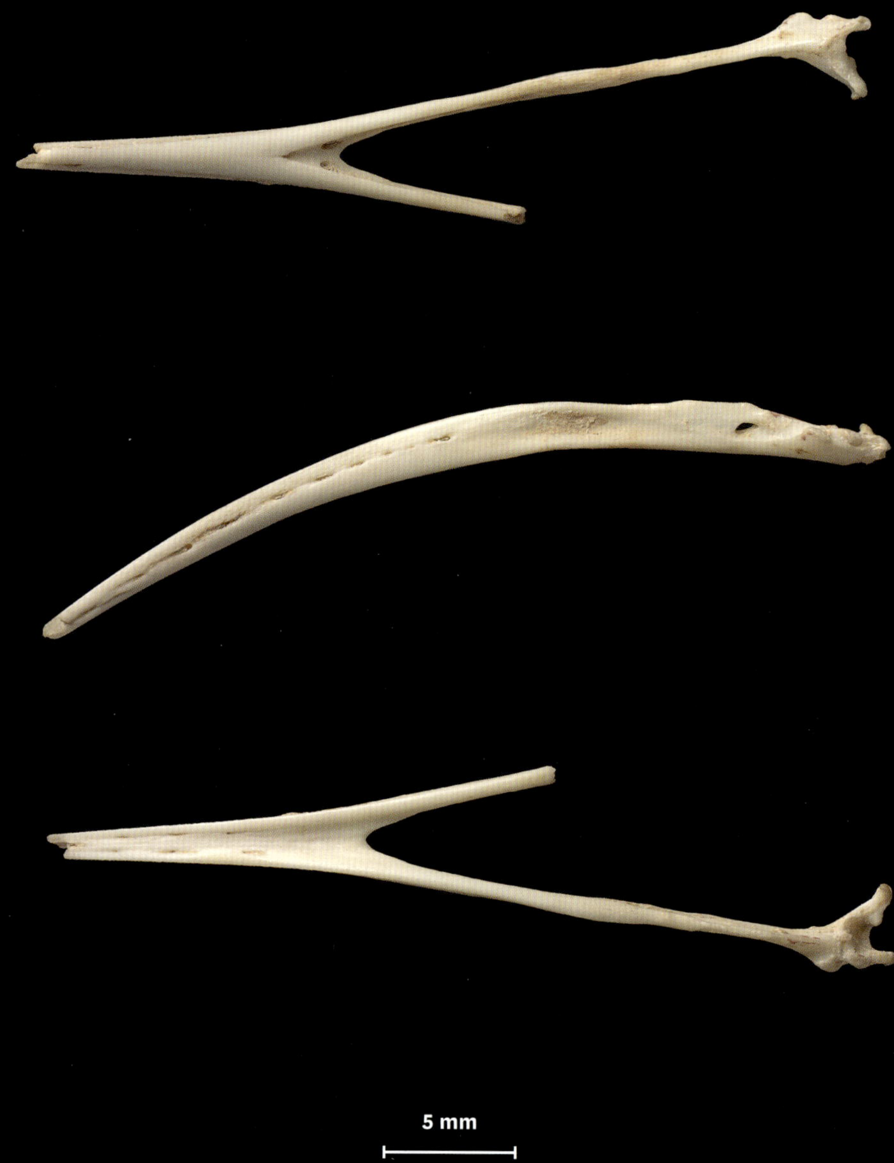
5 mm

Long-billed wren mandible

Dendroscansor decurvirostris

While few would argue that moa are New Zealand's most famous extinct fossil birds, dozens of other species met the same fate as moa as a result of human settlement. At the other size extreme from moa are New Zealand's wrens. The rifleman, or tītipounamu, is the country's smallest bird at 6 grams, and is still found at widespread localities. Its relative, the rock wren, survives in alpine habitats of the South Island but is under serious predation threat from introduced mammals. All of the other five members of the New Zealand wren family are extinct, including the bush wren, Lyall's wren and the North and South Island stout-legged wrens.

New Zealand's wrens are considered to be on the earliest surviving evolutionary branch of the world songbird tree, raising the possibility that songbirds originated in proto-Australasia before spreading to become the most diverse bird group globally. The most poorly known of these wren species is the extinct long-billed wren, represented by literally only a small handful of bones. This species was first described in 1991, when it was formally named *Dendroscansor decuvirostris* by Te Papa palaeontologist Phil Millener and Te Papa Research Associate Trevor Worthy. At 30 grams, it was a chunkier bird than the rifleman, and it had a distinctive long, down-curved beak. It was also flightless, so its extinction was almost certainly due to predation by kiore (Pacific rats): a small bird that was incapable of flight would have had no defences against such a predator. AT

Long-billed wren (*Dendroscansor decurvirostris*) mandible, S.22940, paratype, collected by P Millener, Honeycomb Hill Caves, Oparara, 25 April 1983.

054 / 100

Magneto snail

Chrysomallon squamiferum

Metals have always been attractive to humans – in our minds, their lustre and durability evokes something akin to eternity. Even our history, with its Bronze and Iron ages, is partly defined in terms of metals. In Egyptian myths, the gods were said to have flesh made of gold and bones made of silver (silver was the rarest metal then); the strongest of the gods, Seth, had bones made of iron. Now, metallic skeletons are something we would only see in comic book heroes and villains, right? Well, not quite.

A deep-sea snail that lives around hydrothermal vents on the bottom of the Indian Ocean incorporates iron compounds from the vents into its exoskeleton. The outer layer of its shell is made of iron sulphides (mainly the minerals pyrite, FeS_2, and greigite, Fe_3S_4), giving it a bulky, plate-armour look. The remaining shell layers are made of calcium carbonate and proteins, as is usual for snails. Even more astounding, however, the exposed portion of the snail's soft body (called its foot) is covered with tiny iron spikes (known as sclerites). Once again this looks like armour, albeit a scale armour in this case.

Researchers, not always known for their vivid imaginations, call the species scaly-foot gastropod. The media, however, adopted the more glamorous name Magneto snail, after the famous metal-controlling mutant from Marvel Comics. The snail's scientific name, *Chrysomallon squamiferum*, is slightly more poetic: *Chrysomallon*, from the Greek, means 'golden haired' and alludes to the golden pyrite sclerites; *squamiferum*, from the Latin, means simply 'scale-bearing'. RS

Scaly-foot gastropod *Chrysomallon squamiferum*, M.317895, paratype, shell width 27.5mm, collected by ROV Kiel 6000 on board RRS *James Cook* at depth of 2785 m, Longqi vent field, Southwest Indian Ridge, 29 November 2011.

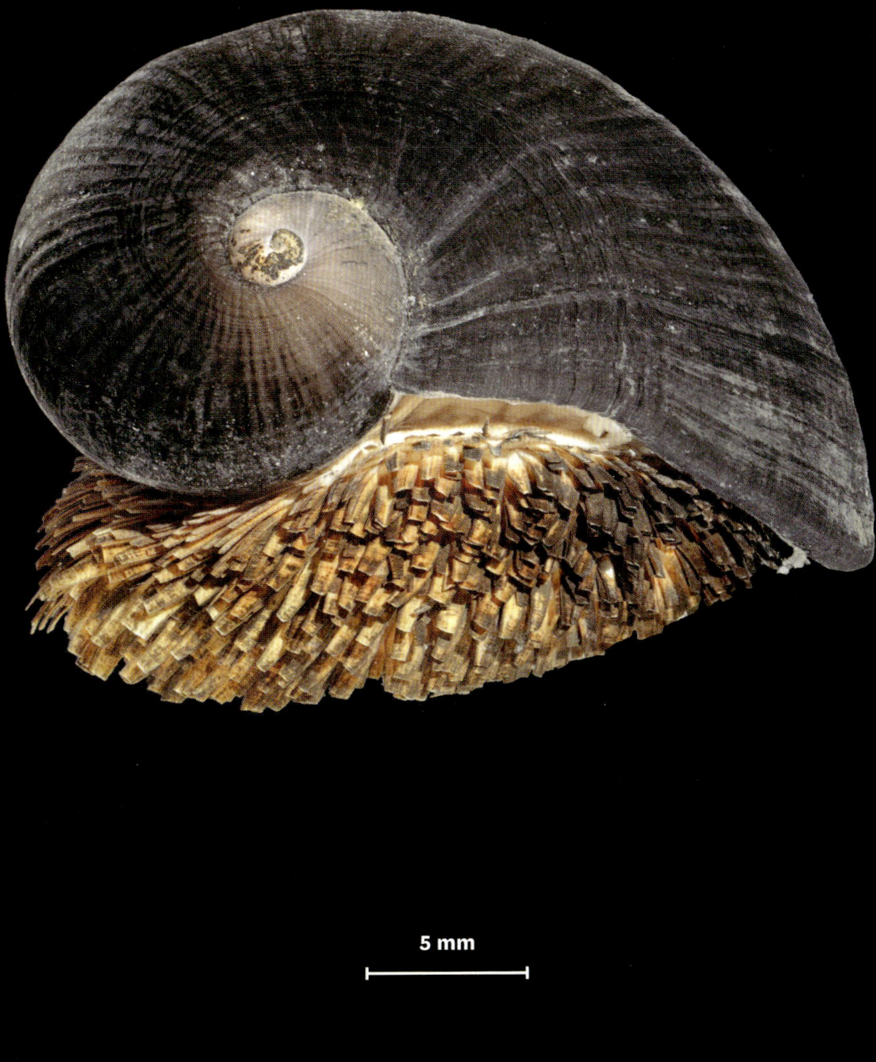
5 mm

Malawi dance garment

Unknown maker

Museum collections have long been an important source of material for scientific research. Material is available for rare, extinct and difficult-to-obtain species and can provide a historical perspective. For example, a study of peregrine falcon eggs held in museums was able to link egg-shell thinning to the introduction of the insecticide DDT.

However, some research techniques, such as many DNA extraction methods, involve destroying part of the specimen. A leaf might be removed from a herbarium specimen, or a piece of bone cut from a fossil and ground up. Curators, who are tasked with caring for museum collections, need to balance specimen preservation with their use for research. Consequently, I have been working to develop DNA extraction methods that minimise damage to specimens, such as ‘soaking’ DNA out of bones.

The item pictured here is a dance garment from Malawi, which was donated to Te Papa’s predecessor in 1936 and is now stored in Te Papa’s International History Collection. Te Papa conservator Anne Peranteau and I were interested in learning more about the construction of this garment. Its label described it as being ‘made from monkeys’ tails’, but we suspected that there was more than just monkeys involved, given the diversity of the fur and skins. Wanting to avoid damaging the garment, we trialled a new method of obtaining DNA – first rubbing the surface of the skins with a PVC eraser and then isolating DNA from the eraser fragments.

Using this method, we did indeed find monkey DNA (from vervet monkeys), but also a much more diverse range of animals, including African wildcat, yellow baboon and slender mongoose. LS

Dance garment, GH024606, unknown collector, Malawi (Ngoni culture), c. 1900.

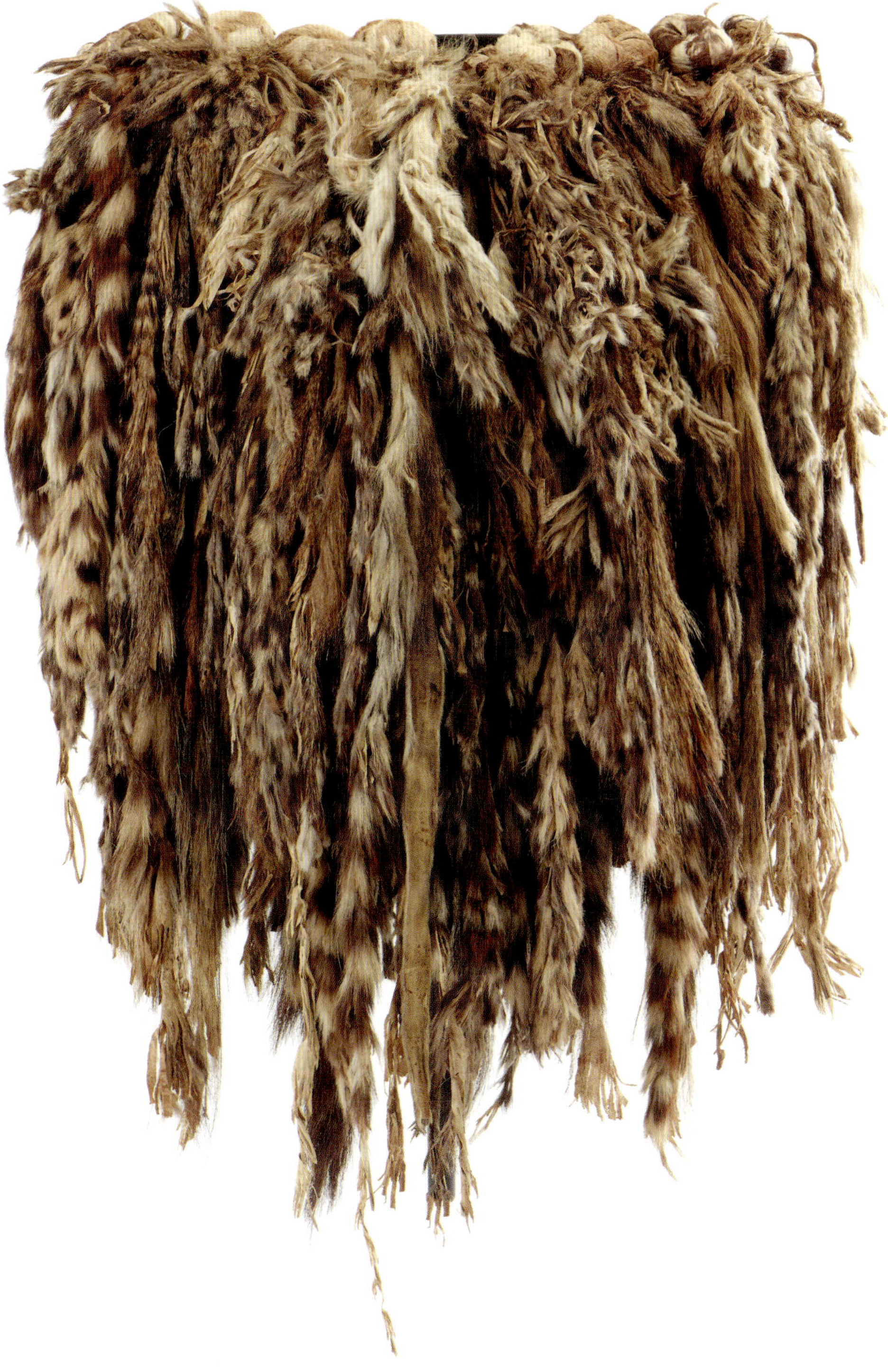

Male-eating bivalve

Parvamussium maorium

When I began studying deep-sea bivalves of the genus *Parvamussium* in preparation for taxonomic revisions in the 1980s, I happened to mention to Geoff Hicks, our crustacean curator at that time, that they were carnivores and that minute crustaceans had been found in their stomachs. Geoff extracted gut contents from a number of *Parvamussium maorium* specimens, collected at a depth of 1009 metres on the Challenger Plateau, and found that all had eaten copepod crustaceans (Harpacticoida) – and all of these copepods were male. This was extremely intriguing, since females are predominant in the total darkness of the deep sea and males are rare – even unknown for some species.

This presented us with two questions: Why only males, and how did the bivalves catch them? We speculated that female copepods attract males using pheromones or light emission, and that the bivalves attract the males by mimicking female copepods. These bivalves have never been studied alive, but their shells are translucent and the mantle edge of the soft parts within are fringed with tentacles, so any visual cue would be visible from the exterior.

Later, during shallow-water field work in New Caledonia, I was the first to observe and record living specimens of a *Cyclopecten* species. Related to *Parvamussium*, these bivalves are also translucent and are known to be carnivores. Viewed in a dish of sea water under a microscope, *Cyclopecten* crawled rapidly on a slender, worm-like foot, attached itself to a coral fragment with a hair-like strand, orientated itself and partly opened its two valves. Between these, the withdrawn foot could be seen wriggling about, presumably mimicking a nematode worm to attract prey. This observation suggested that movements of the foot in *Parvamussium* may be the lure, if indeed the attractant is physical. BM

***Parvamussium maorium*, M.274125, collected by RV *Tangaroa* at depth of 1005–1009 m, Challenger Plateau, 18 April 1980.**

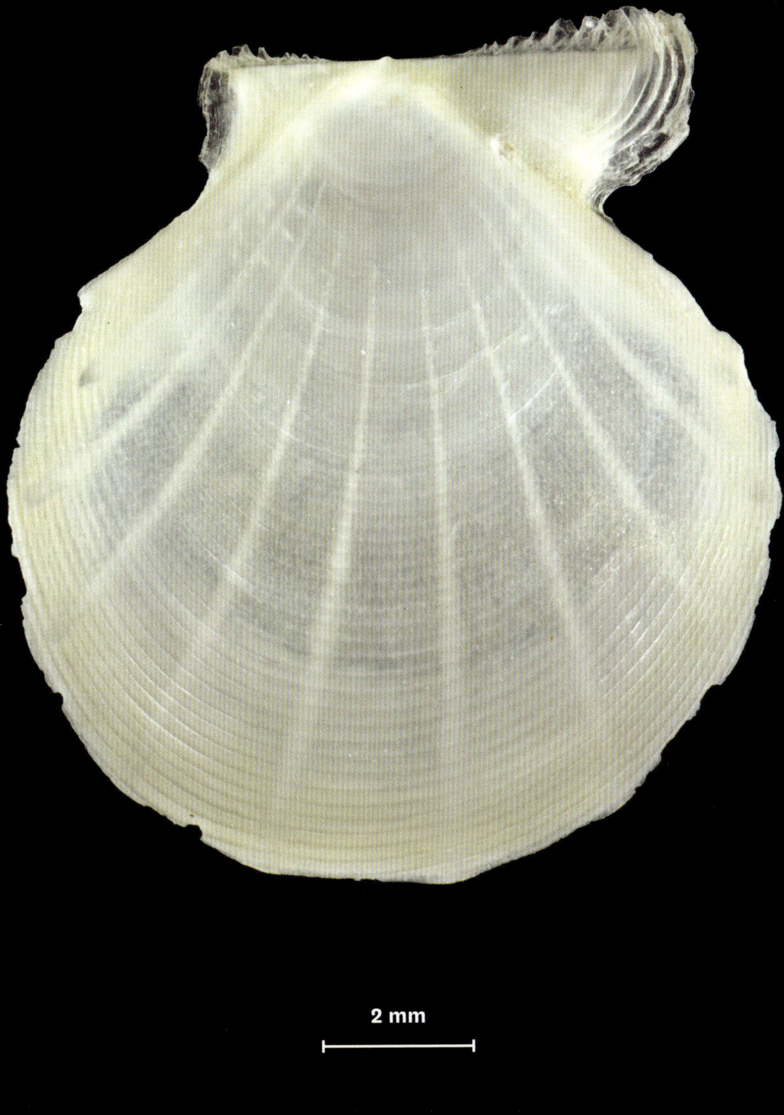
2 mm

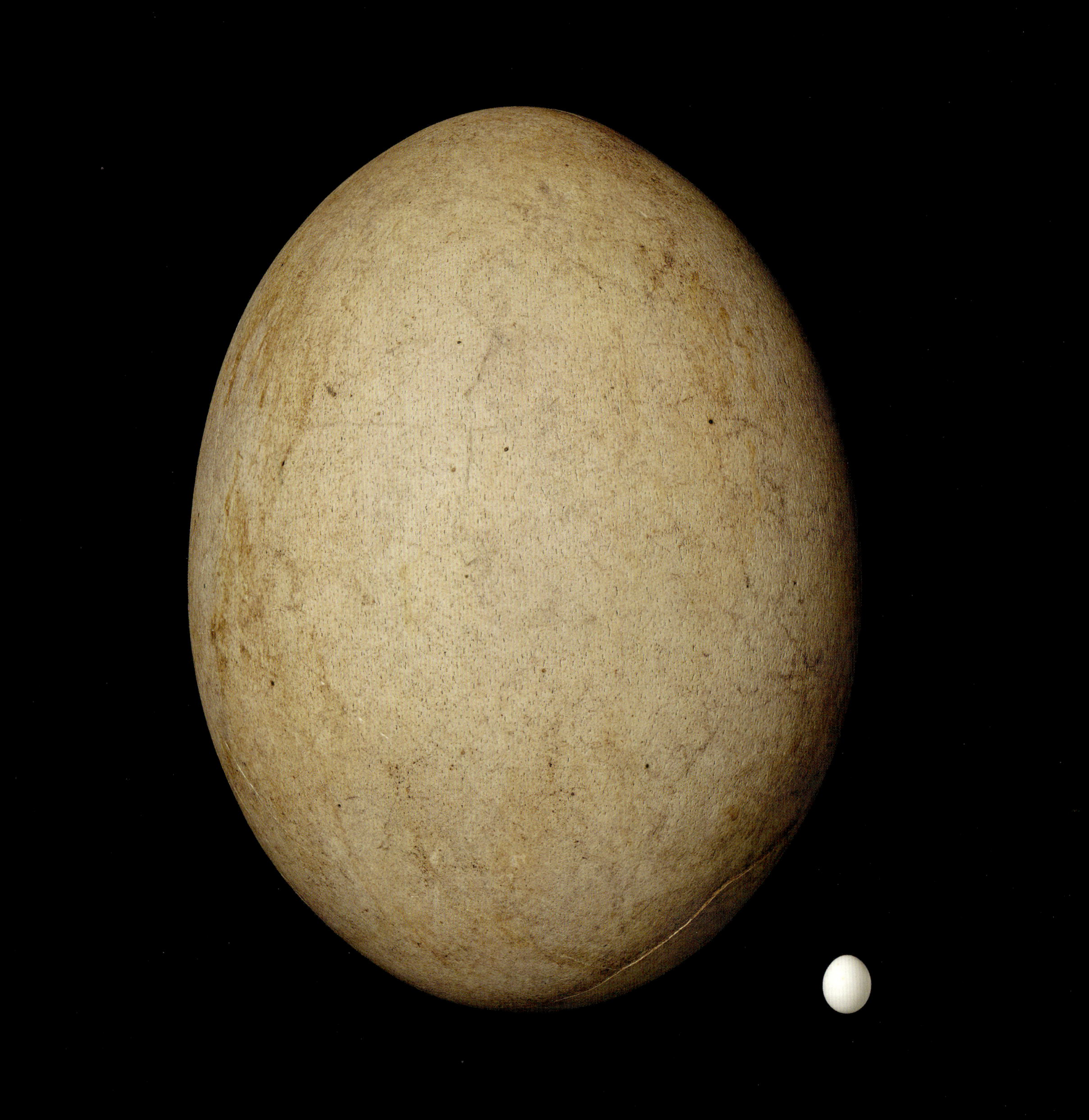

Moa egg

Dinornithiformes order

Moa eggs are almost mythically rare. Of all bird eggs, they had the thinnest shells in relation to their size and so are incredibly fragile. As a consequence, while fragments of moa egg shells are common there are fewer than three dozen eggs that are sufficiently intact for their size and shape to be appreciated.

Most people think of moa as being a single species, but this is far from the truth. They represent the greatest diversification from a single ancestor of any New Zealand bird group, with three families, six genera and nine different species. Their sizes reflected this diversity, with estimated weights ranging from 12 kilograms to 242 kilograms and their eggs ranging in volume from about 0.5 litres to 4 litres.

The largest of all moa eggs is held at Te Papa, and measures 240 millimetres x 178 millimetres. It is assumed to have been laid by a South Island giant moa, and was found at Kaikōura in the late 1850s while house foundations were being dug. The egg was from an early Māori burial site and was probably used to hold or carry water, as shown by the perforation made in it.

This enormous egg travelled the world after its discovery. It was displayed at the New Zealand Exhibition in Dunedin in 1865, then taken to England later that year. It was sold into private ownership at auction in London for £200 in 1865, and was exhibited at the Colonial and Indian Exhibition in London in 1886. This national treasure returned to New Zealand in 1966 when it was purchased by the Dominion Museum. CM

South Island giant moa (*Dinornis robustus*) egg, ME012748, length 240 mm, width 178 mm, unknown collector, Kaikōura, late 1850s; rifleman (*Acanthisitta chloris*) egg, OR.007264, length 16 mm, width 13 mm, collected by L Vangioni, Akaroa, no date.

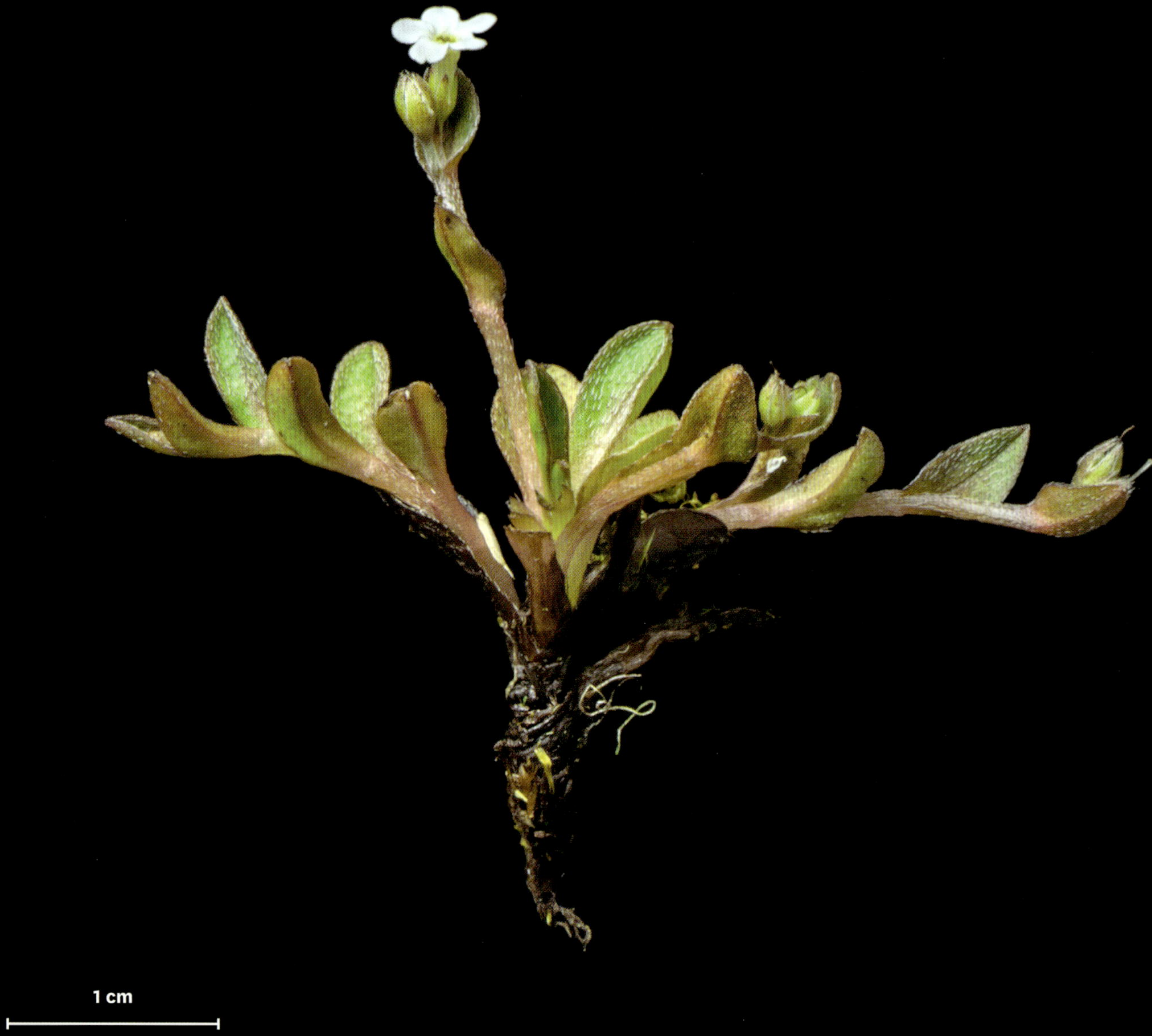
1 cm

Moss-dwelling forget-me-not

Myosotis bryonoma

Discovering new species is, essentially, detective work using techniques in the field, herbarium, laboratory and library. The key is having an expert eye to discern the new and the different by comparing different plant specimens, and scrutinising all available information. Field work is vital to see the plants alive and in their natural habitats.

The specimen shown here is one of my favourites, not just because it is the type specimen of a new species, but also because of the story behind its collection. Like most New Zealand forget-me-nots, it is found in a very specific habitat: the tiny plants grow only in a few high-lying bogs in central Otago. Getting to the remote location where my colleague Antony Kusabs and I collected this specimen was a special treat. In January 2016, we were flown by helicopter (an exhilarating first for me!) to the top of the Garvie Mountain Range, about 1800 metres above sea level, where we spent the entire day walking and searching. Over an exhausting yet productive day, we found, photographed and collected several specimens now housed at Te Papa. Before we pressed this particular plant, former Te Papa researcher Phil Garnock-Jones took this stunning photograph.

Some of the photographs we took that day were used in a 2016–17 Te Papa exhibition highlighting taxonomic research at the museum. The public was invited to suggest names for the new species, and the one we eventually chose – *Myosotis bryonoma* – was inspired by more than 500 submissions. The name is based on Greek, and means 'moss-dwelling forget-me-not' in honour of its unique habitat. HM

Moss-dwelling forget-me-not (*Myosotis bryonoma*), SP104478 (whole plant), holotype, collected by HM Meudt and A Kusabs, Garvie Mountains, Central Otago, 11 January 2016.

Objects

N—P

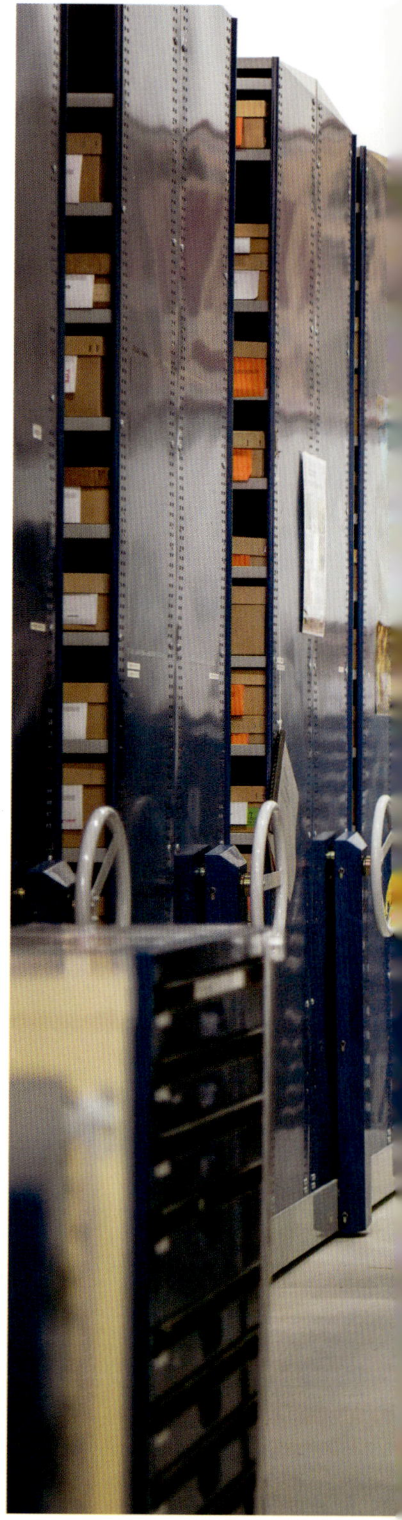

Botany Researcher Heidi Meudt, far left, Curator Botany Carlos Lehnebach, Curator Botany Leon Perrie and Research Fellow Pat Brownsey in the Botany collection storage room at Te Papa's Cable Street building.

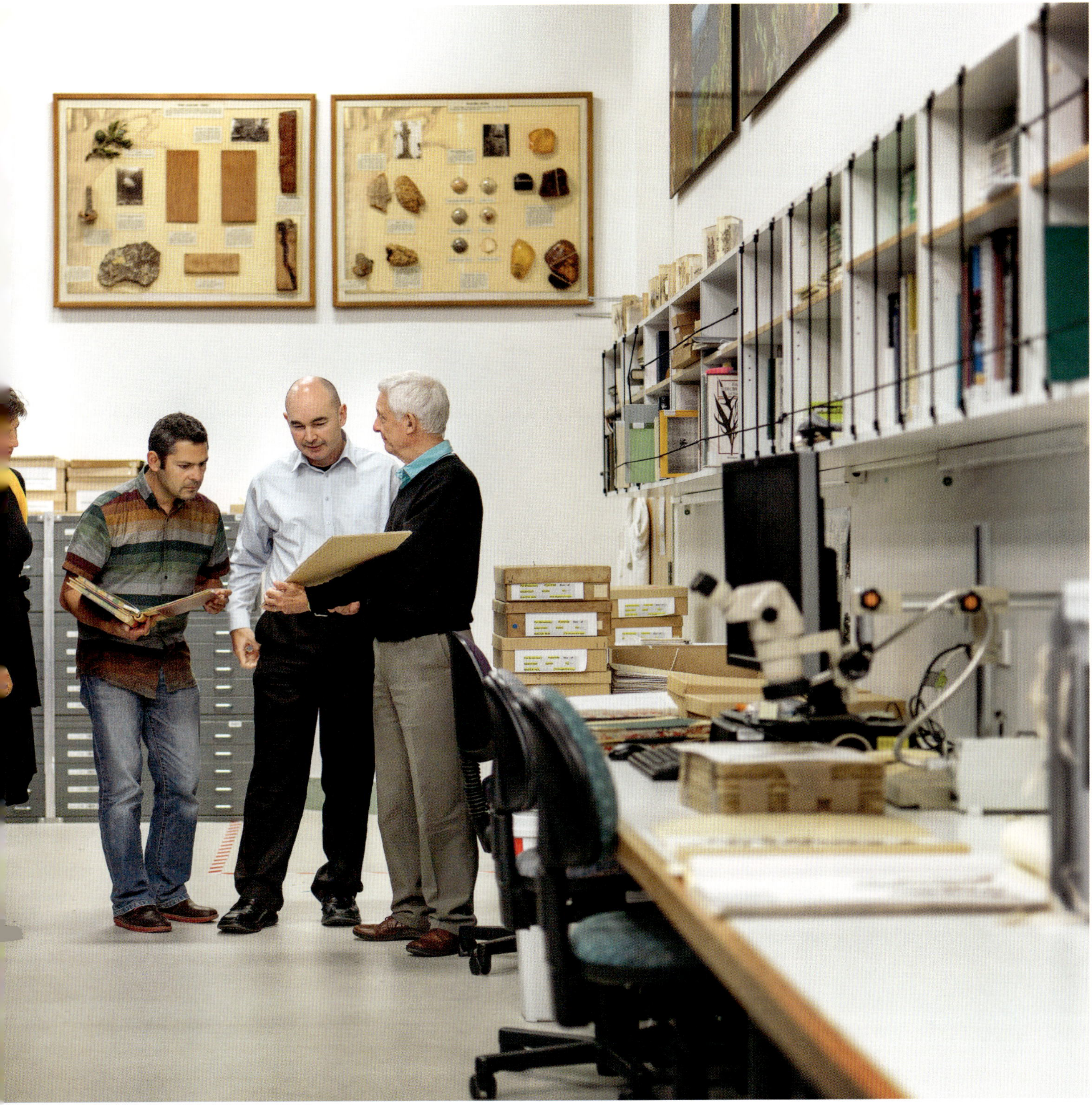

053 054 055

059 060 061

065 066 067

071 072 073

077 078 079

083 084 085

56 057 058

062 063 06

068 069 07

074 075 076

080 081 08

086 087 08

Native forget-me-not

Myosotis lyallii subsp. *elderi*

Lucy Beatrice Moore (1906–1987) was an outstanding botanist. Her expertise and comprehensive knowledge are clearly evident in the first two volumes of the *Flora of New Zealand* series: she wrote many of the sections, including the taxonomic treatment of native forget-me-nots (*Myosotis*), in which she described *Myosotis elderi* as a new species and chose the specimen shown here as the holotype (the specimen upon which the name and description of a species is based).

059/100 The specimen is a favourite, partly because it is complex. In the bottom right-hand corner is a label hand-written by the collector, Donald Petrie (1846–1925), another celebrated New Zealand botanist. It shows that Petrie collected all six plants on the sheet near Arrowtown and identified them as *Myosotis pygmaea*. However, the notes in the bottom left-hand corner, in Lucy Moore's handwriting and dated 19 May 1958, show that her careful eye distinguished three different species: three plants of '*Myosotis pygmaea* var. nov.' on the left, and three plants of her new species *Myosotis elderi* on the right. Plants of the third species were removed to a different sheet.

When I hold this specimen, I feel a tangible connection with Lucy; I can sense her dedication and perseverance in revising the taxonomy of New Zealand forget-me-nots – a task that has now passed to me, over half a century later. Further specimens and data have now been collected, and a new, comprehensive taxonomic treatment is needed. Lucy's specimen notes (like those shown here), her *Flora* sections and her later scientific publications have provided the groundwork for my own ongoing *Myosotis* research. I hope that she would have been pleased to have a fellow woman botanist pick up where she and others left off. HM

Forget-me-nots: **(right)** *Myosotis lyallii* subsp. *elderi*, SP002648, holotype; **(left)** *Myosotis drucei*, SP004733; both collected by D Petrie, mountains near Arrowtown, Otago, no date.

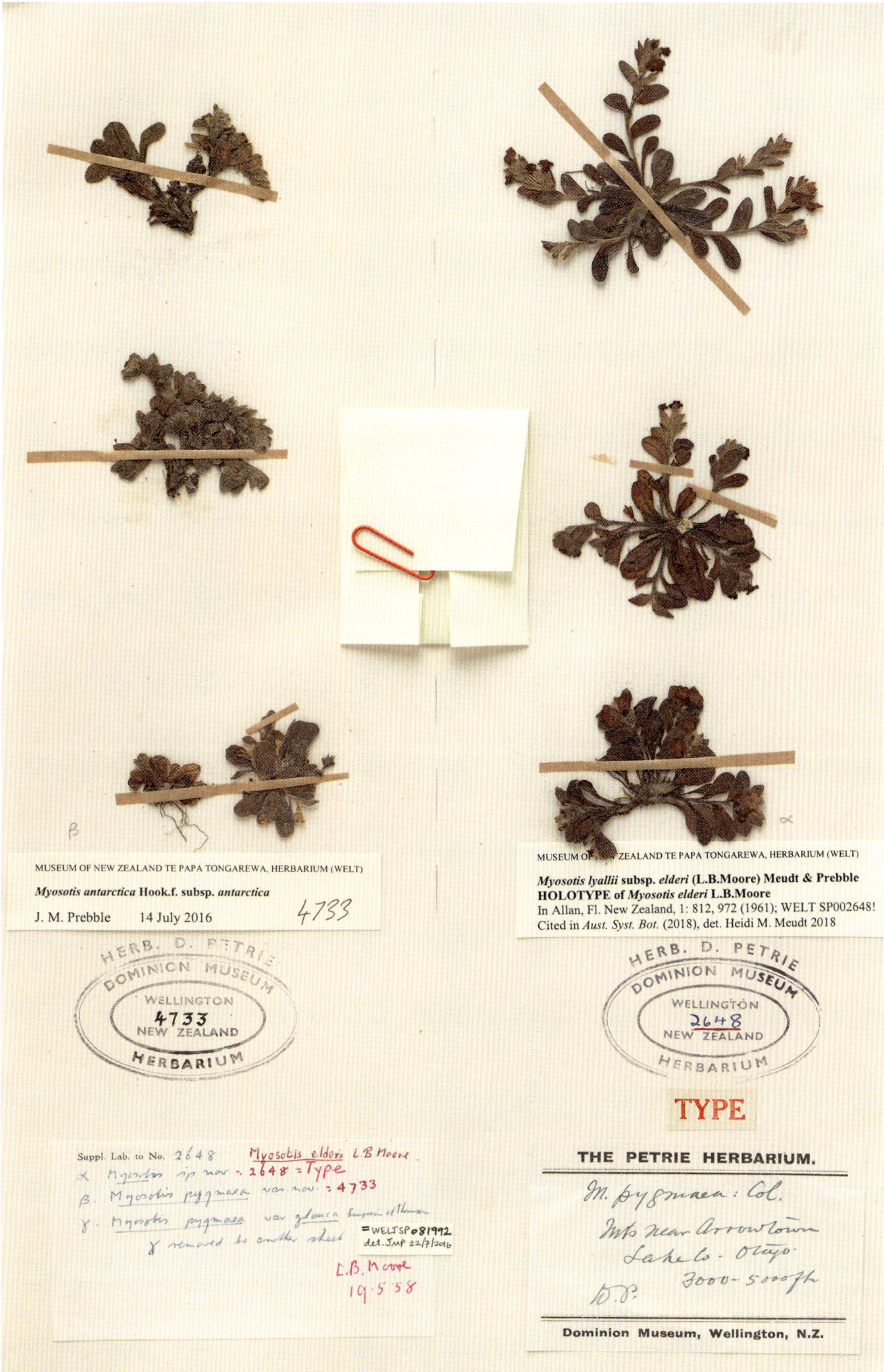

MUSEUM OF NEW ZEALAND TE PAPA TONGAREWA, HERBARIUM (WELT)
Myosotis antarctica Hook.f. subsp. antarctica
J. M. Prebble 14 July 2016
4733
MUSEUM OF NEW ZEALAND TE PAPA TONGAREWA, HERBARIUM (WELT)
Myosotis lyallii subsp. elderi (L.B.Moore) Meudt & Prebble
HOLOTYPE of Myosotis elderi L.B.Moore
In Allan, Fl. New Zealand, 1: 812, 972 (1961); WELT SP002648!
Cited in Aust. Syst. Bot. (2018), det. Heidi M. Meudt 2018
HERB. D. PETRIE
DOMINION MUSEUM
WELLINGTON
4733
NEW ZEALAND
HERBARIUM
HERB. D. PETRIE
DOMINION MUSEUM
WELLINGTON
2648
NEW ZEALAND
HERBARIUM
TYPE
Suppl. Lab. to No. 2648
Myosotis elderi L.B Moore
= 2648 = Type
β. Myosotis pygmaea var. nov. = 4733
γ. Myosotis pygmaea var glauca
γ removed to another sheet
=WELTSP081992
det. JMP 22/7/2016
L.B. Moore
19.5.58
THE PETRIE HERBARIUM.
M. pygmaea: Col.
Mts near Arrowtown
Lake Co. Otago.
3000-5000 ft
D.P.
Dominion Museum, Wellington, N.Z.

Native galaxiids

Galaxiidae family

Galaxiidae is New Zealand's largest and most iconic family of freshwater fishes. Of eight genera and fifty-seven species worldwide, twenty-two species occur in New Zealand: seventeen in genus *Galaxias* and five in *Neochanna*. Twenty of those species are endemic, with several confined to a single river catchment or group of dune lakes. Taxonomic and genetic research has increased the known diversity by eight species since 1990, and at least another three are still to be formally described.

Adult galaxiids are mostly solitary and secretive. Several spend part of their life cycle at sea, migrating into freshwater as schools of juveniles – most often encountered by New Zealanders as 'whitebait'. Whitebait comprises five species; in decreasing order of abundance, these are inanga (*Galaxias maculatus*); kōaro (*Galaxias brevipinnis*); banded kōkopu (*Galaxias fasciatus*); giant kōkopu (*Galaxias argenteus*); and shortjaw kōkopu (*Galaxias postvectis*, shown here). Their adults are the largest in the family, growing to 190–580 millimetres in length. The remaining species live in freshwater all their life and are relatively small (67–175 millimetres,mostly below 150 millimetres).

Many populations are in decline due to loss of habitat, reduced water flow, drought, pollution, and predation by introduced fishes. Several are categorised as Endangered by the International Union for Conservation of Nature (IUCN), including the first galaxiid species described by a New Zealander, the shortjaw kōkopu (*Galaxias postvectis*). This is the second-largest species in our waters, growing to 350 millimetres. The juveniles are the rarest in the whitebait catch, while the adults are nocturnal and very rarely seen. They prefer coastal streams with extensive forest cover and complex bottom structures. Their conservation status is of concern, as few populations of significant size are known. CR + AS

Top: Shortjaw galaxias (*Galaxias postvectis*), P.000981, standard length 155 mm, collected by B Osborne, Queen Charlotte Sound, no date. **Middle:** Inanga (*Galaxias maculatus*), P.005625, standard length 76 mm, collected by RM McDowall, Lake Waiparera, North Auckland, 25 March 1965. **Bottom:** Koaro (*Galaxias brevipennis*), P.002070, standard length 400 mm, Lake Waikaremoana, June 1928.

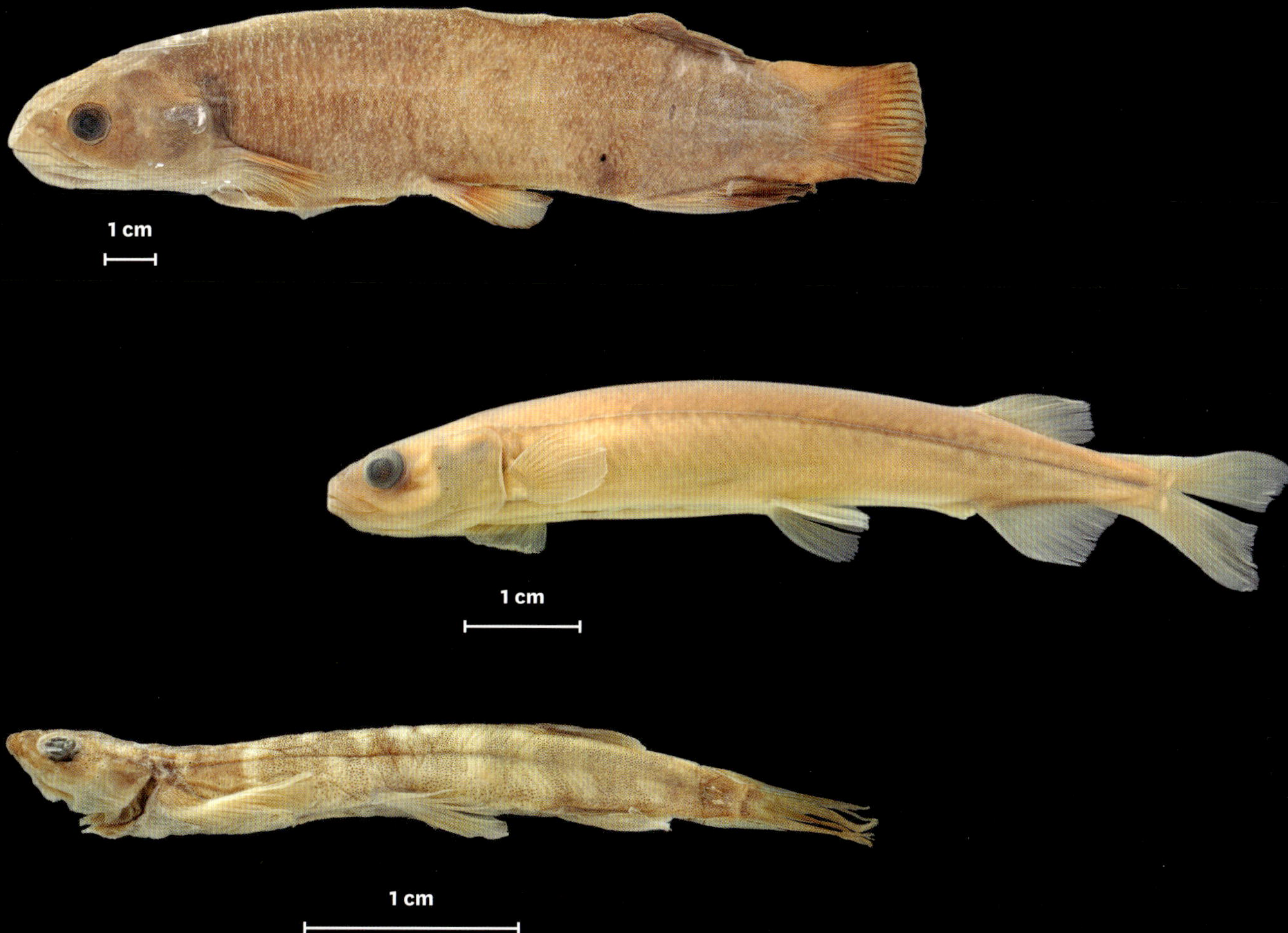
1 cm
1 cm
1 cm

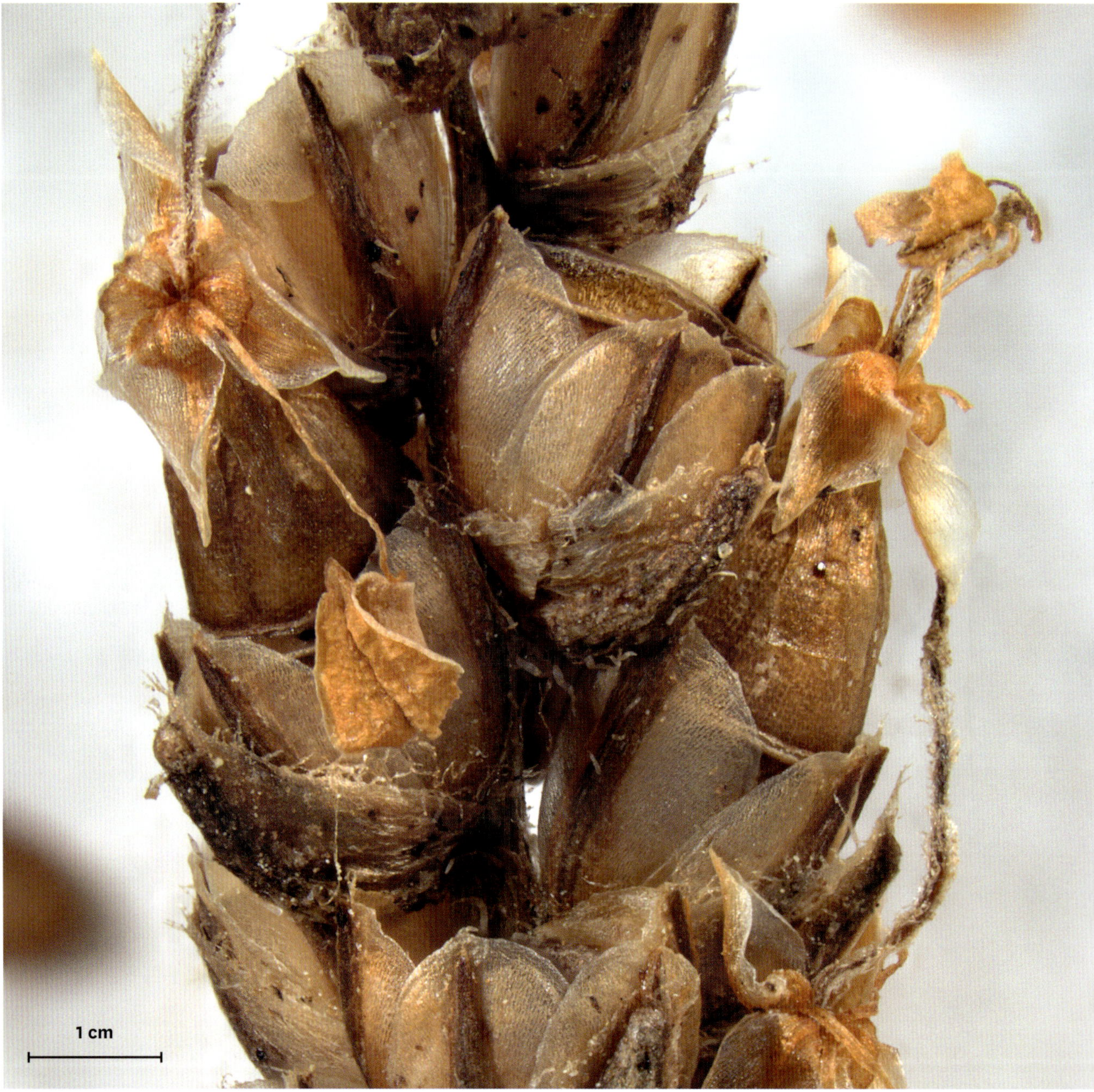
1 cm

Native plantain

Plantago udicola

The specimen shown here demonstrates the importance of three aspects of taxonomic research, which we used to discover and describe the new species *Plantago udicola*: collaboration, collections-based research, and integration of different data types.

We discovered that plants of this species, including this one originally collected by Te Papa researcher Barry Sneddon, have ninety-six chromosomes, a unique (and rather high) number. Counting chromosomes is an important but painstaking taxonomic skill that is disappearing as colleagues such as Brian Murray, who prepared and imaged the chromosomes of this specimen, retire. Collaboration was paramount in the collecting and study of this specimen of *Plantago*, involving many colleagues from several New Zealand institutions and including the co-supervision and training of MSc student Mei Lin Tay.

By comparing numerous specimens from Te Papa and other national herbaria, we found that this specimen, and others like it, had this unique chromosome number coupled with distinctive leaf, bract, seed, flowering scape (stem) and genetic characteristics, as well as a particular habitat preference (*Plantago udicola* means 'plantain of damp places'). This close-up image shows the distinctive hairy bracts at the base of the capsules (fruits) and flowers of the plant. The combined evidence from multiple sources of data led us to describe it as a new species. The species' description is an example of the kind of meticulous, comprehensive, collections-based research that is critical to taxonomy and biodiversity knowledge, and highlights the significance of each and every specimen at Te Papa that might just be a new species waiting to be discovered! HM

Native plantain of damp places (*Plantago udicola*), SP090378/B (flowers and capsules), collected by BV Sneddon, Tableland near Balloon Hut, Kahurangi National Park, western Nelson, 10 April 2008.

062 / 100

Nelson cave spider

Spelungula cavernicola

The specimen shown here is the holotype (name-bearing representative designated when the species is formally described) male of *Spelungula cavernicola*; the common name for this uncommon species is the Nelson cave spider. At around 15 centimetres across, adult females of this species are our largest spiders by leg-span. Long limbs are a feature of many cave-dwelling species – if you meet something predatory in the dark, it's better to lose a bit of leg far away from your main body. These spiders are themselves predators and are known to feed on cave wētā.

Nelson cave spiders have a number of odd anatomical features. First, most spiders can be divided into two major groups, but this species combines elements of both groups. Mygalomorph spiders include tarantulas and tunnelwebs. They have fangs that are aligned with the long axis of the body (i.e. they do not cross) and have four book lungs for respiration (book lungs are named for their appearance, with alternating air pockets and fluid-filled tissue resembling the pages of a book). Most other spiders are araneomorphs. These never have more than two book lungs and their fangs run across the line of the body. *Spelungula cavernicola* mixes araneomorph fangs with the two pairs of book lungs seen in mygalomorphs. Second, the front two pairs of legs of *S. cavernicola* each have one relatively normal-sized claw and one massive one.

As the common name suggests, these spiders are confined to a few Nelson cave systems. Their distinctiveness and very limited habitat range is why they are one of only two species of spider protected under the Wildlife Act 1953. PS

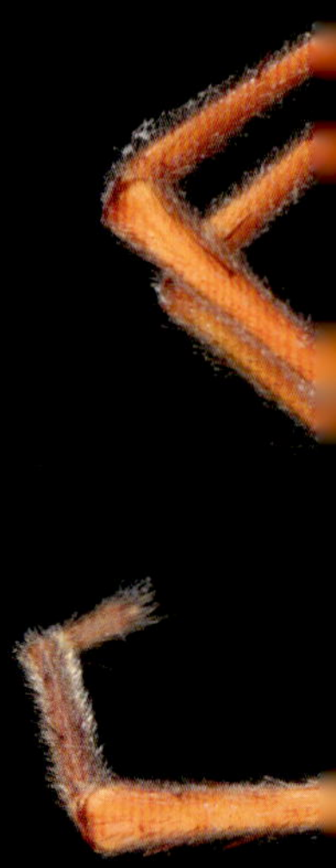

Nelson cave spider (*Spelungula cavernicola*), AS.000014, collected by P Wood, Wonder Sump Cave, Oparara River area, northwest Nelson, 7 February 1983.

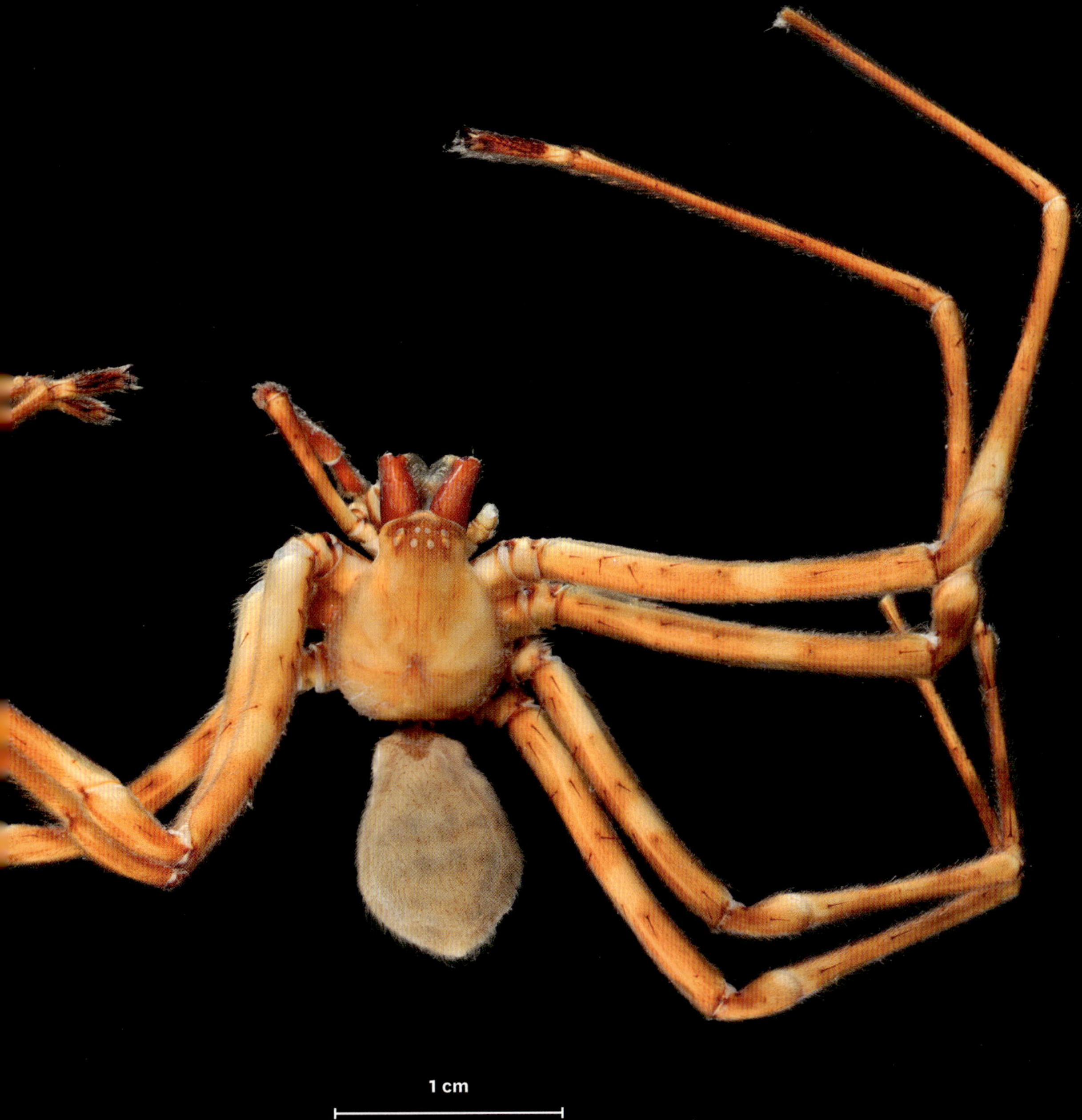
1 cm

New Zealand Ferns, first edition

Herbert Dobbie

For many years, the little book shown here was an enigma. Produced in 1880 by Herbert Dobbie, it was an affordable identification guide to New Zealand ferns, then very much in fashion. Albums of pressed ferns were popular with those wanting to send something uniquely New Zealand to family and friends overseas. However, naming the ferns was difficult because contemporary books were rather technical, with no illustrations to help.

Dobbie used the cyanotype process to produce images of 148 varieties of ferns. After coating sheets of paper with potassium ferricyanide and ferric ammonium citrate, he laid dried fern fronds
063/100 on the sheets, covered them with glass to hold them tight to the paper, and exposed them to sunlight for a predetermined length of time. Lettering and marginal decorations were written on the glass in Indian ink before exposure. After exposure the sheets were washed in water, the exposed parts turning a deep Prussian blue and giving a satisfying white-on-blue silhouette. Although criticised for not showing reproductive structures, the silhouettes were adequate to identify most species of native fern.

Three slightly different editions of Dobbie's book were produced, but it is not known how many were sold. Their fragile nature means that few have survived; only a handful now exist, most in public institutions such as Te Papa.

Over thirty years later, Dobbie published his better-known book *New Zealand Ferns*, which ran to five editions. The book was immensely popular because of its simple descriptions, chatty text and clear photographs of every species. The 1921 edition is intriguingly inscribed 'Second Edition, revised and greatly enlarged', but it was another fifty years before bibliographers realised that Dobbie's blue book was the elusive first edition of his more traditional book. **PB**

Hen and chicken fern (*Asplenium bulbiferum*), 0.043967, cyanotype, 278 mm x 212 mm, from HB Dobbie, *New Zealand Ferns, 148 varieties* (Auckland, 1880).

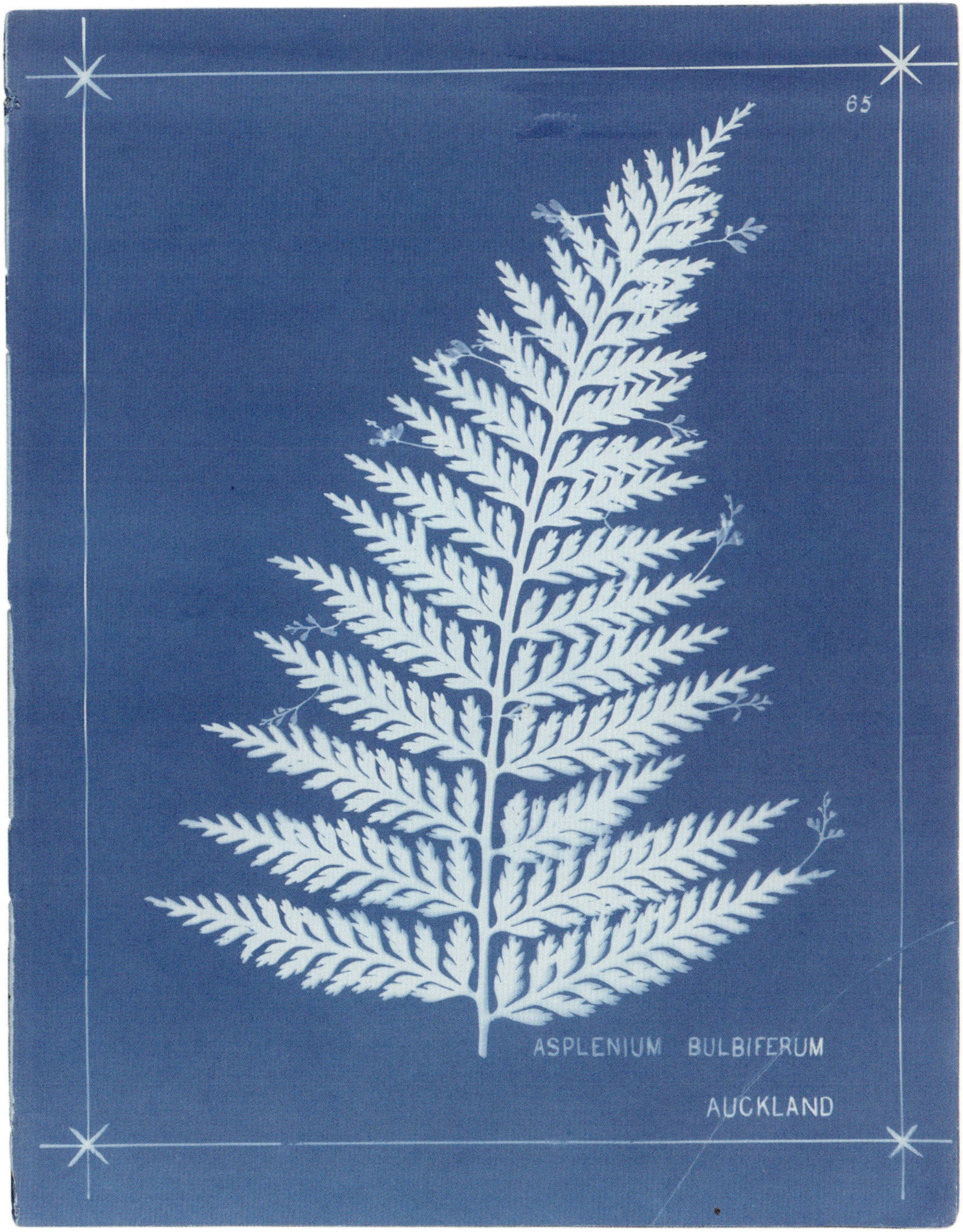
65
ASPLENIUM BULBIFERUM
AUCKLAND

New Zealand grayling

Prototroctes oxyrhynchus

Known to Māori as upokororo, the endemic New Zealand grayling (*Prototroctes oxyrhynchus*) was scientifically described in 1870. Once so abundant that it was used by the cartload as fertiliser, it was extinct by 1930 – a warning to us to take better care of our freshwater life and habitats. Just four voucher (representative) specimens are held at Te Papa: one dry and three preserved in alcohol.

064/100

The grayling was widely distributed around New Zealand, from Bay of Plenty to Southland, mostly in lowland streams and rivers that opened into the sea. Grayling lived at sea for part of their lives, migrating upstream to spawn in immense numbers. Described as one of the most attractive native fishes, its colour varied from a general silvery hue and brown back to a rich, red-brown back speckled with grey and a yellow to golden belly.

Grayling grew up to 450 millimetres long and 1.5 kilograms in weight. They were highly valued both for food and for angling. It is believed that they ate algae and small insects, scraped off rocks with specialised, comb-like, upper-jaw teeth that bit against a horny pad on the lower jaw.

In late 1870, the grayling was noted as being in serious decline – but it wasn't until 1930 that the Marine Department officially recorded it as being on the verge of extinction, and only in 1951 was the fish given legislative protection. Its extinction was never explained satisfactorily, but was most likely due to a combination of the introduction of European trout and the deterioration of the freshwater environment, especially as the result of clear-felling of forest for pastureland. CR + AS

Grayling (*Prototroctes oxyrhynchus*), 1992-0035-2278/1, only known colour image of fresh specimen, drawn by FE Clarke, 1889.

B.   D. 11   P. 15   V. 6   A. 17   C. 17.   H : B :: 1.3 :: 7.2 (1 :: 5½)

Depth : Length : 1.35 :: 7.2 (1 :: 5½)

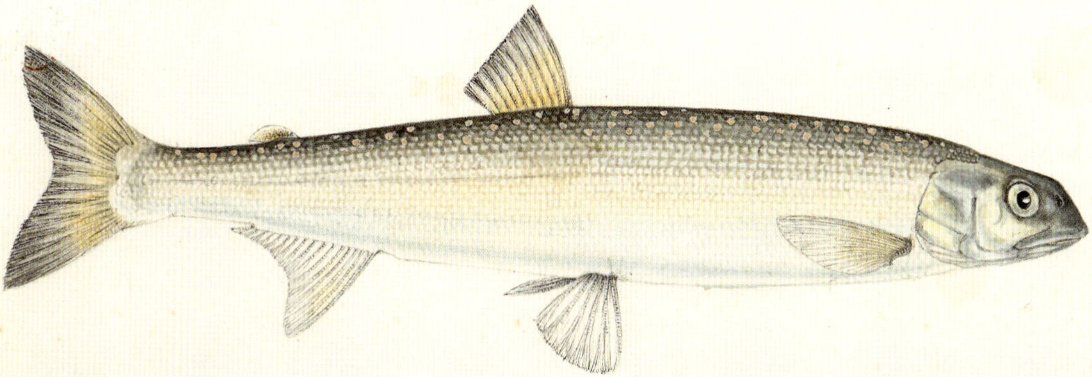

White Spotted Greyling. (Nat: Size)

(Prototroctes Saleii)

Hokitika R. 26th Octr 1889.

F. E. Clarke del:

New Zealand harvestman

Prasmiola unica

This is the only photograph of the endemic New Zealand harvestman *Prasmiola unica*, and to our knowledge it shows the only known specimen. It was collected by Ray Forster, New Zealand's foremost arachnologist, from Days Bay near Wellington in 1942. Despite a number of attempts to find this species again, nobody has yet succeeded. We know very little about this animal's biology, including what the male of the species looks like. Many New Zealand species, particularly invertebrates, are known from a single specimen and this is a perfect example.

Looking like a bizarre spider, this is actually another kind of arachnid called a harvestman (order Opiliones); more precisely, it is a member of the shorter-legged harvestmen (suborder Laniatores). Like many other Laniatores, *Prasmiola unica* is covered in spines and pustules. This may give them the appearance of an aggressive killer that's strayed from a science-fiction movie, but they are typically scavengers rather than predators.

Some harvestmen also have more than one form of male. In long-legged harvestmen (suborder Eupnoi), the chelicerae (the centremost appendages on the front of the animal) on the males can be short and sturdy-looking or long and slender. In females they are much shorter. Males use these appendages in shoving contests, much like stags, and the contrasting forms seem to pit power against reach.

New Zealand has more than 200 species of harvestmen that are found nowhere else on Earth. As a group, New Zealand's endemic harvestmen are most commonly found in native forest habitat. They are usually most active at night and can often be found hiding under logs during the day. PS

New Zealand laniatorid harvestman (*Prasmiola unica*), AH.000081, holotype, collected by R Forster, Days Bay, Wellington, 19 February 1942.

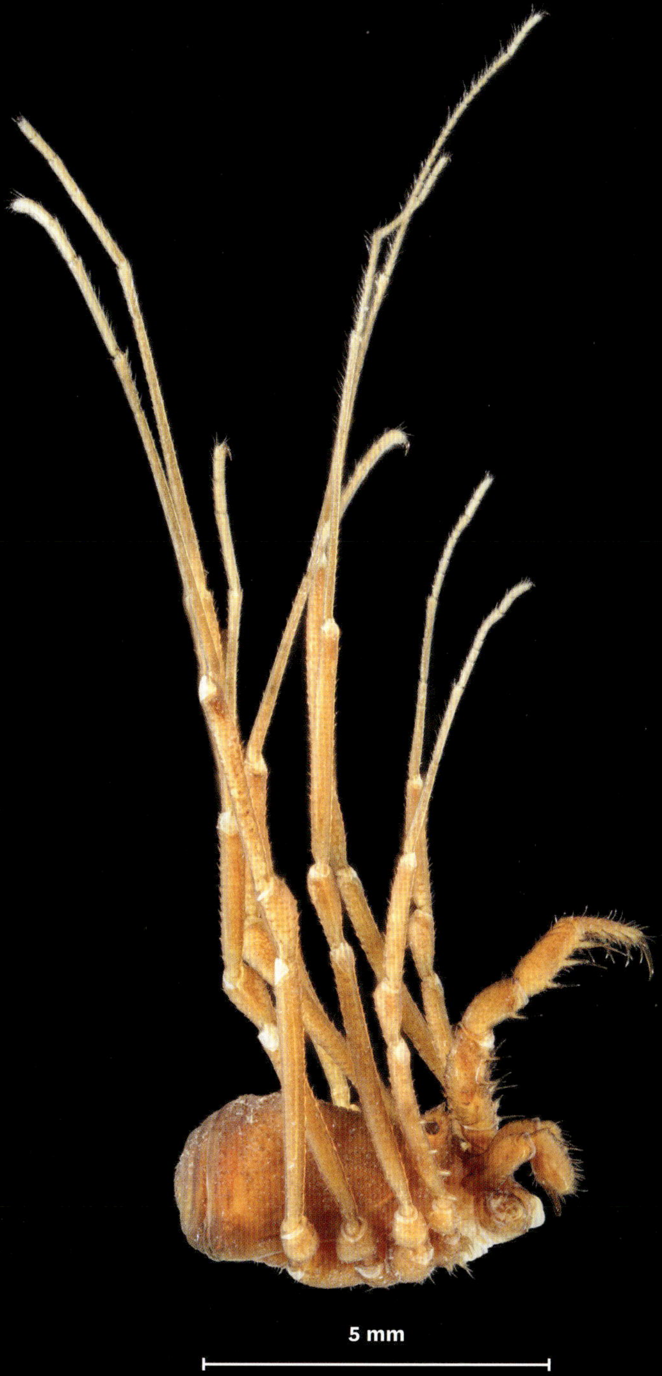
5 mm

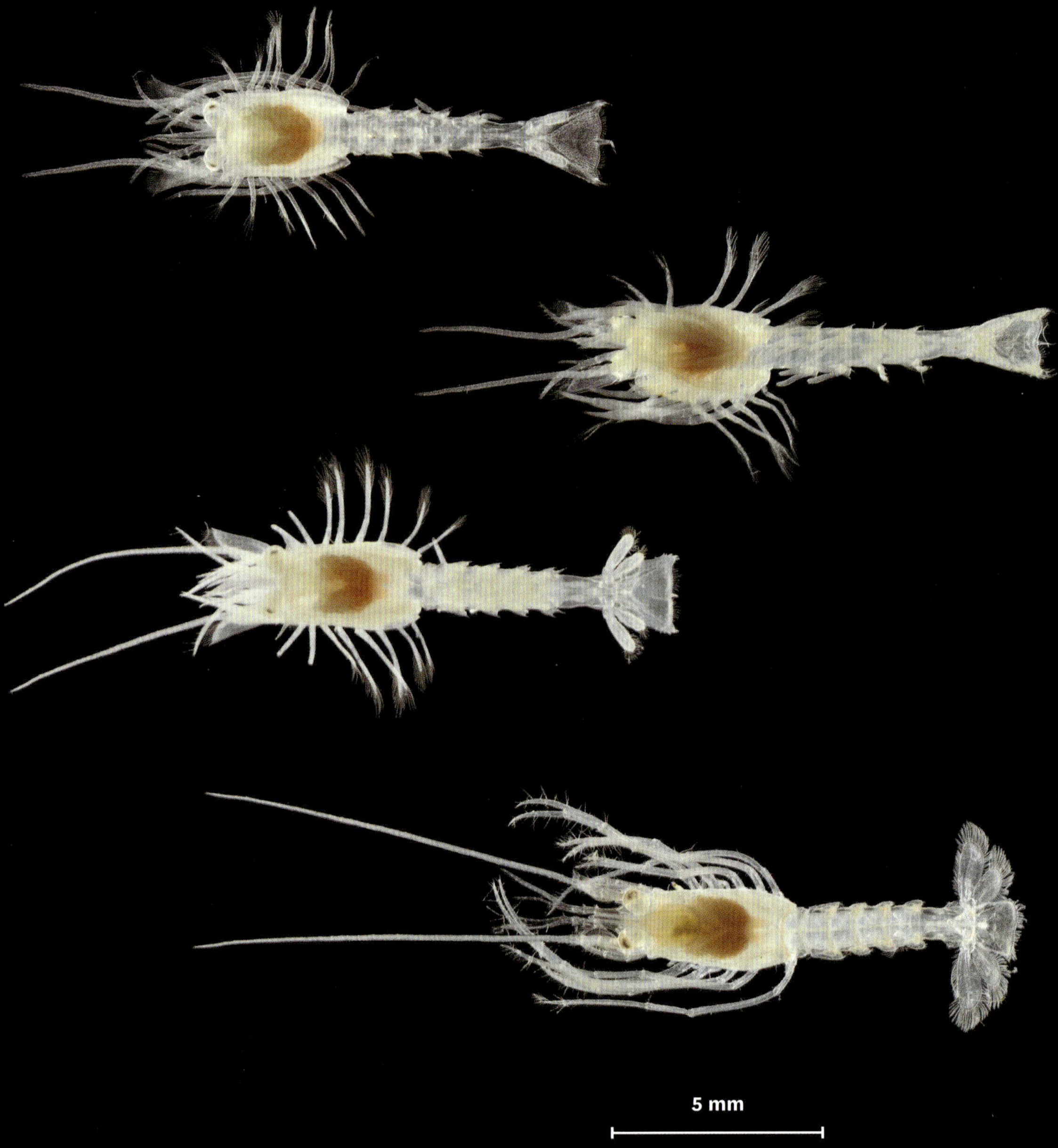
5 mm

New Zealand scampi

Metanephrops challengeri

New Zealand scampi are clawed lobsters that live between 200 and 600 metres deep on sandy mud bottoms suitable for burrowing. Scampi are luxury food; some say the best crustaceans of all. Harvesting them by trawling damages habitat, so catch numbers are limited. Research into the possibility of farming these scampi is therefore under way. An essential but difficult first step was to work out how to rear scampi larvae in laboratory tanks; this has now been achieved by Kevin Heasman of the Cawthron Institute in Nelson.

Shown here are living specimens of the four larval stages. From top to bottom they are: a pre-zoea, first zoea, second zoea and post-larva. While similar in size, changes occur at each moult: antennae get longer, the feathery appendages at the sides (called exopods) gain more setae (bristles), and the tail develops paddle-like limbs. The post-larva has lost most of its exopods and developed three pairs of claws and a tail fan like the adult lobster.

Pre-zoeas hatch from eggs held beneath the female's abdomen. They don't swim or eat, but moult within about an hour to become first zoeas, which still don't eat but do swim. Using their exopods for propulsion, first zoeas position themselves head upwards in the water column and are then dispersed by near-bottom currents. Within two to four days they moult to become second zoeas, which eat live food, and continue to swim and disperse above the bottom for about seven days.

While lobster-like, the post-larva is still semi-transparent and able to swim; it is the transitional stage between swimming and seabed-living. A busy little landscaper, when it finds the right place to burrow it will moult into a small pigmented lobster – on its way to adulthood ready for harvesting! RW

066 / 100

New Zealand scampi (*Metanephrops challengeri*), developmental stages: pre-zoea (CR.023961), zoea 1 (CR.023958), zoea 2 (CR.023959) and post-larva (CR.023960), all body lengths approx. 10 mm, laboratory-hatched by K Heasman, Cawthron Institute, Nelson, 2016–2017.

New Zealand spider orchid

Corybas genus

Orchids in New Zealand? Yes – we have more than one hundred species. Most of our orchids grow in the ground, with just a few species found perching on trees and rocks. Unlike their tropical relatives, New Zealand orchids are small plants, not very colourful, and their flowers are only a few millimetres across. The showiest species have dull green, brown, dark purple or white flowers.

067
100

Some of our most bizarre and fascinating orchids bloom in the middle of winter. The flowers are half-buried in the forest leaf litter and their long, thin sepals and petals point forward, looking almost like spiders ready to attack. These orchids are, unsurprisingly, known as spider orchids (genus *Corybas*), and there are about twenty species in New Zealand. Just how these orchids fertilise their flowers has fascinated scientists and orchid enthusiasts alike since the late 1800s, as spider orchids do not produce nectar to attract and reward their pollinators.

After almost six years of research, we have gathered strong evidence to confirm that spider orchids are pollinated by tiny flies called fungus gnats (genus *Mycetophila*). The orchids appear to fool the gnats by mimicking the aroma of a mushroom, the preferred mating spot of fungus gnats. Following courtship and mating, the female seeks a spot inside the flower to lay her eggs, thereby acting as a pollinator. After hatching, the larvae simply die.

While studying the pollination of these orchids, we also came across five new species of spider orchid; some have been in Te Papa's botanical collections for years, but their distinctiveness had remained unnoticed. Little is known about New Zealand fungus gnats, so there might well be a few species awaiting discovery. CL

Spider orchid (*Corybas acuminatus*) and fungus gnat (*Mycetophila* sp.) loaded with orchid pollen, observed in Waitakere Ranges, Auckland, 2016.

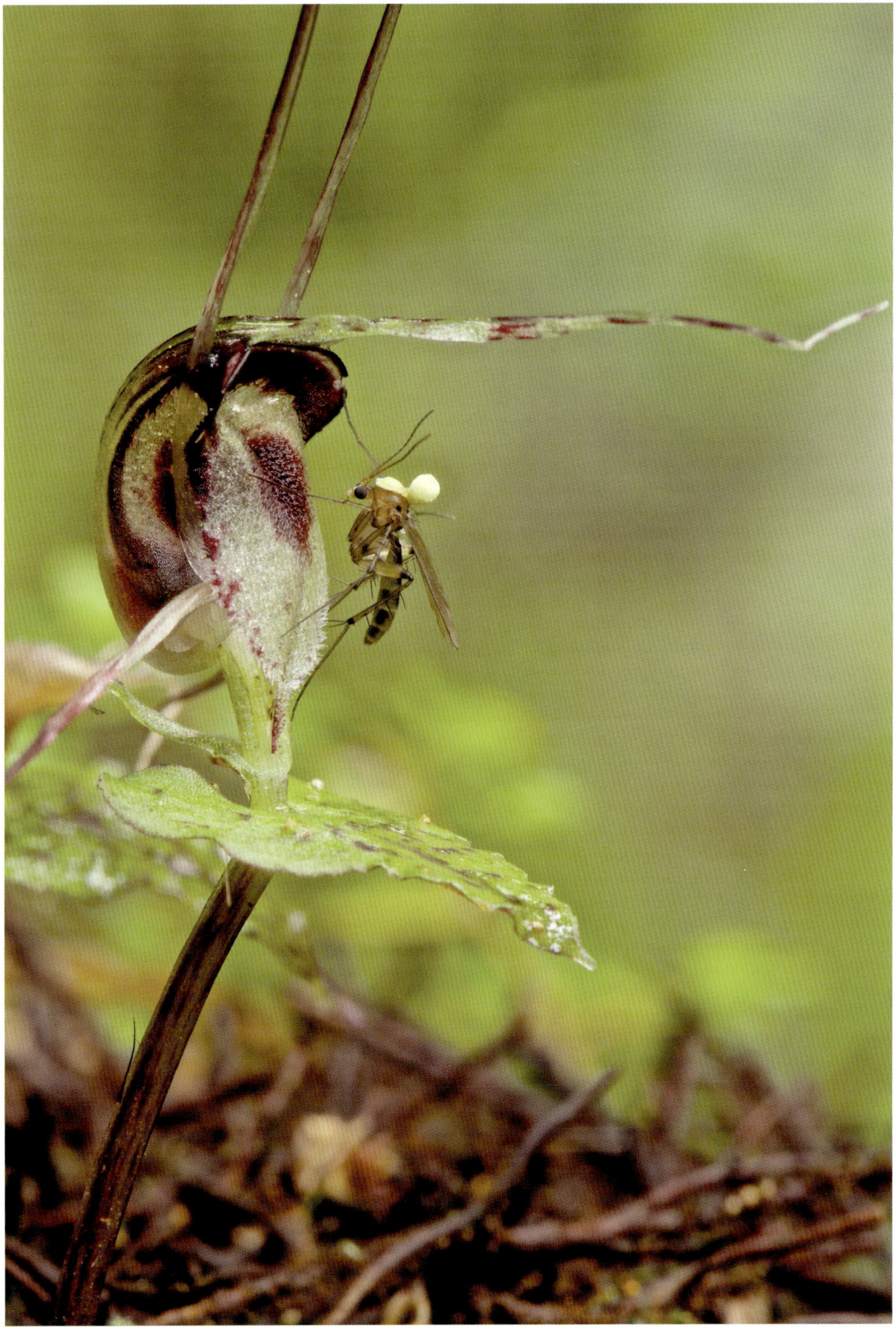

Parasitic wasp

Certonotus fractinervis

Many wasp species are parasitoids of a wide range of insects and spiders – the larvae of a parasitoid develop in or on the host, feeding on its tissues. Not only may this help to control insect pests, but the free-living adult wasps are also very important pollinators.

The host of New Zealand's largest parasitic wasp, *Certonotus fractinervis*, are the larvae of the elephant weevil (*Rhynchodes ursus*), which cause dramatic damage to native trees by creating tunnels in their trunks. How the wasp finds the weevil's larvae, buried deep in their tree-trunk lairs, is quite remarkable. The wasp can bore down into the wood by fully rotating the last segments of its body; as it tunnels, it secretes a substance that destroys wood fibres.

And there's more! The egg-laying organ (ovipositor) of this wasp is longer than the wasp itself and is highly flexible, as demonstrated by *Megarhyssa nortoni*. Introduced into New Zealand to control horntail wasps, the tip of its ovipositor contains metal (ionised manganese, zinc and/or copper), which helps the wasp insert the ovipositor through several centimetres of wood. The female of *C. fractinervis* is also able to detect the presence of the host by smelling with its antennae and feeling for vibrations in the wood.

Once the egg has been injected into the host, the parasitic larva hatches and devours its host over the winter, before pupating. The emerging adult wasp escapes from the tree trunk by chewing itself out; its mandibles are also hardened with metals. Around the time the females emerge, male wasps wander over logs and tree trunks, tapping with their antennae and searching for chewing females with which to mate. JK

Parasitic wasp (*Certonotus fractinervis*), AI.034831, collected by L Sivyer, Owhango (5 km south of Taumarunui), Taupō, April 1989.

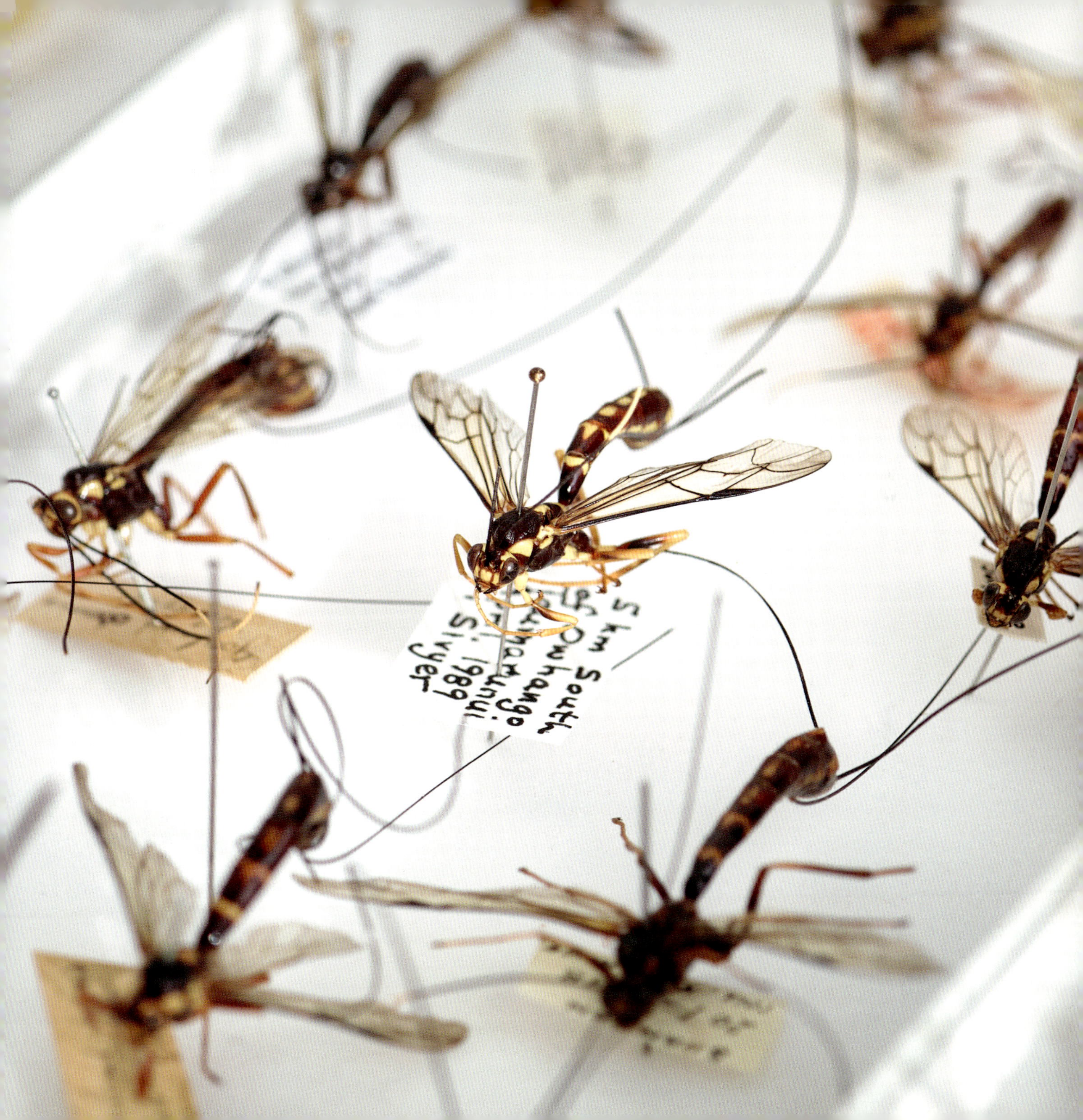
5 km South
Owhango
1989
Sivyer

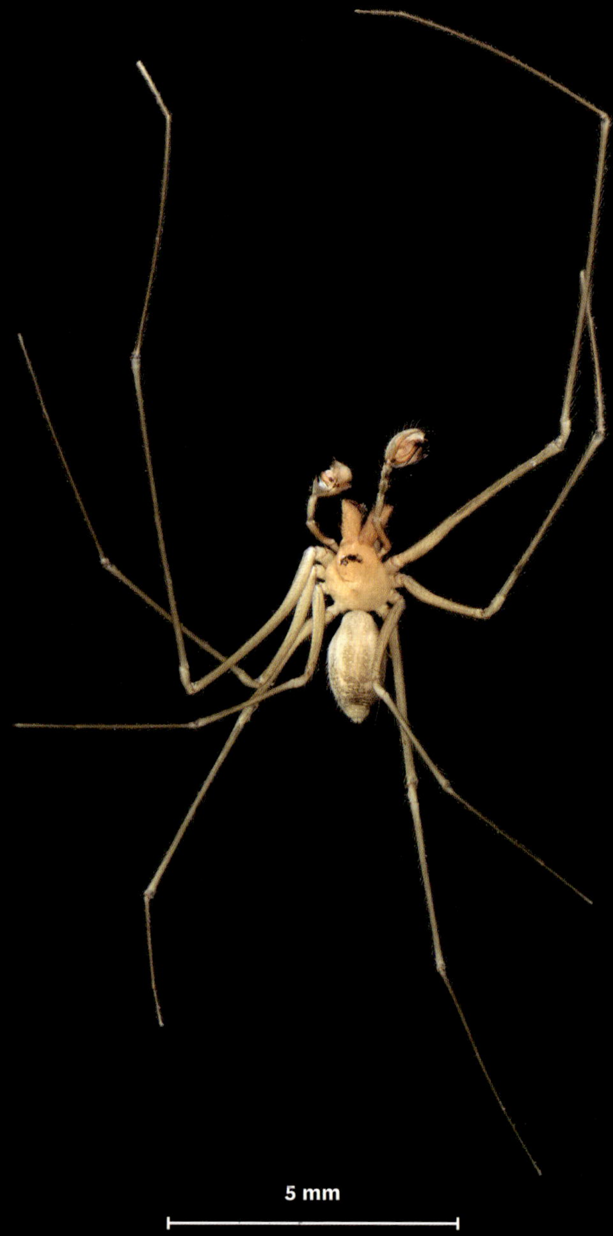
5 mm

Physoglenid spider

Zeatupua forsteri

New Zealand is home to a rich and varied spider fauna. We have over 1100 named species so far, and it would not be out of the question for that total to rise to around 2000 or so. Even more astounding is that about 95 percent of them are endemic. This means that nearly all of our spider species are as unique to New Zealand as more-famous animals such as kiwi and tuatara.

While we do have some fairly large species, much of our spider diversity is at the smaller end of the size scale. One example of this is the family Physoglenidae (formerly part of the family Synotaxidae). These spiders are found only in Australia, Chile and New Zealand, with New Zealand being home to fifty-eight of the seventy-two known species.

To the naked eye, physoglenines resemble a miniature daddy longlegs spider (*Pholcus phalangioides*), but they are not closely related. Many species show strong sexual dimorphism, with males often exhibiting bizarre anatomical features. For example, males of the genus *Nomaua* have their eyes on mounds topped with a row of hairs that looks like a Mohawk haircut. Physoglenines live in native forests and their webs can often be found in low-growing ferns above damp ground.

The species shown here is *Zeatupua forsteri*. It is named for Ray Forster, who did more work on the New Zealand members of the family than anyone else. It is the only member of its genus, and so far has only been found in the beech forest of the Orongorongo Valley near Wellington. This area has been subject to much collecting effort, yet we have never found the female of this species. PS

Physoglenid spider (*Zeatupua forsteri*), AS.001487, holotype, collected by BM Fitzgerald, Orongorongo Valley, Wellington, 1 November 1992.

Pitted tangle fern

Gleichenia inclusisora

Tangle ferns of the genus *Gleichenia* get their common name from the way their repeatedly forking fronds tangle with one another and with other vegetation. Back in 2012, their taxonomy was also in a tangle. I hoped to resolve this by examining Te Papa's 250 or so specimens. As I checked specimen after specimen under a microscope, I was becoming confident that I could attribute each to one of three previously described species. Although superficially similar, fine details consistently separated them. But then I checked the specimen pictured here.

I physically recoiled from the microscope. It was the only time in my life when I did not believe what I was seeing. Strikingly, the spore capsules were embedded in a 'pit' in the underside of the frond segments, rather than sitting more or less flush on its surface. Shocked amazement quickly gave way to realisation that this specimen was so different that it must be a new species! We later named it *Gleichenia inclusisora*, after the 'included' position of its spore capsules (known in ferns as sori).

This specimen was collected in 1948 by the eminent naturalist WRB Oliver. He was not alone in overlooking its significance; I even have to admit to collecting other specimens of *Gleichenia inclusisora* prior to my 2012 investigation and not realising what I had at the time. The long gap between collection and identification clearly demonstrates that recognising the 'new' depends on having a solid understanding of the 'old'. Collections like those of Te Papa help us do both parts of this discovery process: they document what we currently understand about nature's diversity, while simultaneously being a treasure trove of discoveries yet to be made. LP

Pitted tangle fern (*Gleichenia inclusisora*), P005512, collected by WRB Oliver, Mawheraiti, West Coast, 19 April 1948.

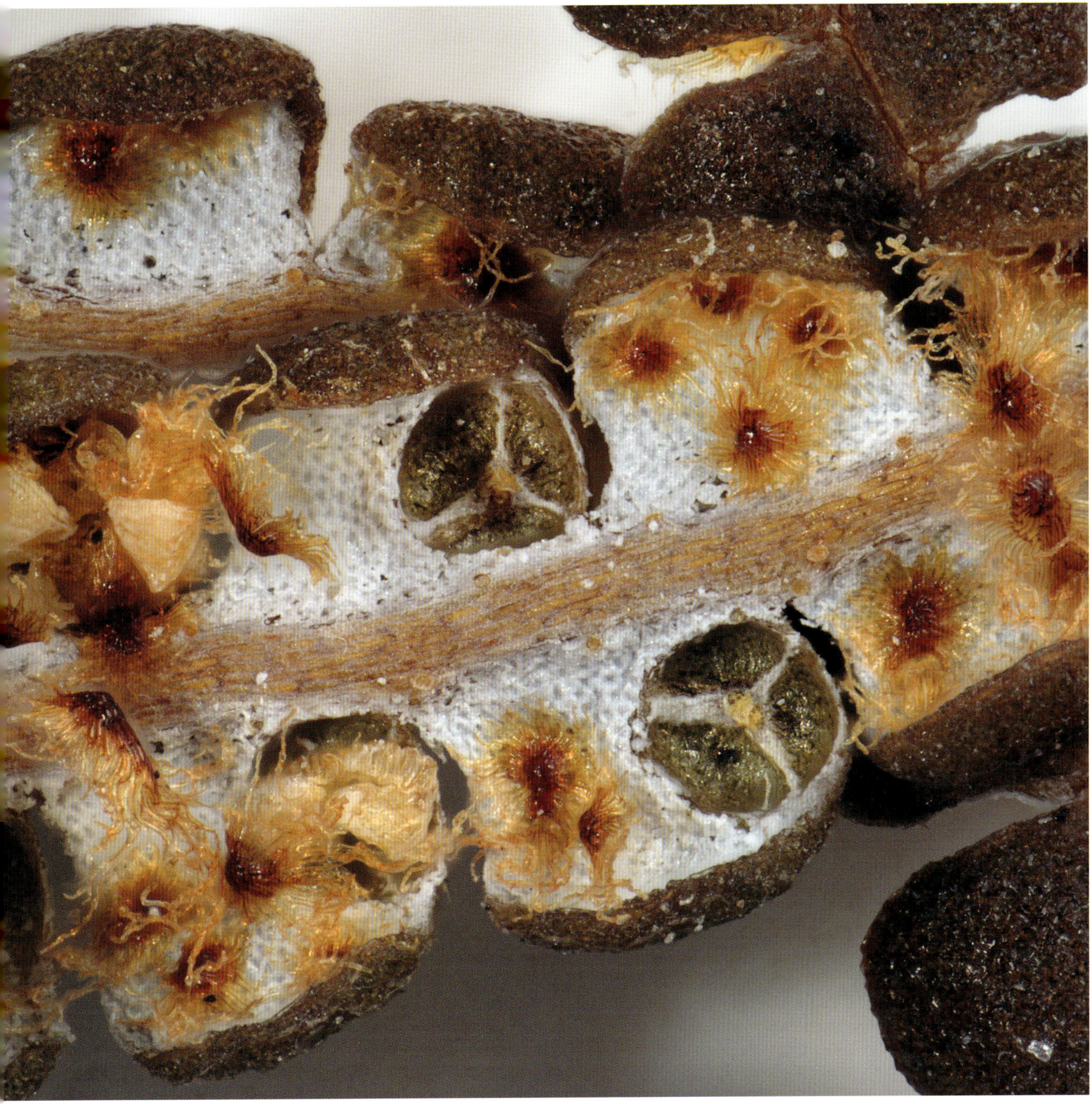

Prickly shield fern

Polystichum vestitum

Printer of the Treaty of Waitangi, missionary, explorer, advocate for Māori, politician, and botanist … that's quite a roll-call for the controversial William Colenso. In 1841, in the Bay of Islands, Colenso met Joseph Hooker, then naturalist on JC Ross's Antarctic Expedition and later to become Director of Kew Gardens and the British Empire's leading botanist. They corresponded for many decades, with Colenso sending New Zealand plants to Hooker for scientific appraisal. Colenso's specimens in Te Papa are largely duplicates (secondary collections) of these exported specimens.

Colenso fancied himself as a plant taxonomist and described numerous new plant species – many of which, in Hooker's view, didn't warrant recognition. The ferns shown here are the type specimens for the names *Aspidium waikarense* and *A. pulcherrimum*, described by Colenso after his 1841 journey through the Urewera region. They are part of Colenso's so-called fern folio, which is, unusually, kept separate from the main botany collection.

I first came across these specimens around 2000, as part of my PhD on *Polystichum* shield ferns. To my eye, both are obviously the same species, *Polystichum vestitum*. It's understandable that Colenso may have been unaware that the species had already been described by Georg Forster after Captain Cook's second expedition, but why Colenso thought his two specimens were two different species escapes me. I take this as a useful lesson that the names I apply to specimens are on display forever, and the specimens themselves are permanent vouchers of both New Zealand's flora and our (hopefully improving) understanding of it. However, I can't completely dismiss the possibility that future analysis may prove Colenso right! LP

Prickly shield fern/pūnui (*Polystichum vestitum*), P003200 and P003201, type specimens of *Aspidium waikarense* and *A. pulcherrimum*, respectively, collected by W Colenso, Lake Waikaremoana, East Coast, December 1841.

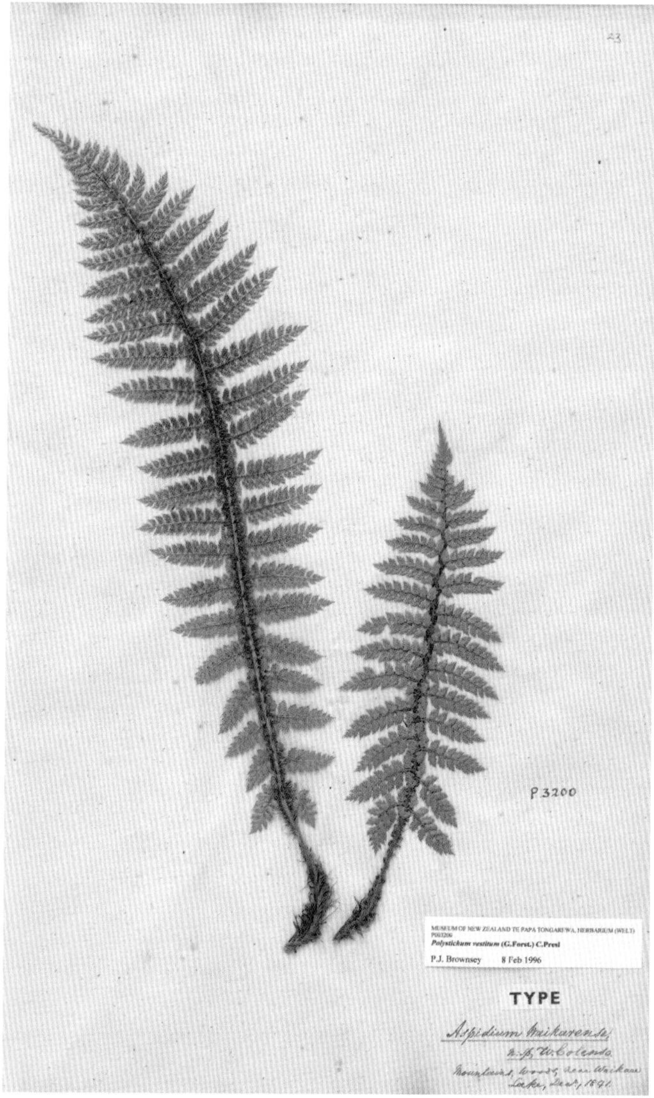
P 3200
MUSEUM OF NEW ZEALAND TE PAPA TONGAREWA, HERBARIUM (WELT)
Polystichum vestitum (G.Forst.) C.Presl
P.J. Brownsey 8 Feb 1996
TYPE

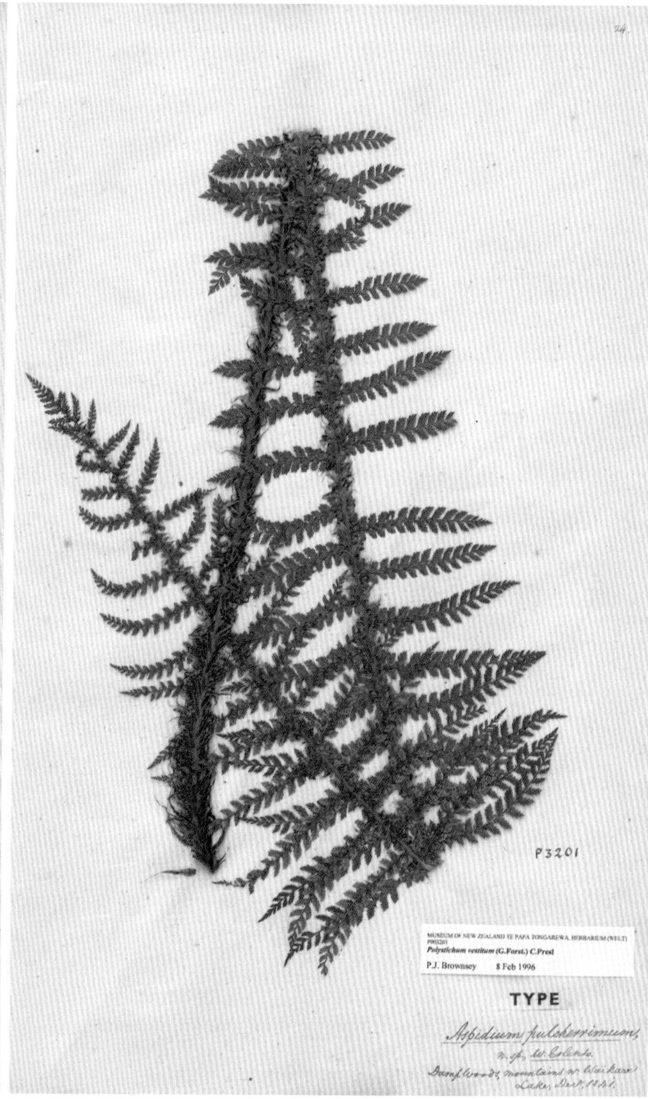
P3201
MUSEUM OF NEW ZEALAND TE PAPA TONGAREWA, HERBARIUM (WELT)
Polystichum vestitum (G.Forst.) C.Presl
P.J. Brownsey 8 Feb 1996
TYPE

Pseudoscorpions

Pseudoscorpiones order

Every now and then, someone contacts Te Papa thinking they've found a baby scorpion, only it doesn't have a stinging tail. What they've found is a pseudoscorpion. Like the scorpions, these animals are arachnids, but they belong to a different order (Pseudoscorpiones). They are also a lot smaller than scorpions – most are no more than a few millimetres long and the largest reaches the less-than-imposing size of just over 1 centimetre.

A pseudoscorpion's most obvious feature is its two pincer-tipped pedipalps (second pair of appendages). These animals are predators and, like their scorpion cousins, use their pincers to grab prey. While they might lack the stinging tail of their scorpion cousins, pseudoscorpion pincers have venom glands that can deliver toxins to subdue their victims. Their tiny size means that they are no threat to us.

The pincers also have another, rather surprising role – they can help these animals fly. Pseudoscorpions are famous for a behaviour called phoresy. This is where one animal uses another for transport. Using one of its pincers, a pseudoscorpion will grab the leg of an insect such as a fly and hitch a ride. This is how they can end up indoors when they are much more at home in leaf litter and soil.

New Zealand has around seventy known species and subspecies in this group, but there could be at least as many again still awaiting discovery. In contrast, we have no species of true scorpions in New Zealand; should anybody find one, they should contact the Ministry for Primary Industries immediately! PS

New Zealand pseudoscorpion species: (A) ***Apocheiridium validum*****, AF.000010, holotype, collected by M Luxton, Totara Reserve, Pohangina Valley, Palmerston North, 21 August 1965. (B)** ***Tyrannochthoniella foveauxana*****, AF.000031, holotype, collected by HJ Cranfield, Ruapuke Island, Foveaux Strait, 29 January 1965. (C)** ***Sathrochthoniella zealandica*****, AF.000016, collected by R Cooper, Taurewa, Tongariro National Park, 12 May 1965. (D)** ***Austrochthonius inversus mordax*****, AF.000008, collected by M Luxton, Wairarapa, 22 April 1965. (E)** ***Morikawia densedentata*****, AF.000001, collected by M Luxton, Omahuta Forest, North Auckland, 30 June 1965.**

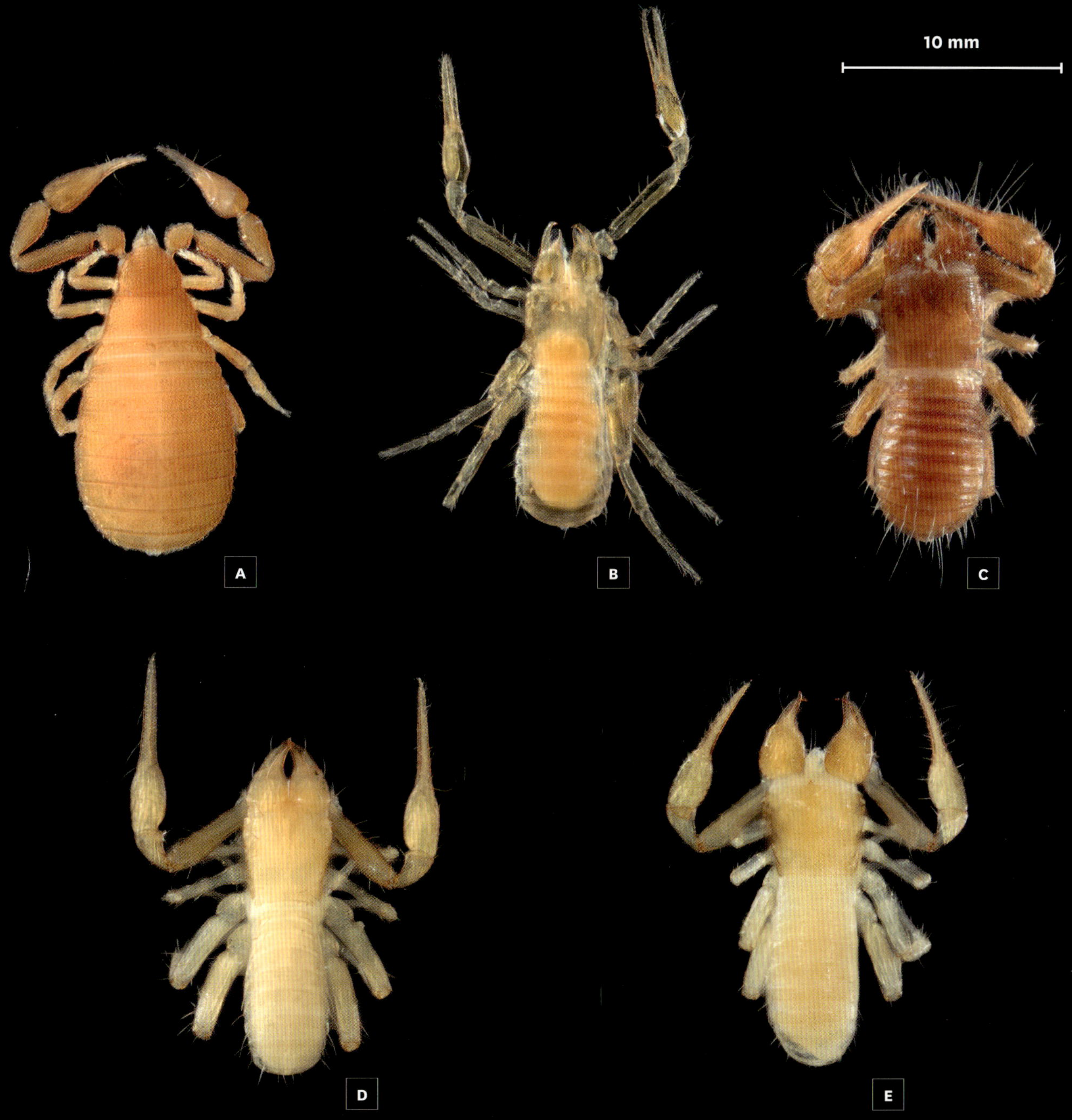
10 mm
A
B
C
D
E

Objects
R—S

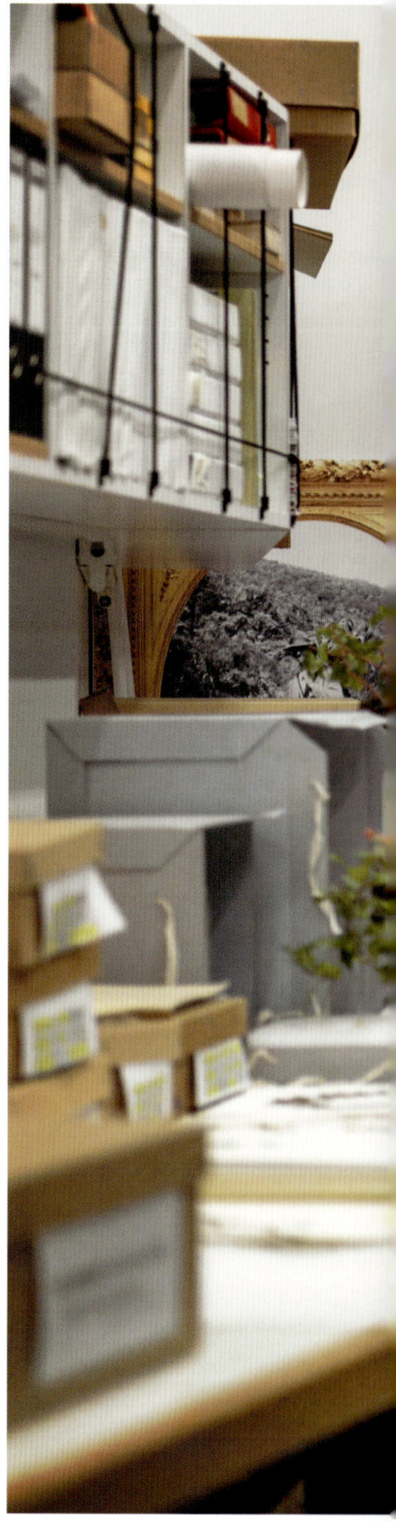

Manager Science and Humanities Collections Jennifer Dalen, far left, Research Technician Botany Bridget Hatton and Kaitiaki Taonga Collection Manager for Botany Antony Kusabs in the Botany collection storage room at Te Papa's Cable Street building.

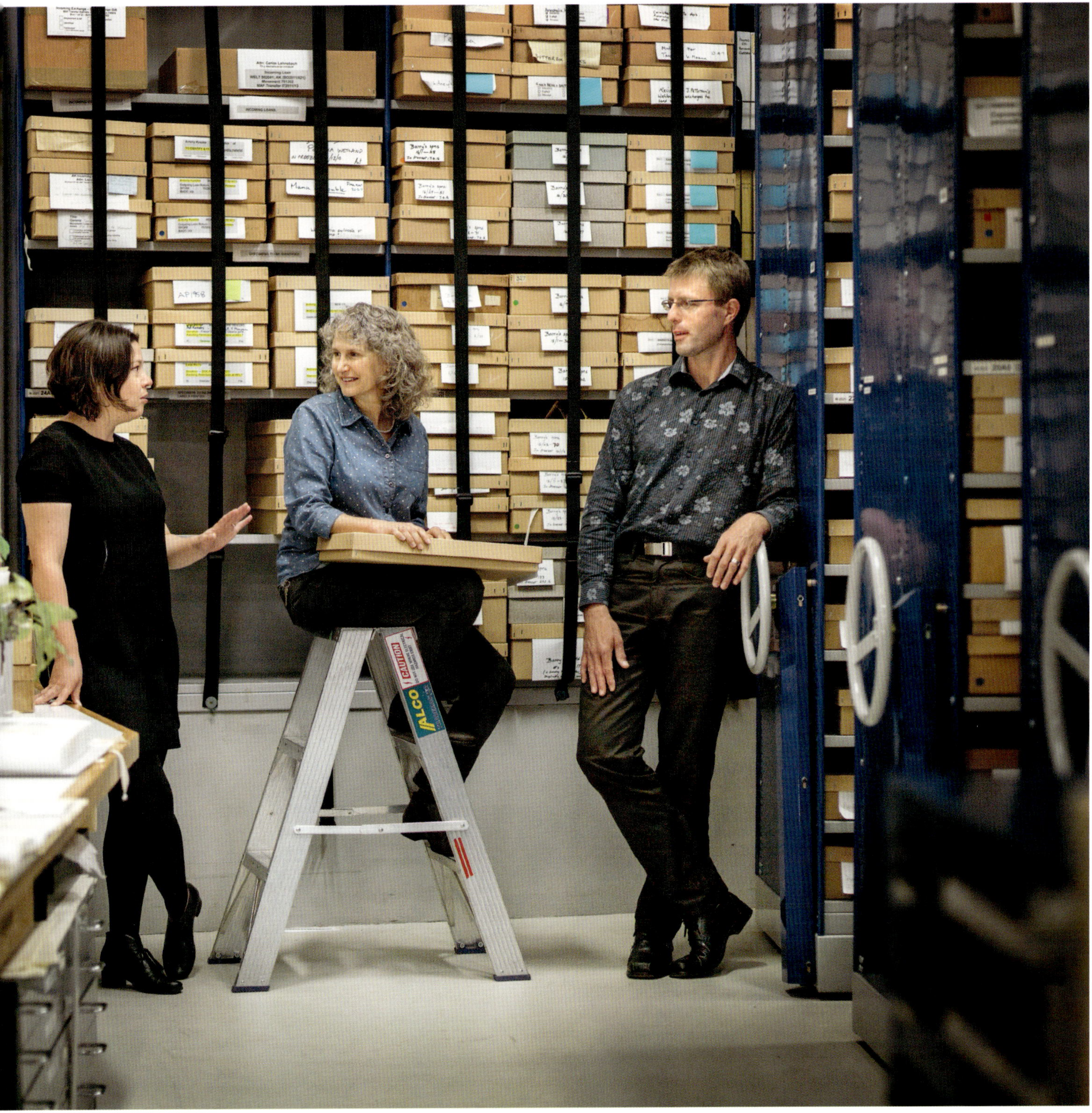
ALCO
CAUTION

064 065 066

071 072 073

077 078 079

083 084 085

089 090 091

095 096 097

067 068 069

074 075 076

080 081 082

086 087 08

092 093 09

098 099 10

Red seaweed

Adamsiella chauvinii

Provenance – especially locality and date of collection – is critical for turning a plant, animal or fungus specimen into a data point for scientifically documenting biodiversity. A specimen without provenance is worthless, although there are rare exceptions: arguably, it would be better to have a unicorn specimen with no provenance rather than no specimen at all.

Nowadays, provenance is recorded in Te Papa's central collections database, including GPS coordinates, habitat description and information about abundance. Changes in the name given to a specimen are also dutifully recorded. Labels summarising these data are computer-printed and fixed to the specimen; the information for each specimen is (or should be) comprehensive and comprehensible.

Before computers, however, specimen data were hand-written, and bad handwriting is neither a modern curse nor restricted to medical doctors. Although I am throwing stones in a glasshouse here, I am regularly bewildered by the scrawl of past collectors and researchers, and I haven't developed the ability of some of my colleagues to intuitively divine my predecessors' meaning. This is why I am always relieved to come across labels written by Nancy Adams, for both her own specimens and those she curated while working at the museum from 1959 to 1987. The care that Nancy took is evident in her neat, legible labels that can be read instantly. A quick survey suggests that about 5 percent of Te Papa's New Zealand seaweed and vascular plant specimens bear a label written by Nancy.

The specimen shown here is a red seaweed from Stewart Island, identified initially by Nancy as *Lenormandia chauvinii*. The species is now known as *Adamsiella chauvinii*, the genus being named after Nancy in recognition of her botanical expertise, particularly with seaweeds. LP

***Adamsiella chauvinii*, A002971, collected by P Cresswell, Paterson Inlet, Stewart Island, 26 January 1970.**

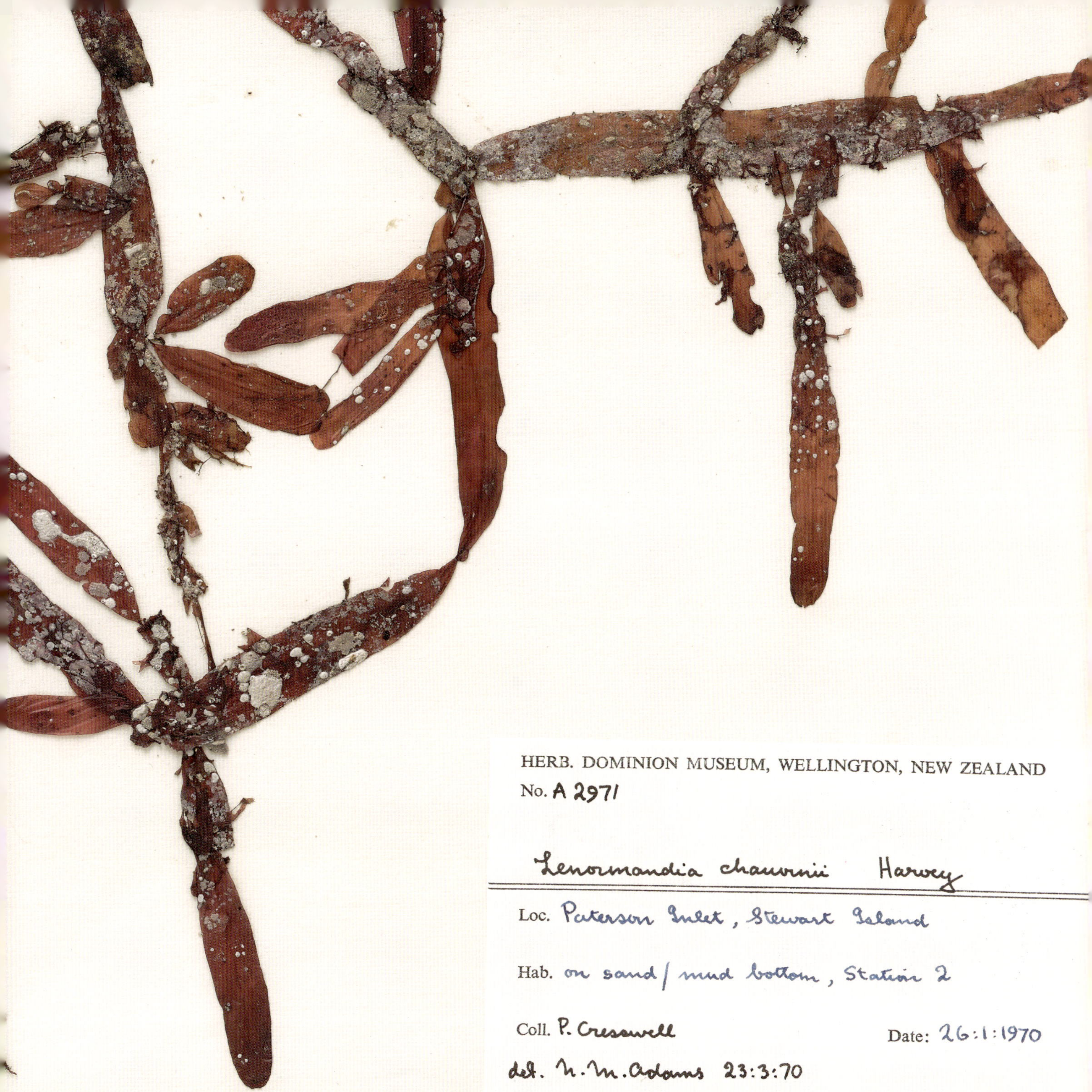

HERB. DOMINION MUSEUM, WELLINGTON, NEW ZEALAND

No. A 2971

Lenormandia chauvinii Harvey

Loc. Paterson Inlet, Stewart Island

Hab. on sand/mud bottom, Station 2

Coll. P. Cresswell

Date: 26:1:1970

det. N. M. Adams 23:3:70

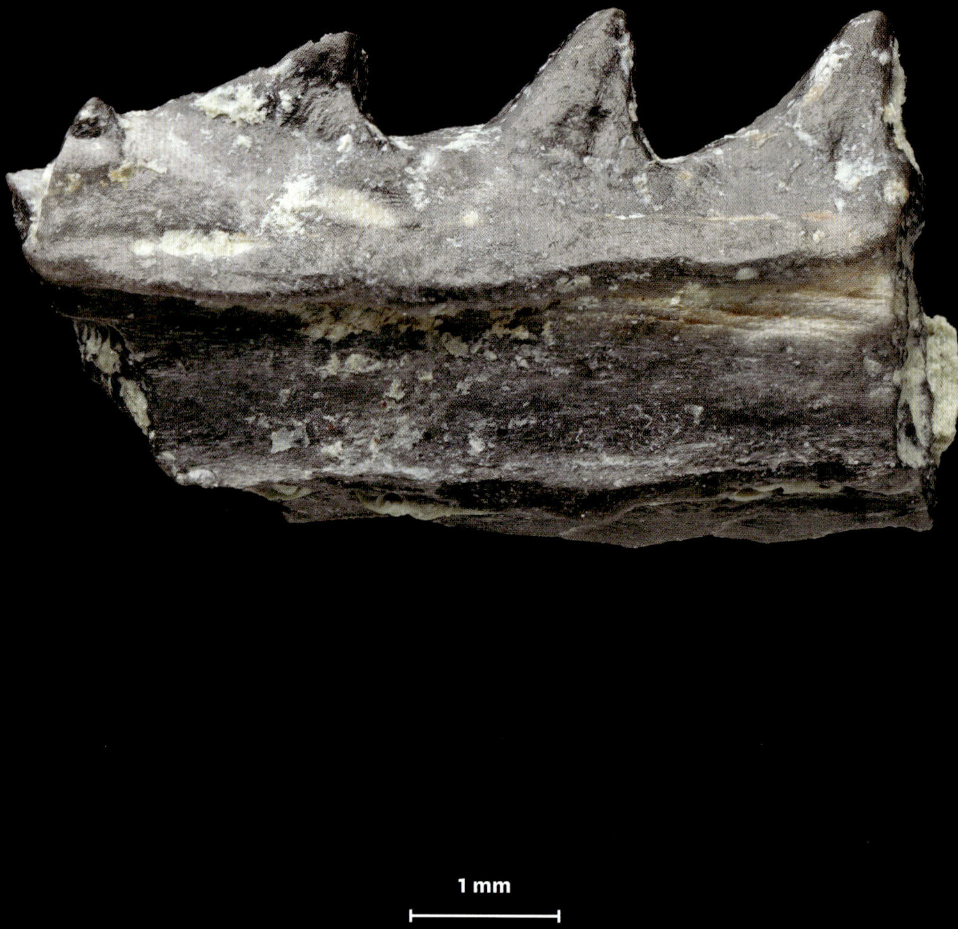
1 mm

Rhynchocephalian mandible

Sphenodon punctatus

New Zealand's tuatara is one of the most unusual animals on Earth. Although often thought of as a type of lizard, it is actually a rhynchocephalian, which is on a unique branch of the reptile tree. The other major groups of living reptiles are turtles, crocodilians, snakes and lizards. Birds are technically in the same tree, being the only surviving descendants of dinosaurs; and, like all mammals, humans have also evolved from an early branch of the reptile tree.

We know very little about the origin and evolutionary history of tuatara. They are the sole survivors of the rhynchocephalians, which flourished globally about 200 million years ago. Rhynchocephalians were presumably widespread on the giant southern land mass Gondwana about sixty million years ago, because their fossils have been found in rocks of that age in Argentina. At about this time, Zealandia (proto-New Zealand) became isolated as Gondwana broke up, presumably carrying the ancestors of tuatara on it. Excavations in rocks aged sixteen to nineteen million years old near St Bathans, in central Otago, have revealed very fragmentary rhynchocephalian fossils that closely resemble tuatara; aside from these, the tuatara fossil record is limited to just the past few thousand years.

Tuatara were widespread on mainland New Zealand until human arrival brought devastation. They are generally unable to breed if rats are present, so the arrival of kiore (Pacific rats) with Polynesian settlers saw tuatara disappear from everywhere except a few rat-free islands and areas only recently colonised by rats. The future for tuatara is now much brighter, however, because rats have been eradicated from numerous islands, allowing both remnant tuatara populations to recover and other populations to be re-established through translocation. **AT**

Rhynchocephalian fossil mandible fragment, S.42282, collected by A Tennyson, T Worthy and C Jones, Vinegar Hill, St Bathans, 13 December 2001.

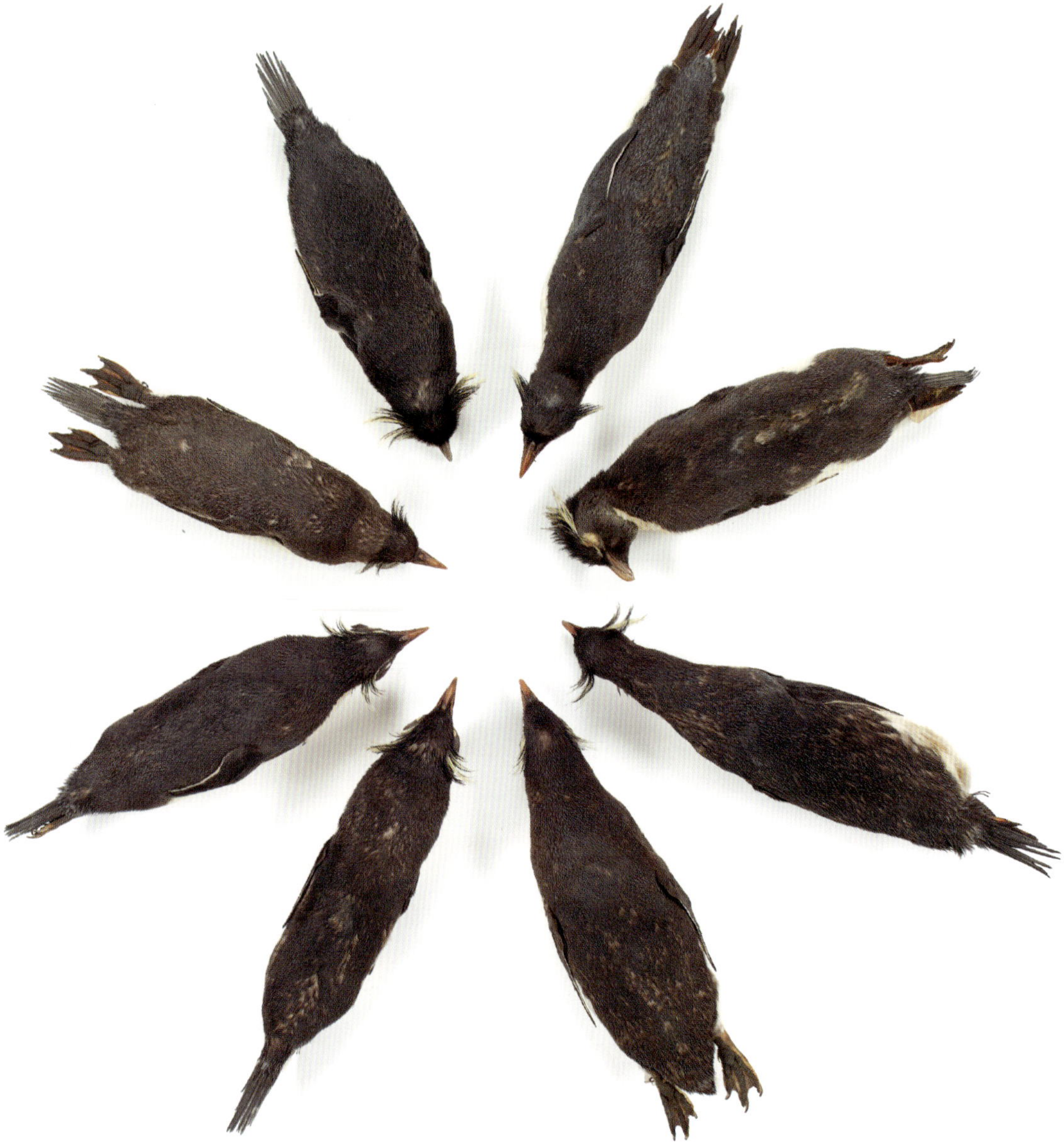

Rockhopper penguin

Eudyptes chrysocome

Sometimes objects collected by a museum a very long time ago, and that may seem otherwise ordinary, can spark new avenues of scientific investigation. Here is one such example.

Rockhopper penguins, found around the Southern Ocean, nest on small subantarctic islands spanning all the meridians of longitude of the globe. Penguin numbers have decreased dramatically at some sites, and as there were no introduced predators or other threats at these sites, the changing oceanic food web was thought to be the cause. Evidence of this was only found in the 2000s, when scientists analysed feathers from museum specimens of penguins dating back to 1861. The chemical composition of the feathers revealed clues about the penguins' food sources at the time each specimen was collected.

Penguins lay down feathers only once in the year, which vary subtly in chemical composition depending on each bird's diet at the time. These feathers, like our fingernails or hair, are formed from what was consumed immediately beforehand. Stable isotopes of carbon and nitrogen from feathers, tissue or blood are used to assess a specimen's position in the food web, and reveal patterns in diet that may change between regions, time periods and populations.

The scientists found that over time, the penguins had consumed prey with lower carbon isotope and nitrogen isotope ratios. This showed that, as time went by, the penguins' prey was coming from marine food webs of lower productivity and the prey was from lower down in the food chain. Simply put, the penguins were less able to find enough rich food near to their nest sites to survive in big numbers.

This example shows how museum collections can form a kind of library or forensic database going back in history. Modern analytical techniques can reveal information that was simply not available when the samples were collected. SW + JS

Rockhopper penguin (*Eudyptes chrysocome*) from Te Papa collections, during different decades from around New Zealand breeding sites.

Rowi's feather louse

Apterygon okarito

'Tell me who you carry and I'll tell you who you are' is one of my former colleague Ricardo Palma's favourite sayings. Ricardo studies bird lice, and each species of louse is usually found exclusively on one or a very few species of bird. Ricardo can often tell you which bird a particular louse came from, and he can help identify decayed, beach-wrecked birds based on lice harvested from their feathers. It's a lot quicker than DNA analysis if you know what you're looking at.

The fact that bird lice are very species-specific helped Ricardo in one of his career highlights. Not only did he describe a new species of louse, but his lousy evidence also helped to reveal a new species of kiwi, which he also described. According to Ricardo, he's the first person since the late eighteenth century to describe both a new species of bird and one of its lice.

Ricardo had a few lice collected in 1979 from a kiwi at Okarito on the West Coast. They looked distinct from other kiwi lice, and this got Ricardo thinking that the bird they came from might also be a new species. DNA analysis confirmed that the kiwi was indeed a new species, and additional specimens helped prove that the louse was a new species too. In 2003, Ricardo co-authored the paper that described the rowi, also called the Okarito brown kiwi (*Apteryx rowi*), and a year later he described its unique louse, *Apterygon okarito*.

Te Papa's louse collection consists of around 70,000 slide-mounted specimens and many more preserved in ethanol. It includes head lice from an ancient Egyptian comb and from extinct species of bird such as huia. **PS**

Rowi's feather louse (*Apterygon okarito*), AI.012588, collected by Department of Conservation staff from rowi kiwi (*Apteryx rowi*) collected at The Forks, Okarito, south Westland.

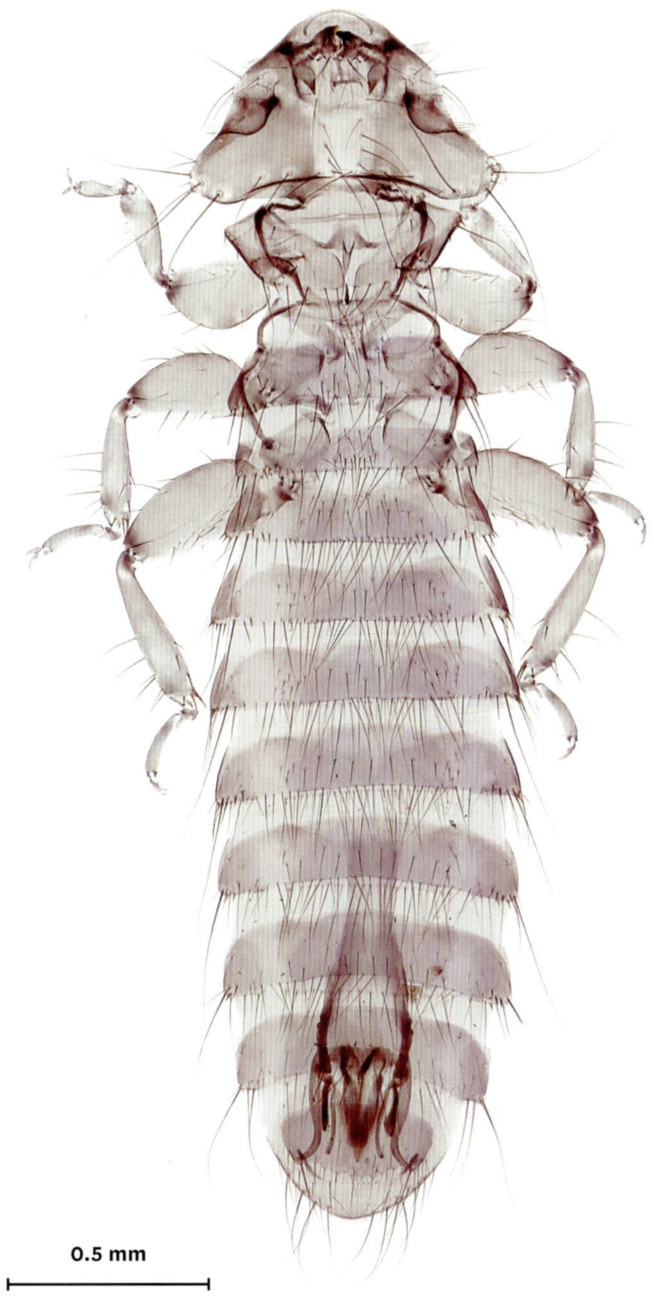
0.5 mm

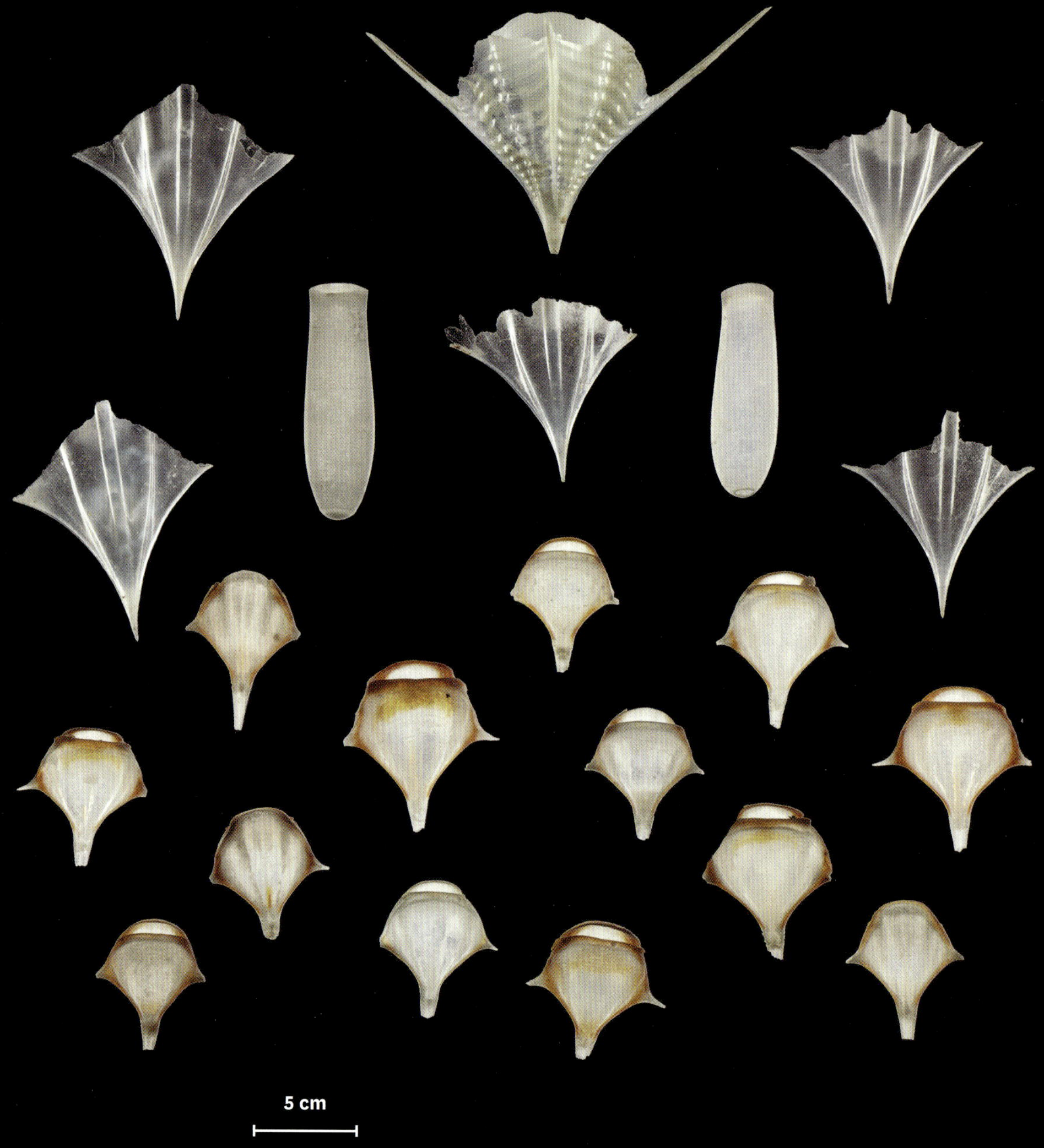
5 cm

Sea butterflies

Pteropoda order

When we think about snails, the image that comes to mind is a soft-bodied animal carrying a bulky spiral shell on its back. However, there is a group of animals so other-worldly in appearance that it is hard to imagine that they are actually snails. They are called pteropods, which comes from the Greek word for 'wing-footed'. In these animals, the typical gastropod foot (the 'belly' of a snail) is divided in two large lobes resembling wings. The animals flap these wings to 'fly' through the ocean, gathering planktonic food particles with the help of a mucous web.

Pteropods tend to be found in warmer waters, but are known around the world, including in polar regions. Some forty species can be found offshore all around New Zealand. They can be roughly divided in two sub-groups, the sea angels and the sea butterflies. Sea angels, like slugs, have lost their shells in the course of evolution. They resemble angel-like Christmas-tree ornaments. Sea butterflies have shells, albeit very weird ones – they may look like globes, ice-cream cones, hollow tusks, test tubes or tiny spaceships.

Because their calcareous shells are more easily preserved, sea butterflies are the pteropods typically found in museum collections such as Te Papa's. Even so, their shells are diaphanous and extremely fragile. Our oceans are becoming more acidic as global temperatures rise, and at some point we expect the water to be acidic enough to dissolve the shells of sea butterflies. Not only would the world be a poorer place without them, but – more alarmingly – their disappearance would also send shockwaves through the food web, contributing to the collapse of entire ecosystems. RS

Sea butterflies, suborder Euthecosomata, M.11049, M.20834, M.215524, M.300309, width 2 cm (top centre specimen), unknown collectors, collected on different occasions in the Bay of Bengal and at several localities around the North Island, New Zealand, 1957–2002.

Sea louse

Gloiopotes huttoni

This streamlined crustacean, *Gloiopotes huttoni*, is well designed for its chosen habitat: it lives on fast-swimming big-game fishes. *Gloiopotes* are fish lice, external parasites of these impressive marine predators.

The photographs show side, top and underneath views of a 13-millimetre-long female *Gloiopotes* from a New Zealand striped marlin (*Kajikia audax*). As the fish swims at high speed, the louse, aided by its flattened and hydrodynamic body shape, clamps the edges of its large dorsal shield to the host's skin to prevent being swept away.

Behind the shield is a large segment with two gold-coloured egg-strings. Each contains roughly one hundred eggs, stacked like plates, which mature as they pass along the string to hatch into the water as microscopic, free-swimming larvae. The larvae then find new hosts, but how, where or when this occurs is unknown.

The appendages or limbs have dark, hardened tips and spines that anchor the parasite to its host. They include mouthparts forward of centre, which penetrate the host to feed on mucus, skin and blood. The parasites mainly congregate on the host's undersides; when numerous, they leave raw, red patches behind.

In New Zealand, *Gloiopotes huttoni* has been collected from striped, black and blue marlin, shortbill spearfish and mako sharks. These hosts all live offshore in oceanic waters and move long distances, even crossing oceans, although little is known about their associations and distribution.

While morphologically similar, *Gloiopotes* species are also variable, making it difficult to distinguish them on different host species. Together with Clive Roberts, Te Papa's Curator of Fishes, we have been collecting *Gloiopotes* specimens to investigate whether DNA analysis can help to identify these parasites found on different host species in New Zealand's waters. RW

Sea louse *(Gloiopotes huttoni)*, CR.0025166, length 13 mm, from skin of striped marlin host *(Kajikia audax)*, caught by R Pattison, in surface water, over 145 m bottom, off Poor Knights Islands, Tutukaka, Northland, 12 January 2013.

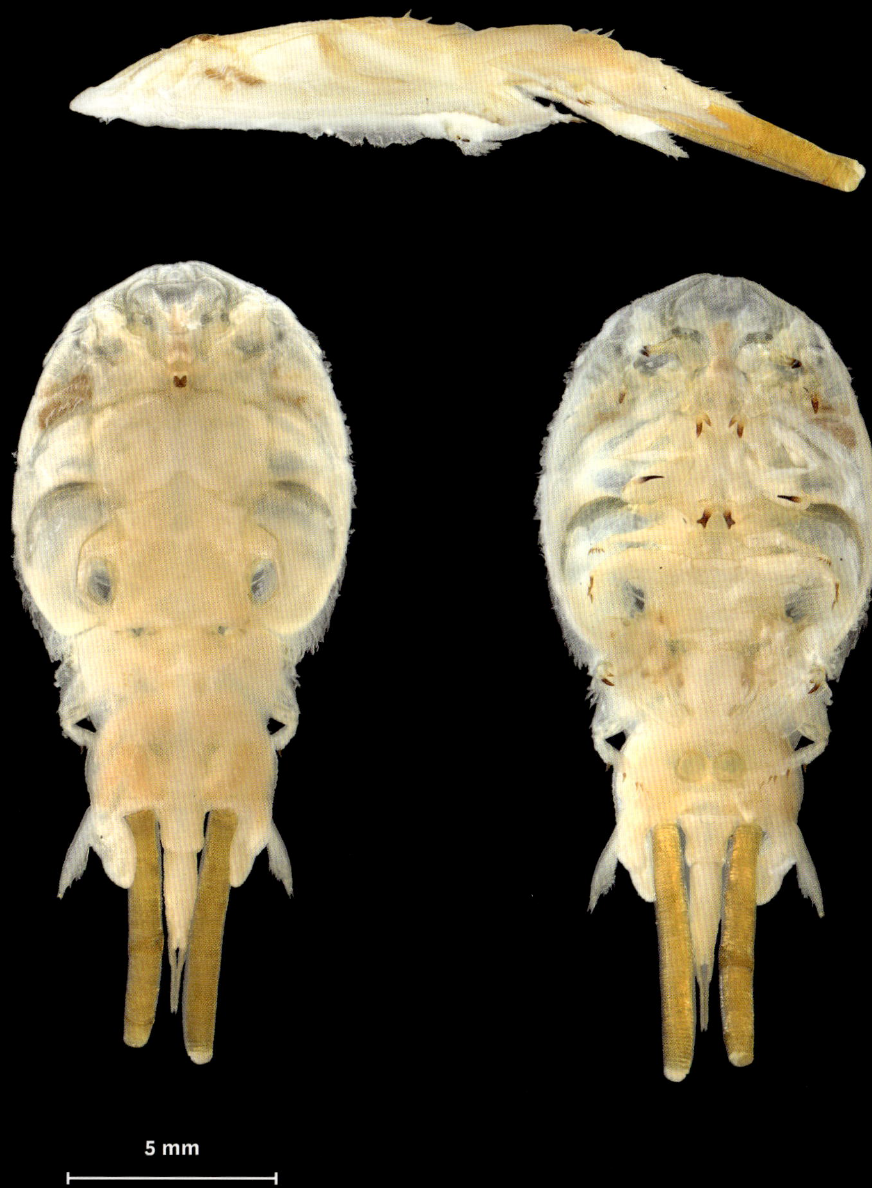
5 mm

079 / 100

Sea spider

Pycnogonida class

The odd, spindly arthropod shown here is a sea spider. These animals are truly in a class of their own, forming the taxonomic class Pycnogonida, meaning 'many knees', and having many unique characters. They're not spiders and are probably not even related to them, but because most species have eight legs – some have ten or twelve – the name has stuck.

The largest sea spiders are deep-sea dwellers. This one, from Antarctica, has a leg-span of 33 centimetres. The record is 70 centimetres, although most of the 1300 species are far smaller and live in shallow water.

Sea spiders have remarkably minimal bodies. At the front is a long proboscis, which the sea spider uses to suck the soft parts from sea anemones and other invertebrates. Most sea spiders have eyes on a turret, but some deep-sea species, including this one, are eyeless. At the back is a minute abdomen. There's no room in the body for major organs, so the stomach extends into all the legs. The body bears a pair of palps (sensory appendages), and ovigers (legs modified to carry eggs) in front of much longer legs. After the female lays its eggs, the male takes over. He holds great pom-poms of eggs on his ovigers until they hatch, and then takes care of the young.

The animal's blood is not confined to blood vessels, and the main job of pumping blood around is performed by peristaltic action (rhythmic contraction and relaxation) during digestion in the stomach extensions in the legs. Pycnogonids lack gills. Given their long legs and tiny bodies, they have a high surface-area-to-volume ratio. Coupled with an exoskeleton of thin, unhardened chitin, this allows sufficient oxygen to diffuse through into the blood. RW

Sea spider class Pycnogonida, PY.000178, maximum leg span 330 mm, head width 6.1 mm, collected by D Barber on the FV *San Aotea II* at depth of 585 m, Ross Sea, Antarctica, 11 January 2004.

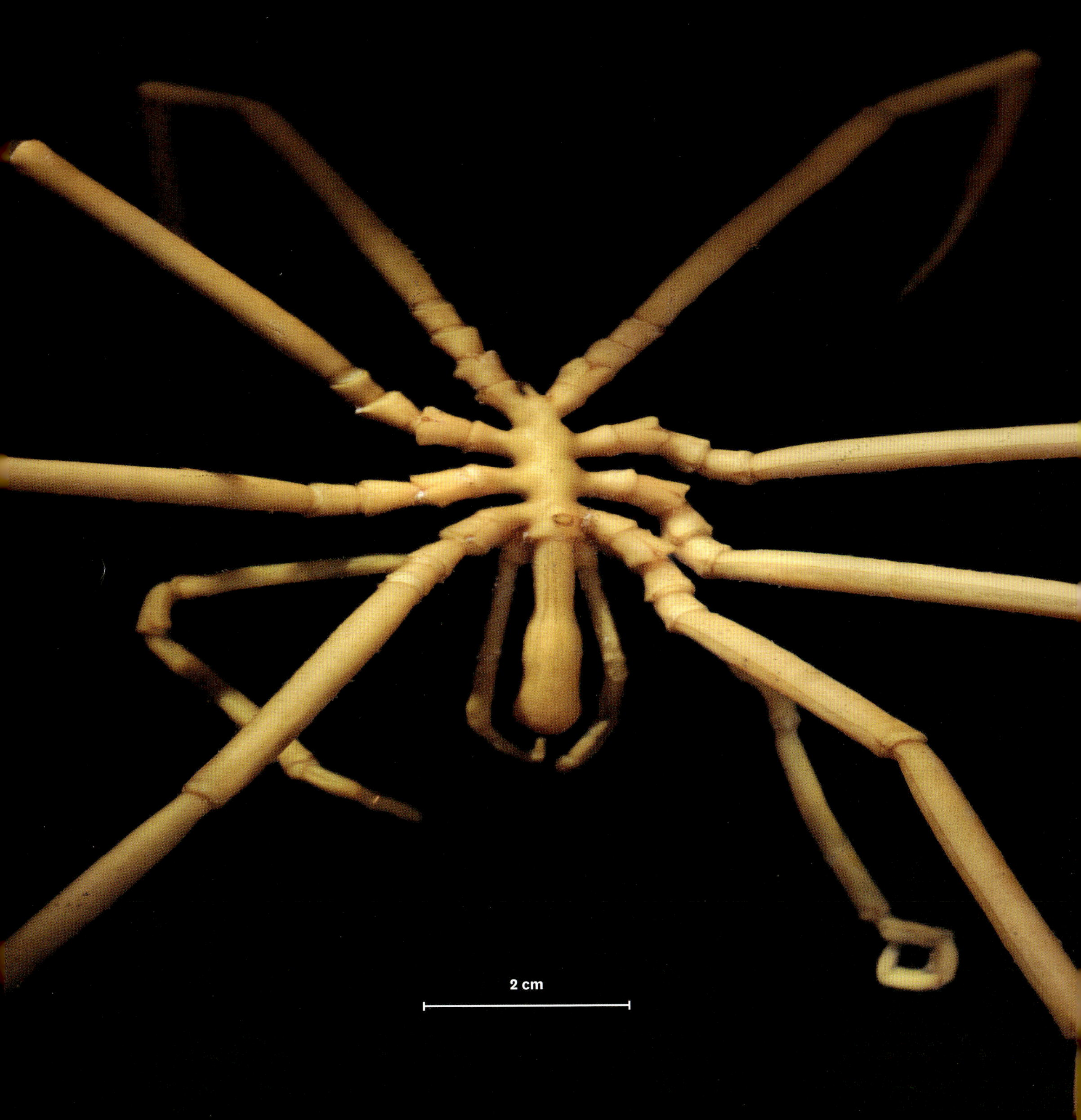
2 cm

Short-tailed bat

Mystacina tuberculata

The short-tailed bat (*Mystacina tuberculata*), known to Māori as pekapeka-tou-poto, is quite unlike any other bat on the planet. It is the sole survivor of a family known only in New Zealand (and previously in Australia), and thought to have diverged from other bats more than forty million years ago. Whereas bats are defined as mammals that fly, this little group of antipodean bats folded up their wings and walked.

Short-tailed bats can, and do, fly strongly – up to 30 kilometres a
080 night – but they use their legs more than any other living bat. Whether
100 scuttling over the forest floor, ploughing through leaf litter or plunging down petrel burrows, this behaviour gives them access to a much wider variety of food than they could catch on the wing. This wide diet is also one of the features that sets short-tailed bats apart from their relatives. They can catch flying insects using echolocation, lick nectar from flowers, crunch on big flightless insects or chew the fat off dead birds.

But such terrestrial living now comes at a price. It was a profitable lifestyle in a land that had few ground-hunting predators, but it makes these bats highly vulnerable to the rats, cats and stoats that people have brought to Aotearoa. Desperate measures are being taken to safeguard the few remaining bat populations, and unfortunately some attempts have failed.

Juvenile short-tailed bats relocated to Kāpiti Island succumbed to a mystery infection that rotted their large external ears (which are vital for echolocation). And the animal photographed died after eating poison placed to protect its colony from rats and stoats. Despite these setbacks, there are still a few healthy populations, including on Hauturu/Little Barrier Island and Whenua Hou/Codfish Island. CM

Short-tailed bat (*Mystacina tuberculata*), LM002498, length 73 mm, collected by G Dennis, Pureora Forest Park, Waikato, 9 January 2009.

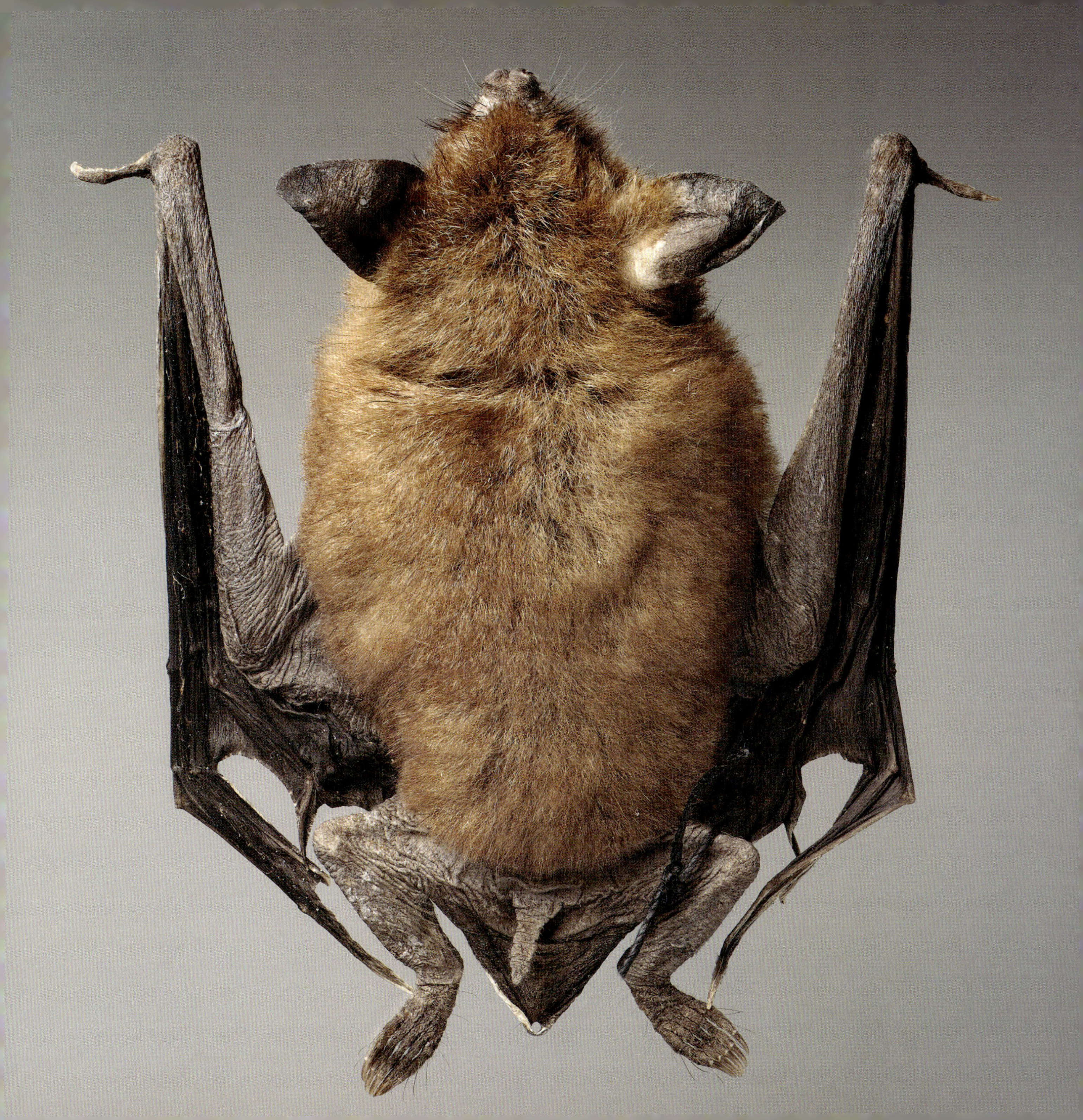

081

100

Snow hebe illustration

Bobbi Angell

Today, we live in a world where digital photos of plants can be instantly created and shared online with a mobile phone. However, there is something about a hand-drawn scientific illustration that feels both 'old school' – like something rediscovered in a large, leather-bound tome from the 1800s – and modern at the same time, bringing a fresh perspective.

The detailed pen-and-ink illustration shown here was drawn by Bobbi Angell, a professional botanical illustrator from the USA who has completed thousands of such line drawings for Floras, books and scientific articles, including some of mine for projects on *Veronica, Plantago, Ourisia* and *Myosotis*. This is one of several of Bobbi's drawings that are held in Te Papa's herbarium, stored alongside some of the dried plant specimens that she studied to create them. It takes time, patient observation and considerable skill to create such illustrations. These high-quality and detailed drawings are extremely effective in helping readers to visualise the plant by bringing to life its important distinguishing characteristics.

This unique illustration shows not one but six different species and subspecies of snow hebe from ten Te Papa specimens. Snow hebes are compact cushion plants in the genus *Veronica* and are native to high-elevation habitats in the South Island. The most important characters for telling them apart – how and where the hairs are distributed on the leaves and bracts – are minuscule and difficult to see with the naked eye. Studying snow hebes requires a hand lens or microscope to identify them. But here the artist has effectively and rather exquisitely revealed these tiny, yet important, characters in her drawing. HM

Pen-and-ink illustration by Bobbi Angell of various snow hebe species on 41 cm x 31.5 cm card.

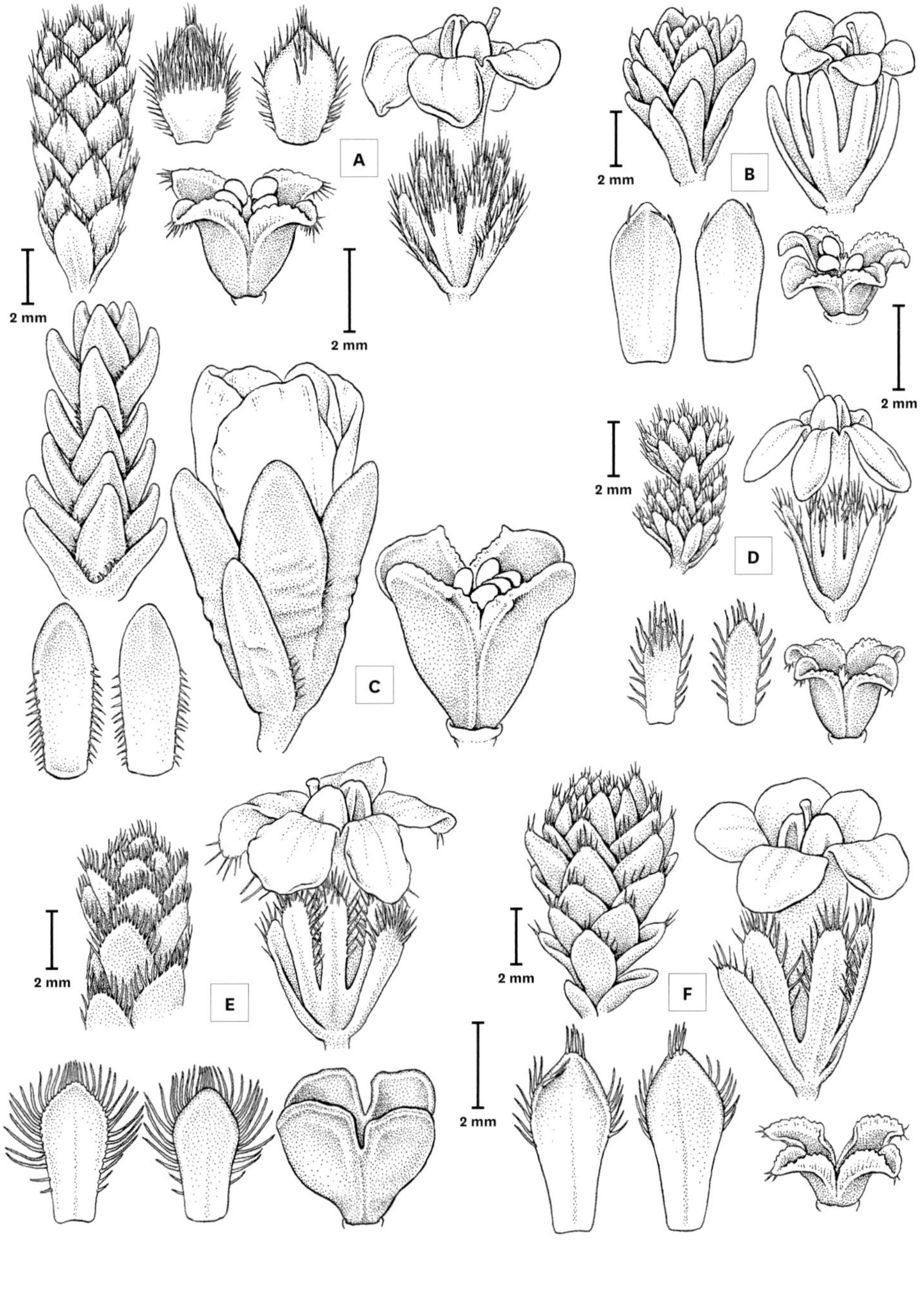
A
2 mm
2 mm
B
2 mm
2 mm
C
D
2 mm
E
2 mm
F
2 mm
2 mm

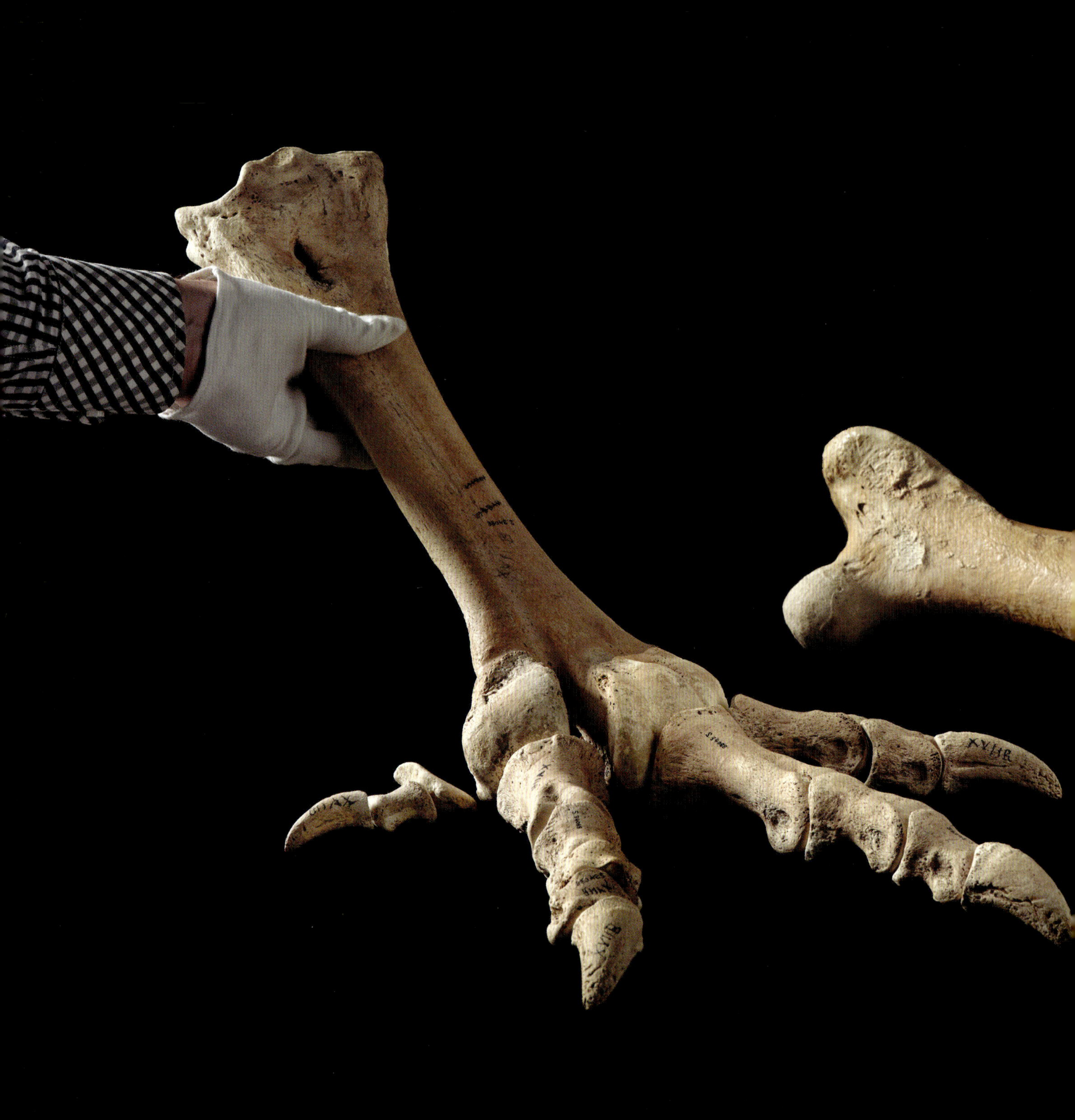

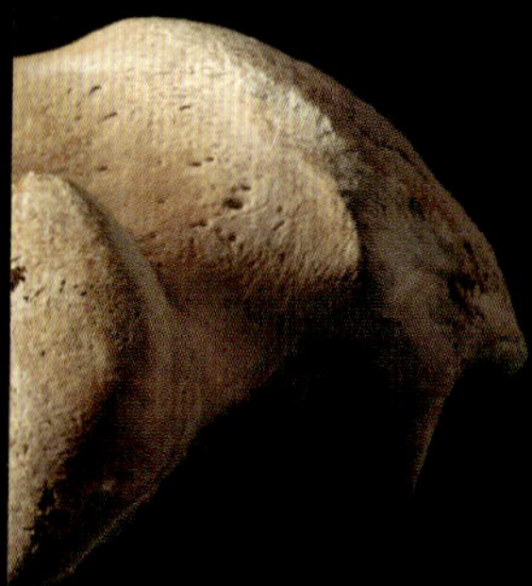

South Island giant moa bones

Dinornis robustus

Seeing a real moa leg bone is a special moment for any New Zealander, but to be able to go one step further and actually hold a moa bone, and fully appreciate its weight, strength and size, is a completely absorbing experience. In one of my first back-of-house tours of Te Papa, I was leading a group of people around our Tory Street facility. Our curatorial team was pre-positioned within the collections and I was there to move people through the facility and talk about the collections we have and the research we do.

We entered the fossil vertebrate collection, where curator Alan Tennyson began speaking about some of the fossils in Te Papa's collection. He talked about the iguanodontid tooth (one of the first fossils identified as a dinosaur, see page 117), our giant penguin fossils (including perhaps the largest species ever to live, see page 93) and other collection items only found at Te Papa.

Then, the leg bone of a giant moa was brought out to show the group. Each of us was allowed to pick up the bone and feel its weight. We quickly realised just how massive and powerful these extinct animals had been. Giant moa are the tallest birds ever to have walked the Earth, and they could weigh as much as 242 kilograms. Encountering objects up close in a museum is an unbeatable experience, and I could see the special connection each person was making as they held the bone. I, too, was completely blown away by this experience. DP

South Island giant moa (*Dinornis robustus*) fossil skeleton, S.34088, collected by J Eyles and R Scarlett, Pyramid Valley, North Canterbury, 1949.

South Island kōkako nest and egg

Callaeas cinerea

Among the poignant treasures in the Te Papa collection is the South Island kōkako nest. It is a reminder of when this enigmatic bird, now probably extinct, was sufficiently common that a wandering naturalist could chance upon its nest. Until 2014, this nest was simply labelled 'Freshwater Basin, Milford Sound'. It was clearly old, but there was no date, no collector's name, and no detail about how or where it was found.

083

100

Part of a museum curator's role is solving historical puzzles, and sometimes we get lucky. While researching another topic, I read an article on 'Rare or Little Known Birds' published by Canterbury-based naturalist and author Thomas Potts in 1880. Potts described a visit to Milford Sound in January 1873, in the company of James Hector, the founding Director of the Colonial Museum (now Te Papa): *whilst exploring the bush that fringes Milford Sound, the writer was so fortunate as to discover five [kōkako] nests, at heights varying from ten to seventeen feet above the ground ... The parent ... was ... covering two nestlings as yet unable to see ... The other nests were found in damp situations (one with a broken egg) in a small patch of bush at Freshwater Basin, close by Lady Bowen Waterfall.*

He went on to describe how James Hector collected one of the nests for the Colonial Museum, with other nests collected for the Canterbury Museum (where Potts was a member of the board of trustees). More than 140 years after the event, we now know that this nest was collected by two giants of New Zealand natural history, and the circumstances of its discovery. CM

South Island kōkako (*Callaeas cinerea*) nest, OR.029158, maximum width 270 mm, collected by T Potts, Freshwater Basin, Milford Sound, January 1873; South Island kōkako (*Callaeas cinerea*) egg, OR.007626, 42 x 27 mm, collected by W Campbell, Hokitika, no date.

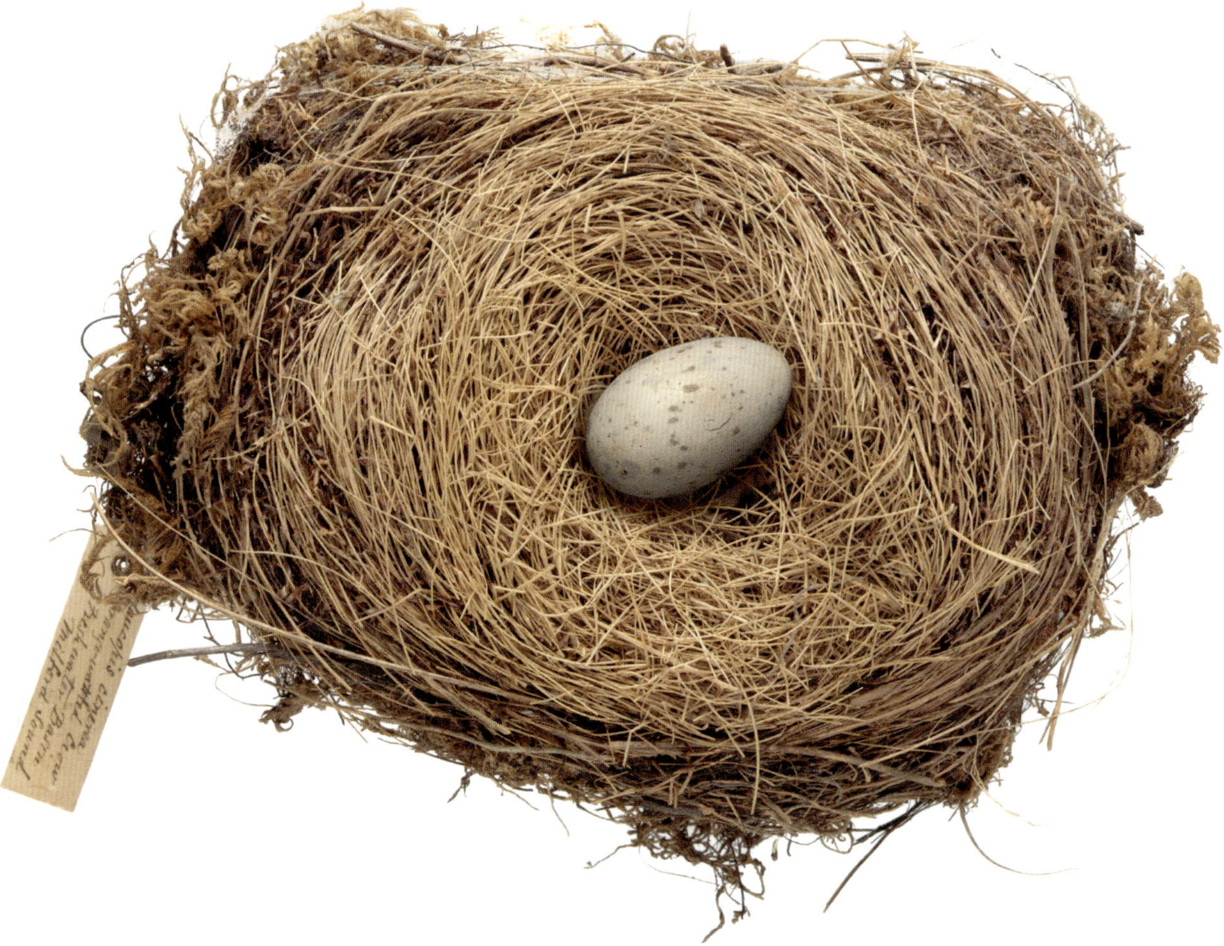

084 / 100

South Island snipe

Coenocorypha iredalei

In mid-March 1964, the first muttonbirders of the season stepped ashore on Taukihepa/Big South Cape Island, the largest of the muttonbird islands that lie off the coast of Rakiura/Stewart Island. Until that year, Taukihepa was an extraordinary time capsule. Its forests teemed with wildlife that had long since disappeared from the mainland. You could hear the strident calls of tīeke/South Island saddleback, see tiny bush wrens hopping through fern thickets, and startle South Island snipe as you walked through the pākihi (low shrubland) vegetation on the summit ridges. At night you could see greater short-tailed bats flitting in your torch beam, and perhaps hear the loud cry of the hakawai, the unseen spirit bird of the muttonbird islands.

But now something was wrong. The forests had fallen silent, many plants had been eaten to the ground, and within the muttonbirding huts everything was a mess. Wallpaper had been stripped off, mattresses ruined and every scrap of food consumed. Somehow rats had got ashore, and had reached plague proportions since the muttonbirders had departed the previous May.

Word of the disaster reached the New Zealand Wildlife Service, and a desperate rescue mission was launched. Miraculously, the tīeke were saved by moving a few to other, rat-free islands. But attempts to rescue the snipe and bush wrens failed, and the rescuers had no knowledge of how to attempt to save the bats. Within a decade, all three species were extinct.

Two snipe (both males) were caught, but died in an aviary during a winter storm. With the extinction of the South Island snipe we also lost the legendary hakawai, as we now know that this was a nocturnal aerial display performed by the snipe. CM

South Island snipe (*Coenocorypha iredalei*), OR.011206, length 200 mm, collected by B Bell, Taukihepa/Big South Cape Island, 1 September 1964.

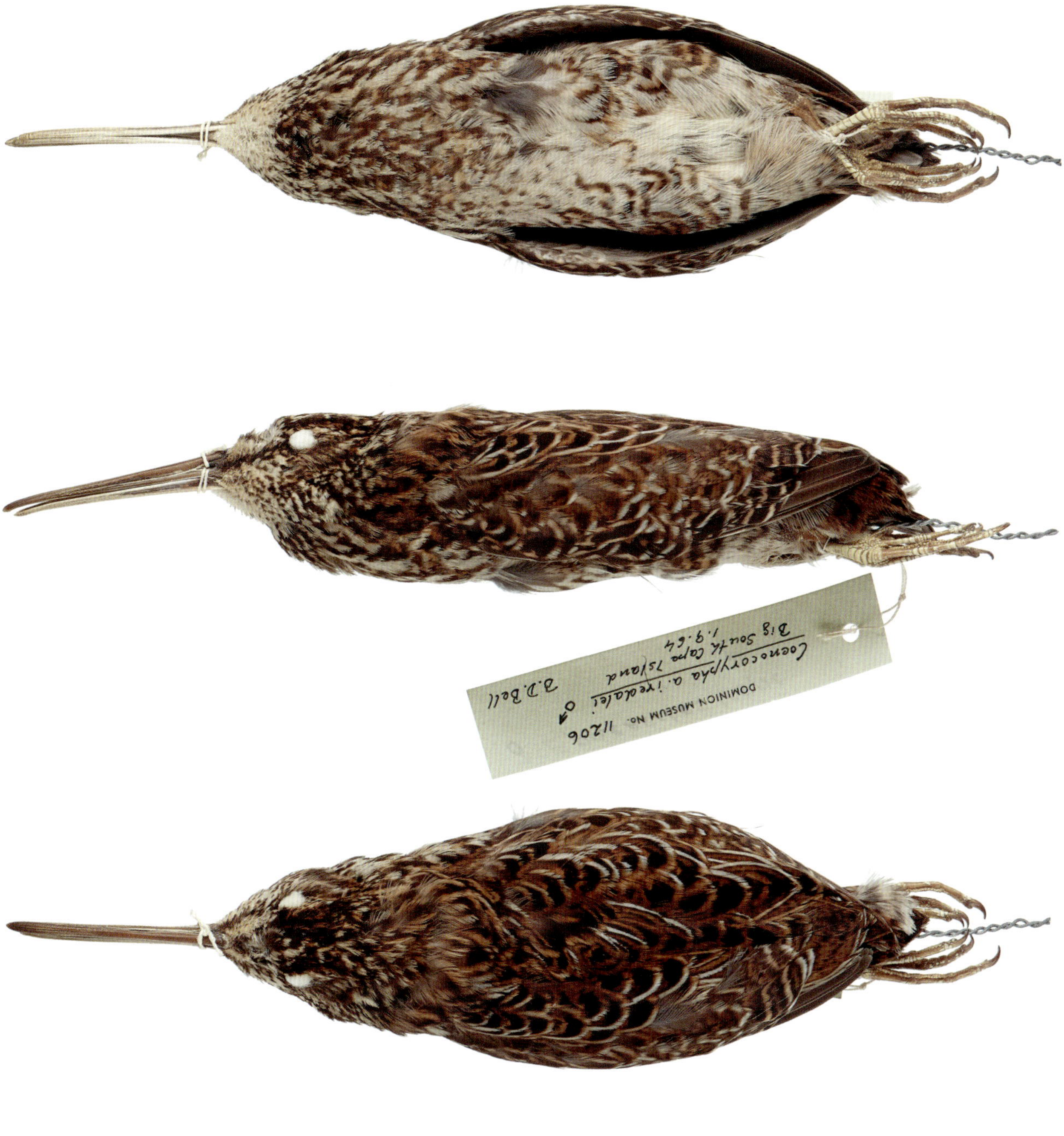
DOMINION MUSEUM No. 11206
Coenocorypha a. iredalei ♂
Big South Cape Island
1.9.64

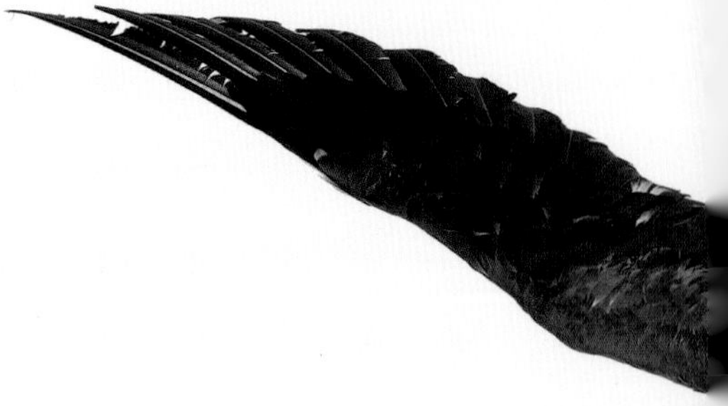

085 / 100

Southern Buller's albatross

Thalassarche bulleri bulleri

New Zealand is home to more than half of the world's twenty-four albatross species, and boasts more endemic species of albatross than any other nation, with ten endemic taxa (species, genus, family, etc.). The southern Buller's albatross is one of these endemic taxa, nesting at two sites near Stewart Island, the Tini Heke/Snares and Hautere/Solander Islands. With its beautiful dramatic plumage and extravagant courting displays, this is one of my favourite seabird species.

In 2016 and 2017, the Te Papa team undertook two research expeditions to Hautere/Solander Island to study where the albatrosses were feeding, and found that they were travelling as far away as Hawke's Bay, the sub-Antarctic Auckland Islands and into the coastal waters of Tasmania. Each bird can travel 100–2000 kilometres during its foraging trips at sea, which last between two and ten days. During the long hours of waiting for GPS-tagged birds to return to the nest, we were serenaded by other birds vying for nesting spots with non-breeding individuals, and showing off their plumage to each other.

These southern Buller's albatrosses nest in the New Zealand winter-time, raising their chicks over several months in the cold and changeable climate of Foveaux Strait. Several albatross species follow this slightly unusual breeding calendar, with their chicks hatching into late summer and being fed by both parents throughout the winter. The cool waters of southern New Zealand's marine ecosystem is productive and dependable in the winter-time, enabling this breeding pattern to take place; although the most numerous of New Zealand's marine-dependent birds, such as sooty shearwaters and white-capped albatrosses, nest in the spring-time and raise their chicks over the summer. SW

Southern Buller's albatross (*Thalassarche bulleri bulleri*), OR.025000, adult male, by-catch, collected by K Bradey (government fisheries observer), at sea in New Zealand at 44°30'S, 166°24'E, 15 June 1995.

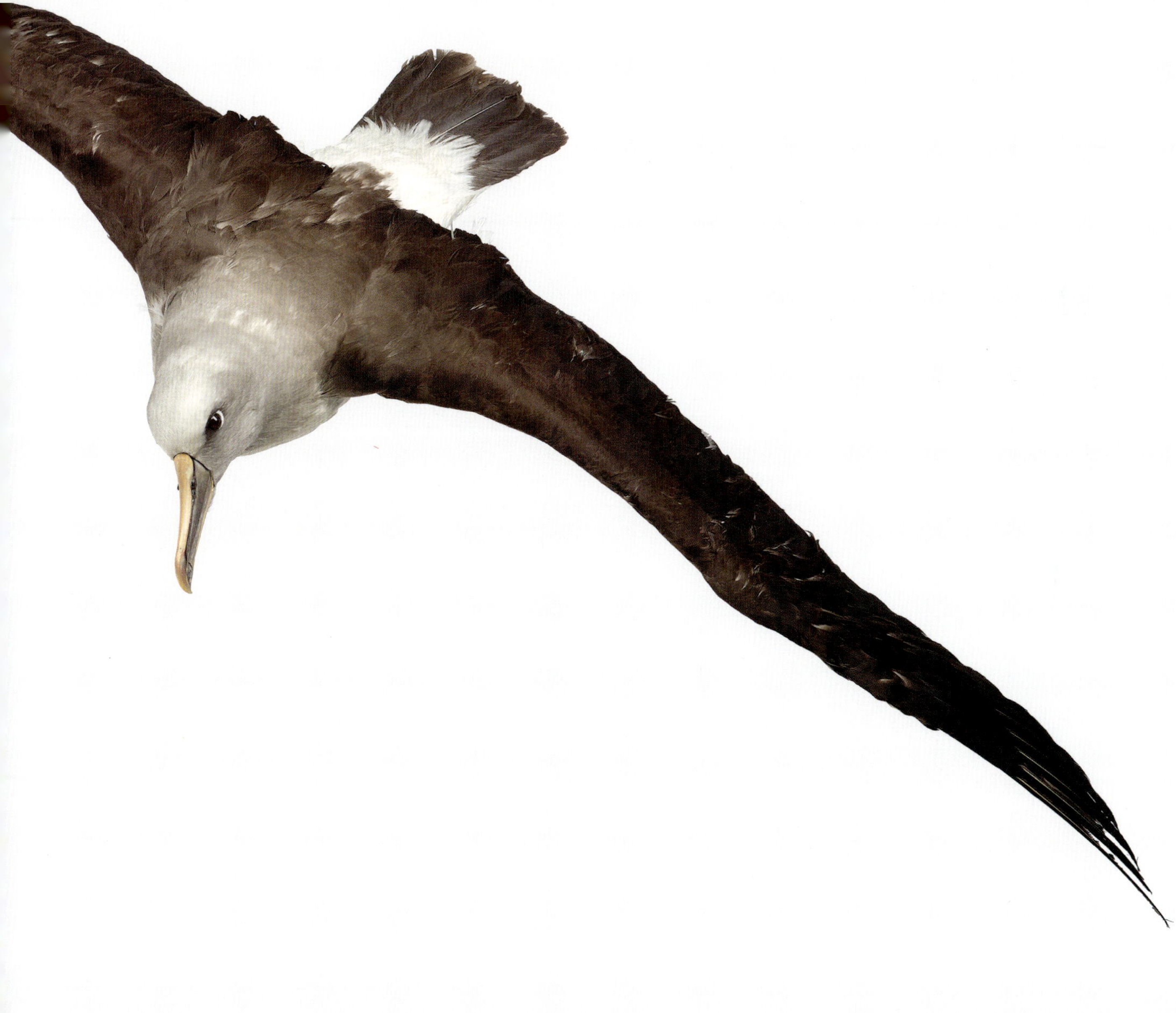

Springtails

Collembola class

John Salmon's name may be familiar to many as the author of the book *The Native Trees of New Zealand*. Less familiar to most is his work on entomology, including on Collembola, a class of arthropods considered to be cousins to the insects and more commonly known as springtails. New Zealand has over 380 springtail species, and around 80 percent of these are only found here. Salmon described more New Zealand species than anyone else, making him a key figure in this field. There are many more species still awaiting discovery or formal scientific naming and description.

Springtails are small, typically around a millimetre or two in length, although there are a few species that get to 10 millimetres long. They also have a great variety of forms and colours, as we can see from the selection of images here. Springtails can be present in huge numbers, with densities in the tens of thousands per square metre. They perform a useful role in recycling nutrients. Their common name arises from their primary defence mechanism. Most have a structure called the furcular located towards the rear. Normally this is tucked away, but it's also held under tension. If a springtail feels threatened, the furcular is released and in about one-fiftieth of a second it hits the surface underneath with enough force to literally spring the animal to safety.

Te Papa holds Salmon's Collembola collection, consisting of many thousands of specimens mounted on slides or stored in vials of ethanol. In recent times it has been used by researchers from as far afield as Brazil and Russia. PS

Type specimens from the Salmon collection of New Zealand springtails, class Collembola, all collected by JT Salmon unless otherwise indicated: **(A)** *Parisotoma picea*, AI.000294, collected by J Sorensen, Tucker Cover, Campbell Island, 12 November 1945. **(B)** *Paronana karoriensis*, AI.000202, collected Akatarawa, Wellington, 29 March 1937. **(C)** *Urewera splendida*, AI.000351, collected Awahuri, Palmerston North, 17 December 1939. **(D)** *Entomobrya opotikiensis*, AI.000281, collected Opotiki Beach, Bay of Plenty, 14 December 1937. **(E)** *Setanodosa tetrabrachta*, AI.000368, collected by R Forster, Red Rocks, Cook Strait, 30 August 1941. **(F)** *Parasalina (Tasmasecta) boldensis*, AI.000116, collected Bold Peak, Lake Wakatipu, 11 February 1943. **(G)** *Entomobrya clitellaria australasia*, AI.000110, collected Karori, Wellington, 7 April 1940. **(H)** *Proisotoma terrigenus*, AI.000367, collected Johnston's Hill, Karori, Wellington, 16 September 1942. **(I)** *Pseudokatianna nigretalba nigretalba*, AI.000255, collected Lake Manapouri, 7 January 1940. **(J)** *Acanthomurus setosus violaceus*, AI.000389, collected Lake Manapouri, 7 January 1940. **(K)** *Acanthomurus alpinus obscuratus*, AI.000272, collected Days Bay, Wellington, 19 July 1942. **(L)** *Acanthomurus rivalis*, AI.000321, collected by K Wise, 2 March 1963.

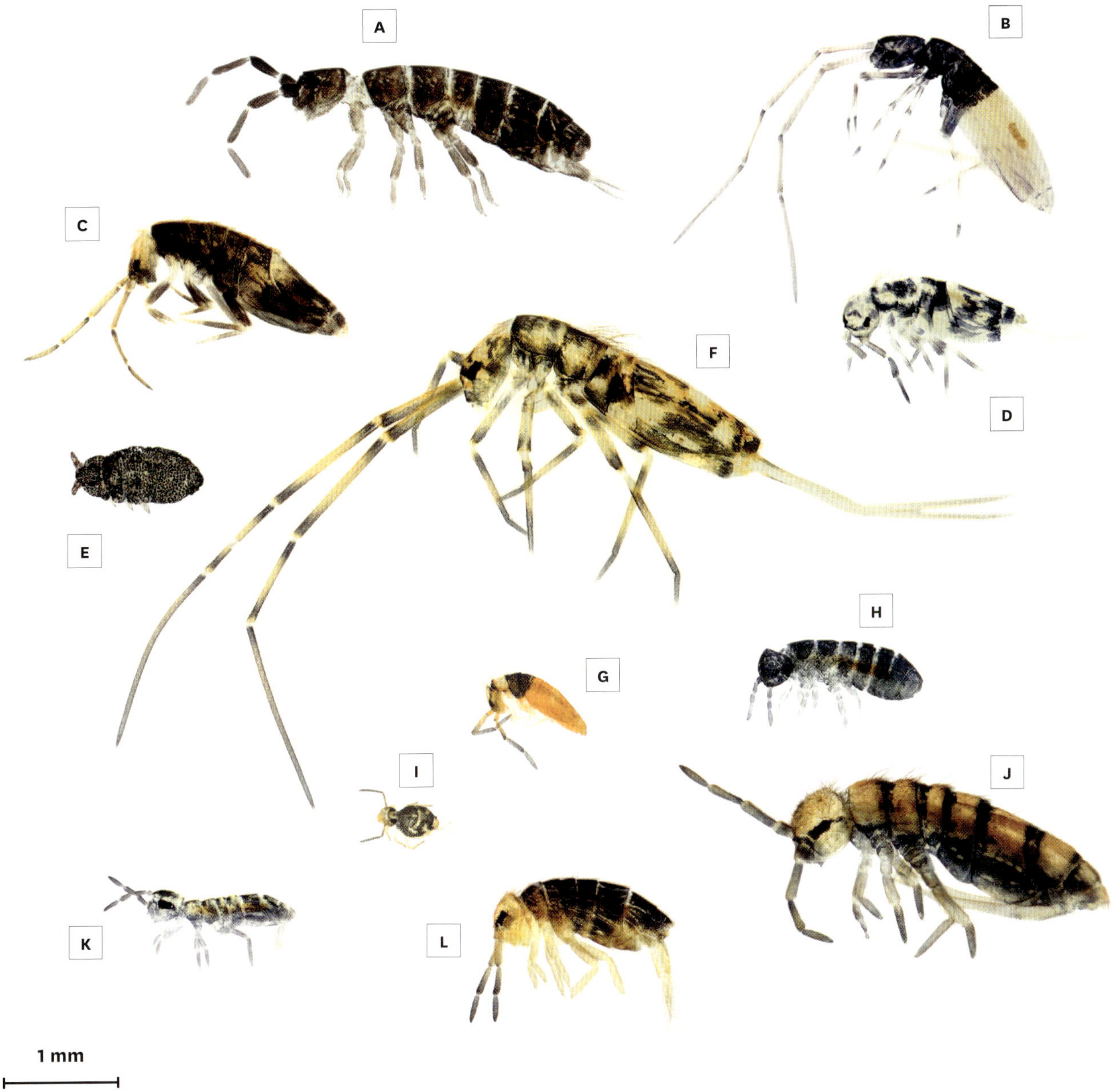
A
B
C
F
D
E
H
G
I
J
K
L
1 mm

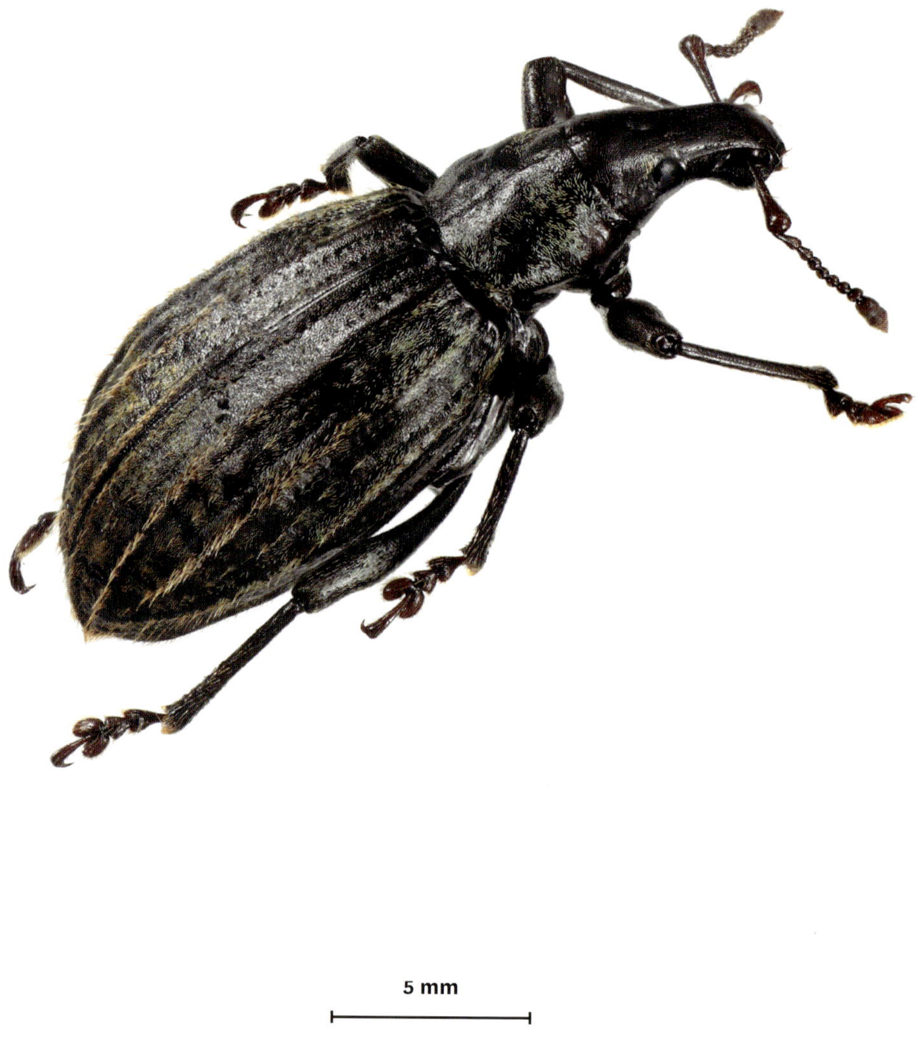
5 mm

Subantarctic weevil

Oclandius species

New Zealand's subantarctic islands are wild, beautiful and of significant ecological importance – many of their species being found nowhere else in the world.

Five island groups are scattered across the Southern Ocean between the Roaring Forties and the Furious Fifties: the Snares Islands, Auckland Islands, Campbell Island, Bounty Islands and Antipodes Islands. Of either volcanic origin or granite outcrops, they sit on the Campbell Plateau. As this plateau is part of the submerged continent Zealandia, the islands have been isolated from other land masses for around 25 million years. This has led to the evolution of numerous endemic species, particularly beetles.

Two species of the extraordinary New Zealand giant weevils are endemic to the subantarctic islands. *Oclandius laeviusculus* and *O. cinereus* were both discovered in 1901 by Captain FW Hutton, who travelled to the island on the famous ship *Hinemoa* to examine the provision depots there and to collect birds for the British Museum.

Studying these beetles helps us understand the origins of the subantarctic fauna, and allows us to compare ecological and evolutionary patterns with tropical or northern temperate groups. More research could reveal why closely related but different species occur in the same microhabitats on one island and whether the Campbell Plateau is a separate piece of Gondwana that is distinct from New Zealand.

Like many other insects on the islands and other giant weevils on the main land, *Oclandius* sp. are flightless. This makes them very prone to attacks by mammal pests, particularly mice. The subantarctic islands have been modified by introduced mammals, including sheep, goats, cattle, rabbits and rats. These pests have now been eradicated from the subantarctic islands, with the exception of the main Auckland Island, where only feral pigs, cats and mice remain. Only ongoing, continuous pest eradication programmes will help these extraordinary weevils to survive. JK

Subantarctic weevil (*Oclandius laeviusculus*), AI.033464, collected by AJD Tennyson, Ocean Island, Auckland Islands, 23 June 1998.

Objects
T—Z

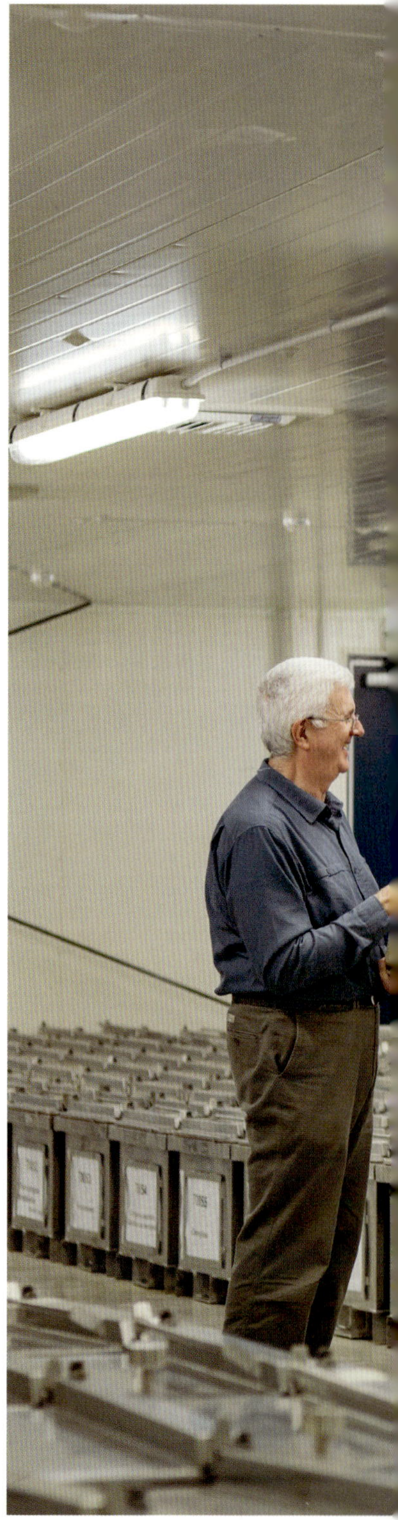

Curator Vertebrates (Fishes) Clive Roberts, far left, Kaitiaki Taonga Collections Manager for Fishes Andrew Stewart, Director Strategy and Performance Dean Petersen and Research and Technical Officer Fishes Carl Struthers in the fish collection storage area at Tory Street, Wellington.

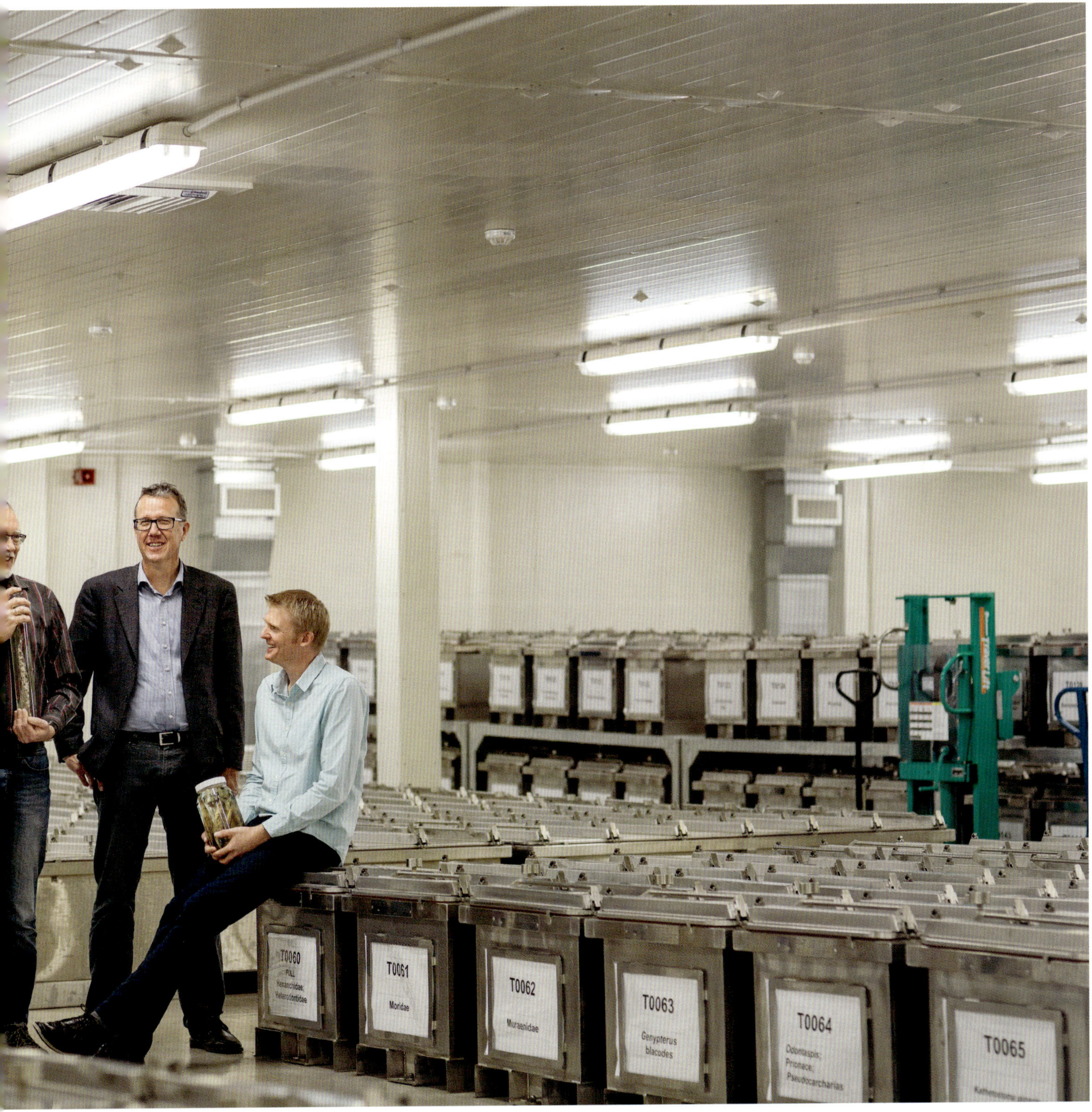
T0060
FULL
Hexanchidae;
T0061
Moridae
T0062
Muraenidae
T0063
Genypterus
blacodes
T0064
Odontaspis;
Prionace;
Pseudocarcharias
T0065

063 064 065

070 071 072

077 078 079

083 084 085

089 090 091

095 096 097

66 067 068 0

3 074 075 0

080 081 082

086 087 088

092 093 094

098 099 100

Takahē

Porphyrio hochstetteri

The South Island takahē, or notornis as it was first known, was New Zealand's original rare bird. At a time when kākāpō, huia and piopio were easy to find, it was the notornis that set pulses racing and sent fortune-seekers into the wilderness.

Part of the mystique of the bird was that it was originally described (in 1848) from bones. These first specimens were found in south

Taranaki, but little more than a year later, a living notornis was captured by a sealing gang in Dusky Sound, Fiordland. This astonishing discovery raised the hope that other 'extinct' New Zealand birds – particularly moa – had also survived in the wilds of Fiordland.

The bird illustrated here was the second notornis caught in colonial times, on Secretary Island in Fiordland, in 1851. These first two specimens ended up in the British Museum, while two further birds caught in 1881 and 1898 ended up in the Dresden Museum in Germany and in the Otago Museum. And that was it. This rarest of all New Zealand birds disappeared again for another fifty years.

The second rediscovery of takahē, in 1948, made headlines around the world, but caused consternation at the Dominion Museum here in Wellington – how could this remarkable story of hope and rediscovery be told without a specimen to display? Then, in 1955, the British Museum generously gifted one of their two takahē specimens back to the people of New Zealand. This bird, captured 168 years ago, spent more than a century in England before returning to its country of origin.

Few specimens in Te Papa have such an astonishing tale to tell, either before or after death. CM

Takahē (*Porphyrio hochstetteri*), OR.022236, height 390 mm, unknown collector, Thompson Sound, Fiordland, 1851.

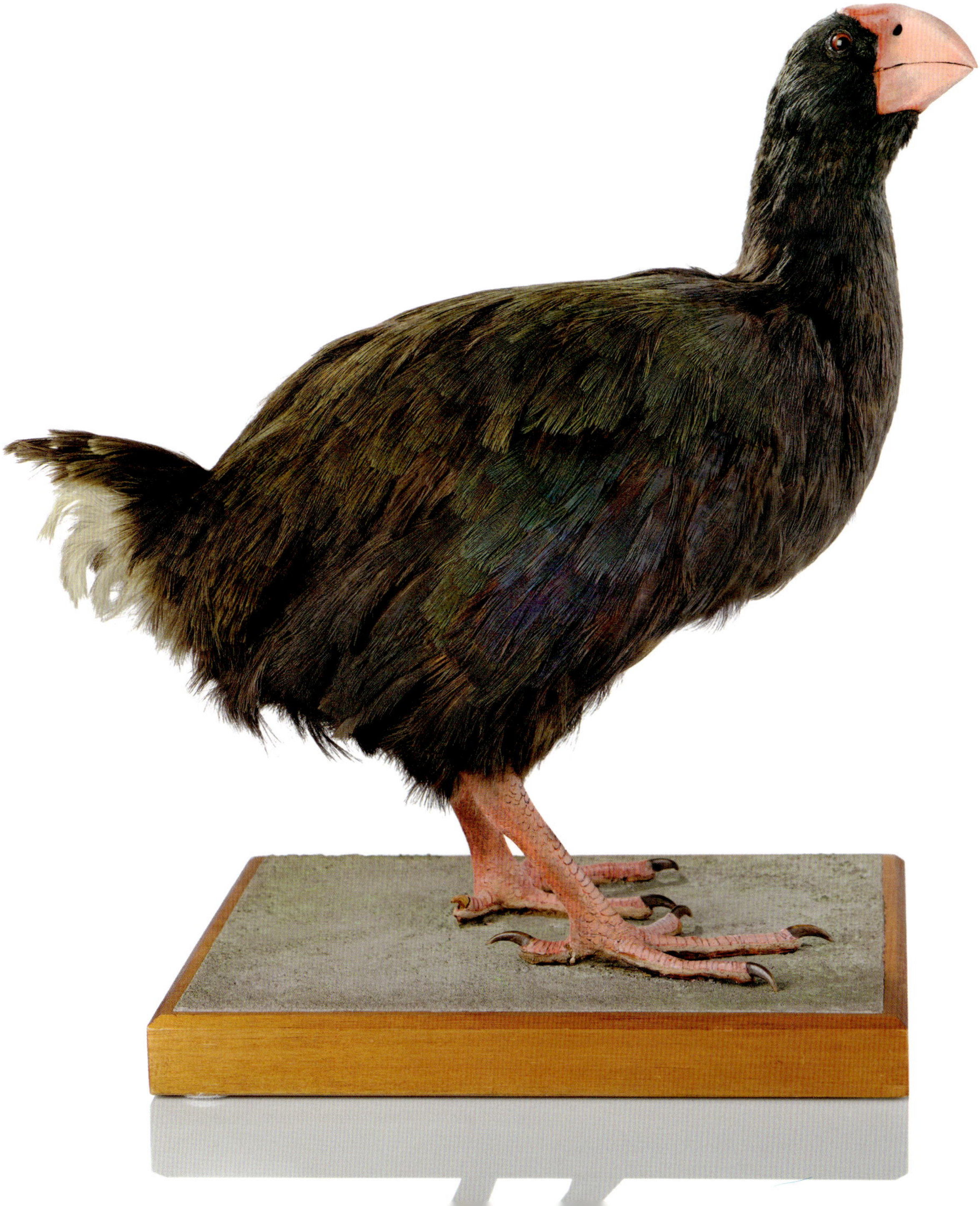

Tateid snails

Tateidae family

The small freshwater snails of the gastropod superfamily Hydrobiodea occur almost everywhere except the polar regions. Found mostly in ponds, seepages, caves and aquifers, a few species occur in lakes and rivers or in brackish conditions. There are vast numbers of species, but until relatively recently only a few were known in the New Zealand region, all of which belong to the family Tateidae.

Potamopyrgus antipodarum, our largest, most widespread and variable species, has spread to Australia, Europe and North America, probably through being sold in aquaria and then escaping into the environment. For many years only a single subterranean (troglobitic) species was known in New Zealand, until *Opacuincola caeca* was discovered in Gorge Creek Cave, near Takaka, and named by National Museum curator Winston Ponder. Further troglobitic species found in artesian bores drilled in the Waimea Plains, near Nelson, in a bore at Roy's Hill vineyard, near Napier, and in streams in caves in the north-western South Island were named by later National Museum curator Frank Climo.

Recent searches by visiting Austrian researcher Martin Haase revealed numerous additional species in caves and seepages in the North and South Islands. One is restricted to the Wellington area, notably the bird-bath spring at the foot of the Wadestown hill, and was named *Potamopyrgus oppidanus*. These findings suggest that many additional species remain to be discovered in unsampled cave systems, and especially (and critically) in subsurface aquifers that have yet to be sampled, notably under the Canterbury Plains and Otago. BM

(Left to right) **Top row:** *Rakiurapyrgus cresswelli*, M.158200, collected by B Elliott, Golden Bay Stream, Stewart Island, 18 January 1970; *Hadopyrgus ngataana*, M.174154, collected by C Mosimann and M Haase, Maitai Cave, south-east of Nelson, 24 November 2001; *Opacuincola cervicesmadentes*, M.174119, collected by G Kuschel, Wet Neck Cave, west of Collingwood, 13 June 1973; *Sororipyrgus marshalli*, M.174081, collected by M Scarsbrook and M Haase, roadside seep east of Pahiatua, 20 November 2003. **Middle row:** *Opacuincola eduardstraussi*, M.174115, collected by G Kuschel, Twinforks Cave, west of Collingwood, 12 June 1973; *Opacuincola johanstraussi*, M.174103, collected by D Roscoe, M Haase and C Mosimann, Ballroom Cave, south of Collingwood, 20 November 2002. **Bottom row:** *Opacuincola josephstraussi*, M.174117, collected by D Roscoe, P Jamieson and B Hazelwood, Aorere Cave, south of Collingwood, 5 June 1977; *Opacuincola permutata*, M.174135, collected by R Bromley, Kennedy's Sink, west of Inangahua, 17 November 2002; *Potamopyrgus oppidanus*, M.174061, collected by M Haase, stream at foot of Wadestown Hill, Wellington, 26 November 2003. All specimens are holotypes.

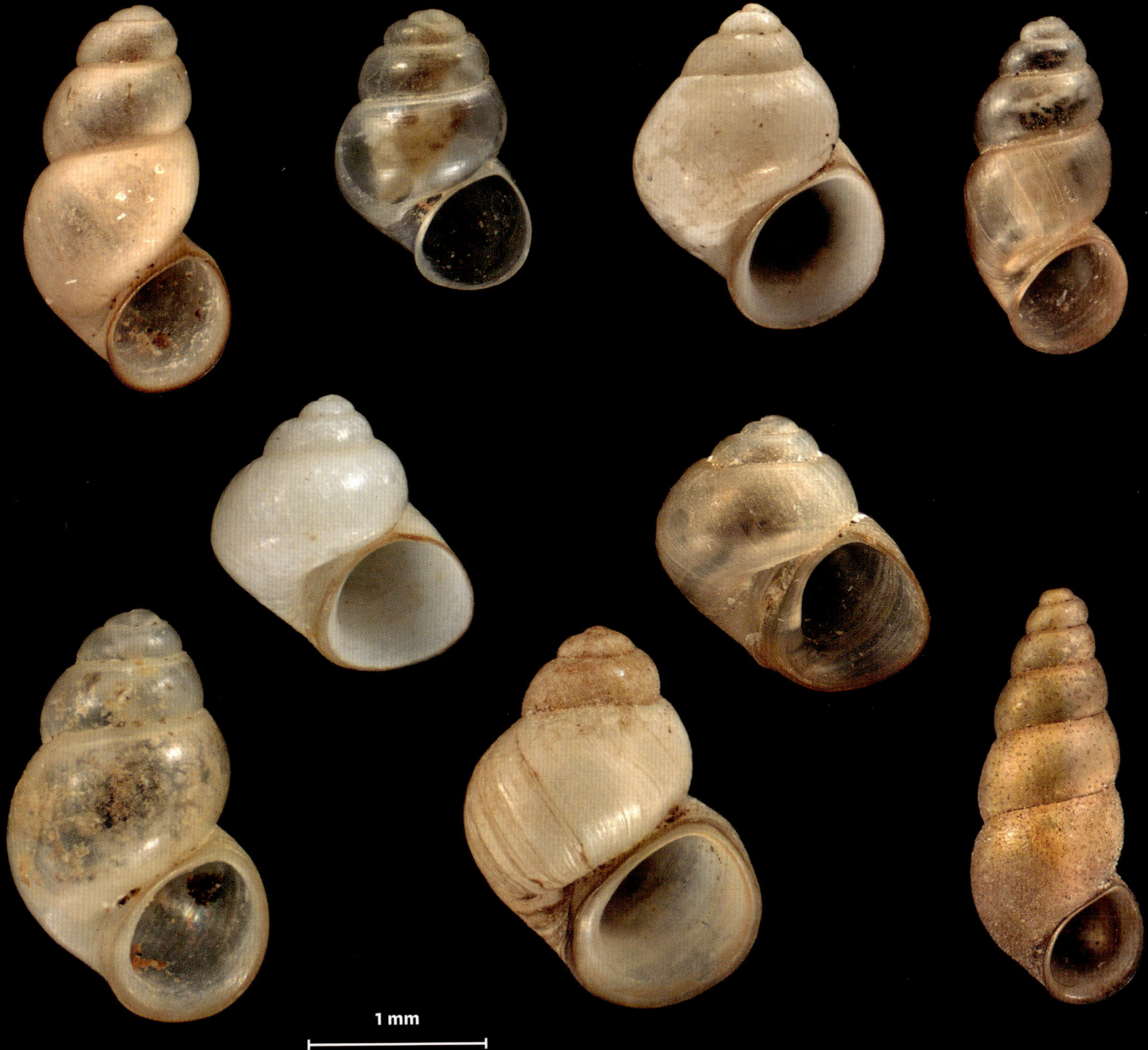
1 mm

090
100

Tree fern vases

Cyathea species

Even before I came to New Zealand, in 1973, I had seen tree fern vases in England. Intriguingly attractive, they hinted at an environment very different from European forests, where tree ferns are unknown. The vases beckoned, and I crossed the world to settle in Aotearoa. As a young researcher, I knew the basic anatomy of a tree fern trunk, but the beauty and exotic nature of the vases were only fully explained when I saw living plants. To the untutored eye, their relationship to tree ferns is hard to comprehend, and they remain objects of mystery and intrigue, provoking excitement and disbelief when presented to a class of school-children.

Tree fern vases are usually made from either the silver fern or ponga (*Cyathea dealbata*) or the mamaku (*Cyathea medullaris*). In cross-section, their trunks are quite different to those of conifers or flowering trees. There are dark brown rings composed of very hard material that give the trunk its rigidity and strength, acting rather like reinforcing rods in concrete pillars. The orange-coloured material inside the dark rings is vascular tissue, which conducts water up the trunk to the fronds, and the pale yellow material is soft pith. The very small dark rings are vascular traces running out into the fronds that grow from the trunk.

To make a vase, the frond bases and fibrous material on the outside of the trunk are stripped away and the soft pith at the centre is hollowed out, leaving a ring of hard tissue. The patterns of yellow, orange and dark brown are created by sculpting the external face, creating a distinctive craftwork that reflects the engineering and physiological needs of native tree ferns. PB

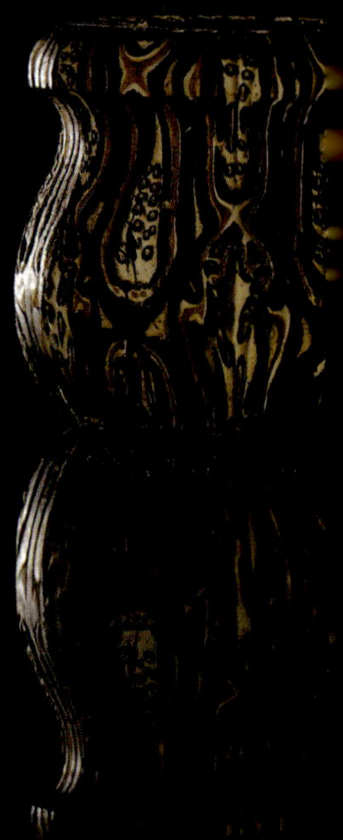

Tree fern vases made from the hollowed-out trunks of ponga (*Cyathea dealbata*) and mamaku (*Cyathea medullaris*), from 100 mm to 280 mm tall.

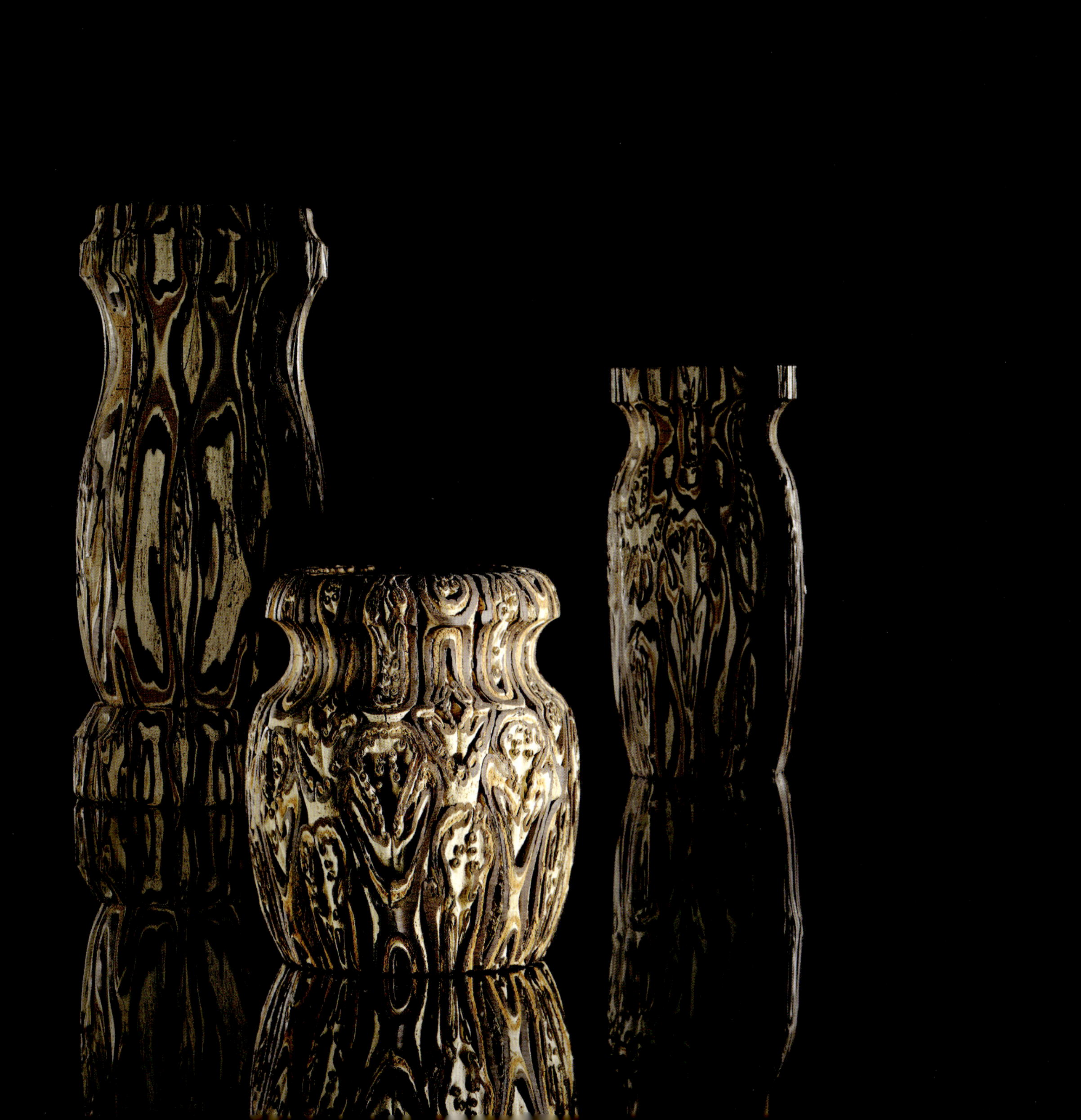

091 / 100

Triplefin

Tripterygiidae family

Triplefins or cockabullies (family Tripterygiidae) are small, benthic (bottom-dwelling) fishes familiar to anyone who has explored a New Zealand rockpool or snorkelled a reef. About 168 species in twenty-nine genera are found worldwide, in shallow tropical to cold-temperate waters, but New Zealand is the capital of triplefin diversity with twenty-seven endemic species in twelve endemic genera. We also have the largest (196 millimetres), deepest-living (to 500 metres), and only estuarine species in the family. There are further species still to be described, particularly from deep reefs near the edge of the continental shelf.

Triplefins are among the most abundant and frequently encountered fishes in our intertidal and shallow subtidal habitats. It is thought that this diversity is due to rapid evolution into different species because of habitat change occurring over the past two to five million years. A wide variety of new habitats arose during periods of rapid climate change, with alternating glaciation and warming coupled with large changes in sea level (by more than 200 metres) and the submerging of land masses.

One of our most attractive species is the yellowblack triplefin, especially the male. It was named *Forsterygion flavonigrum* in 1994, from the Latin *flavus* (yellow) and *nigra* (black), referring to the distinctive head and body colour of mature males. Although they only grow up to 50 millimetres long, they are commonly seen underwater by scuba divers at depths of 10–30 metres, being widespread throughout our coastal waters and as far out as the Three Kings, Chatham, Snares and Auckland Islands. The species inhabits rocky reefs and rocky to stony seabeds to depths of 110 metres. CR + AS

Yellowback triplefin (*Forsterygion flavonigrum*), P.016927, holotype, standard length 40 mm, collected by G Hardy, off Seymour Island, Doubtful Sound, Fiordland, 28 February 1985.

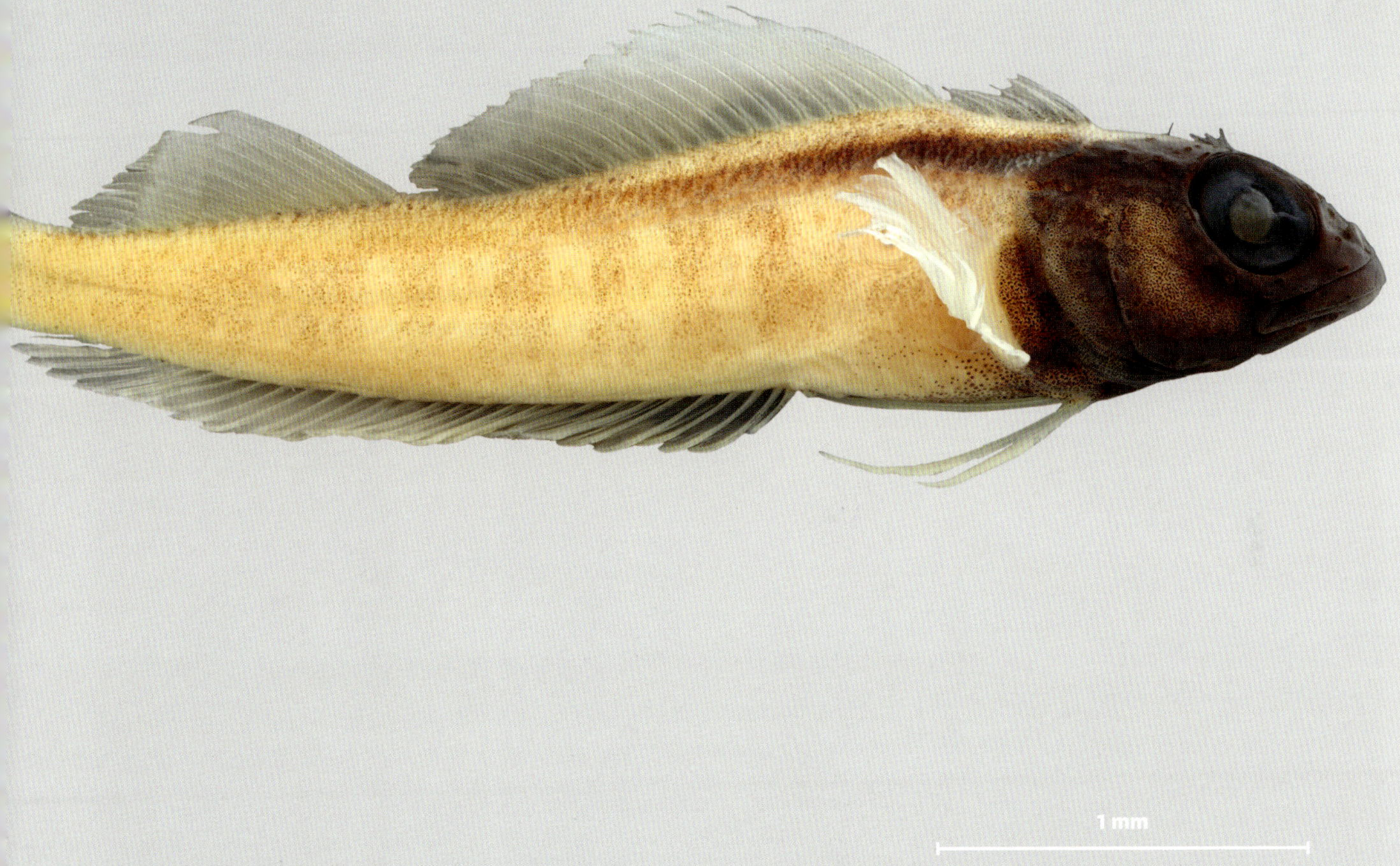
1 mm

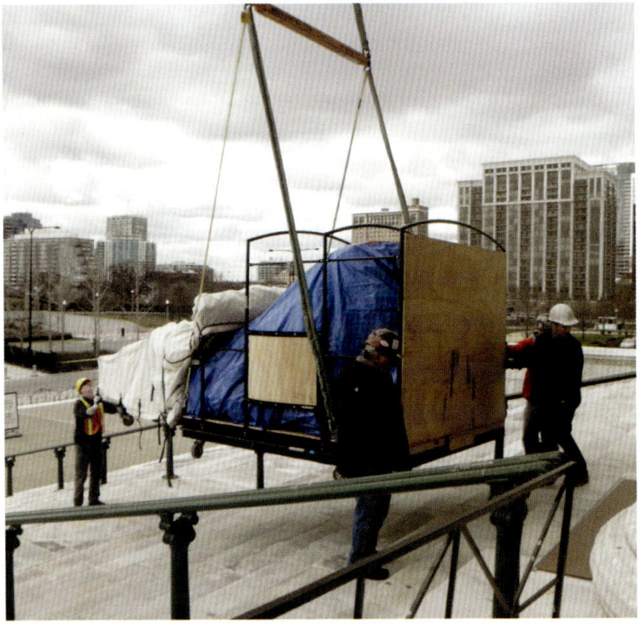
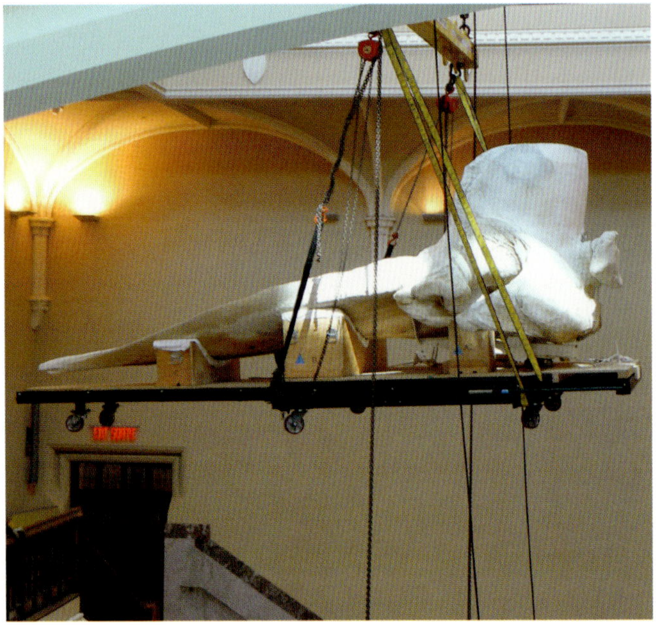
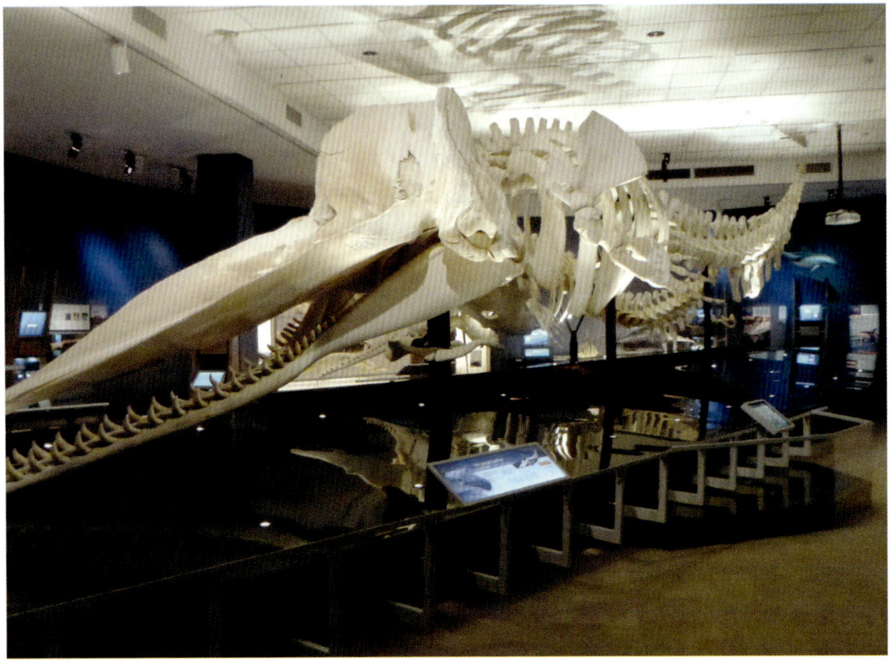
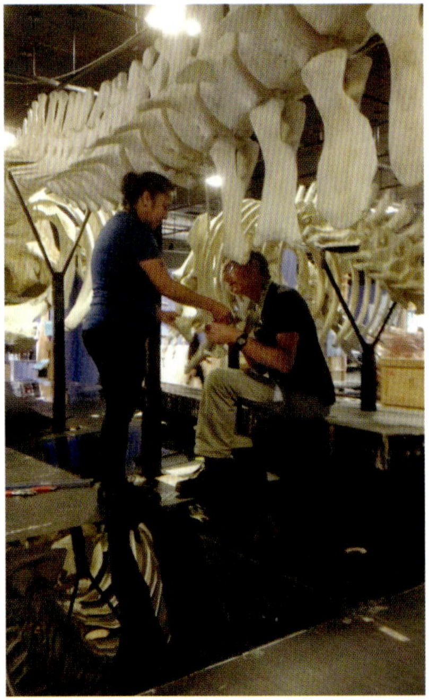

Tū Hononga

Physeter macrocephalus

It's 3AM, and we are somewhere close to the Texas border. It's too dark to see any of the countryside, but I've seen more than enough desert scenery since leaving San Diego yesterday. We're not even halfway to our destination, either. There are still another twenty hours to go until we roll into Grand Rapids, Michigan; still a lot of truck stops, bad meals and not much sleep.

This is all part of touring as a collection manager with Te Papa's *Whales/Tohorā* exhibition and its star attraction Tū Hononga – the skeleton of a 17-metre-long male sperm whale. He was one of twelve whales that stranded on Karekare Beach, West Auckland, in 2003 and was gifted to the nation by Te Kawerau ā Maki. I first met Tū Hononga in 2007, when I helped to put his skeleton together. The following year, the show headed off to North America and has been touring the USA and Canada ever since, covering 25,000 kilometres between sixteen different cities.

When we get to Grand Rapids the hard work will really start: unloading our ten truck-loads of exhibition material and then starting the four-week installation. The biggest challenge is always posed by Tū Hononga's skull – mounted on its travelling trolley, it weighs 750 kilograms and is over 5 metres long. Some very creative solutions have been needed just to get it into some of the exhibition galleries. It takes three days to assemble Tū Hononga, with a large gantry to lift his skull, forklifts to position his thoracic section, and a fair amount of grunting while attaching his ribs and vertebrae. Five months later we'll be back to do it all in reverse and get back on the road. TS

Top left: Moving the skull of *Physeter macrocephalus* into the Field Museum, Chicago. **Top right:** Lifting the skull into the gallery at the Canadian Museum of Nature, Ottawa. **Below right:** Moana Parata and Tom Schultz assemble the skeleton at San Diego Natural History Museum. **Below left:** Tū Hononga on display at the Fernbank Museum of Natural History, Atlanta.

093

100

Type specimen

Clematis petriei

One of the most significant parts of Te Papa's natural history collections is the approximately 5000 type specimens of plants. This is more than the rest of New Zealand's botanical collections put together, although there are even more type specimens of New Zealand plant species in European collections. Type specimens are those specimens on which a scientific name is based. This special status is usually established when a name is first published, although exactly what kind of type a particular specimen is may need further clarification. The large number of types reflects the long pedigree (at least in a New Zealand context) of Te Papa's collections, going back to 1865 with the creation of the Colonial Museum.

In some botany collections, type specimens are stored separately. But Te Papa has so many botany types – about 1.5 percent of its specimens – that they are simply dispersed through the main collection, distinguished by a red folder among the usual beige ones. I find this striking colour contrast a useful reminder when it comes to balancing the everyday utility of these specimens for our ongoing research work with their historical and scientific significance.

The specimen shown here is one of over 500 type specimens collected by Donald Petrie; he collected this one from near Canterbury's Springfield in 1893. He identified it as *Clematis colensoi*, but in 1961 it was used by HH Allan as the type specimen for his description of a new species, *Clematis petriei*. This species grows in the north-east of the South Island.

The skull-and-crossbones stamp in the lower left-hand corner indicates that this specimen was once painted with mercury chloride to deter herbivorous insects, a goal now achieved by quarantine measures and low humidity. LP

***Clematis petriei*, SP025890, collected by D Petrie, Kowai River, Canterbury, January 1893.**

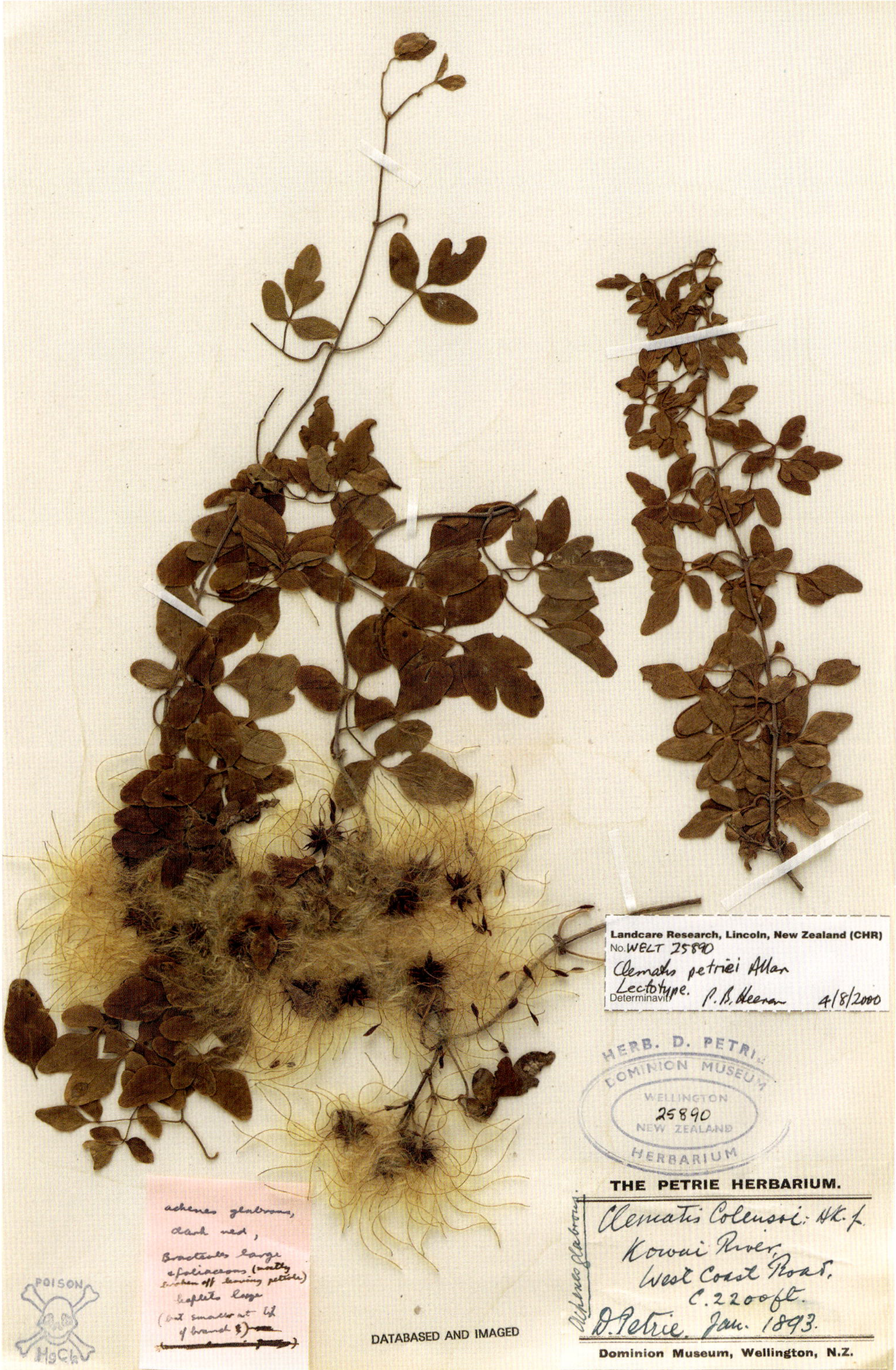
Landcare Research, Lincoln, New Zealand (CHR)
No. WELT 25890
Clematis petriei Allan
Lectotype.
Determinavit P. B. Heenan 4/8/2000
HERB. D. PETRIE
DOMINION MUSEUM
WELLINGTON
25890
NEW ZEALAND
HERBARIUM
THE PETRIE HERBARIUM.
Clematis Colensoi: Hk. f.
Kowai River,
West Coast Road.
C. 2200 ft.
D. Petrie. Jan. 1893.
Dominion Museum, Wellington, N.Z.
achenes glabrous,
dark red,
Bracteoles large
& foliaceous (mostly
broken off leaving petiole)
leaflets large
(but smaller at tip
of branch &)
POISON
HgCl
DATABASED AND IMAGED

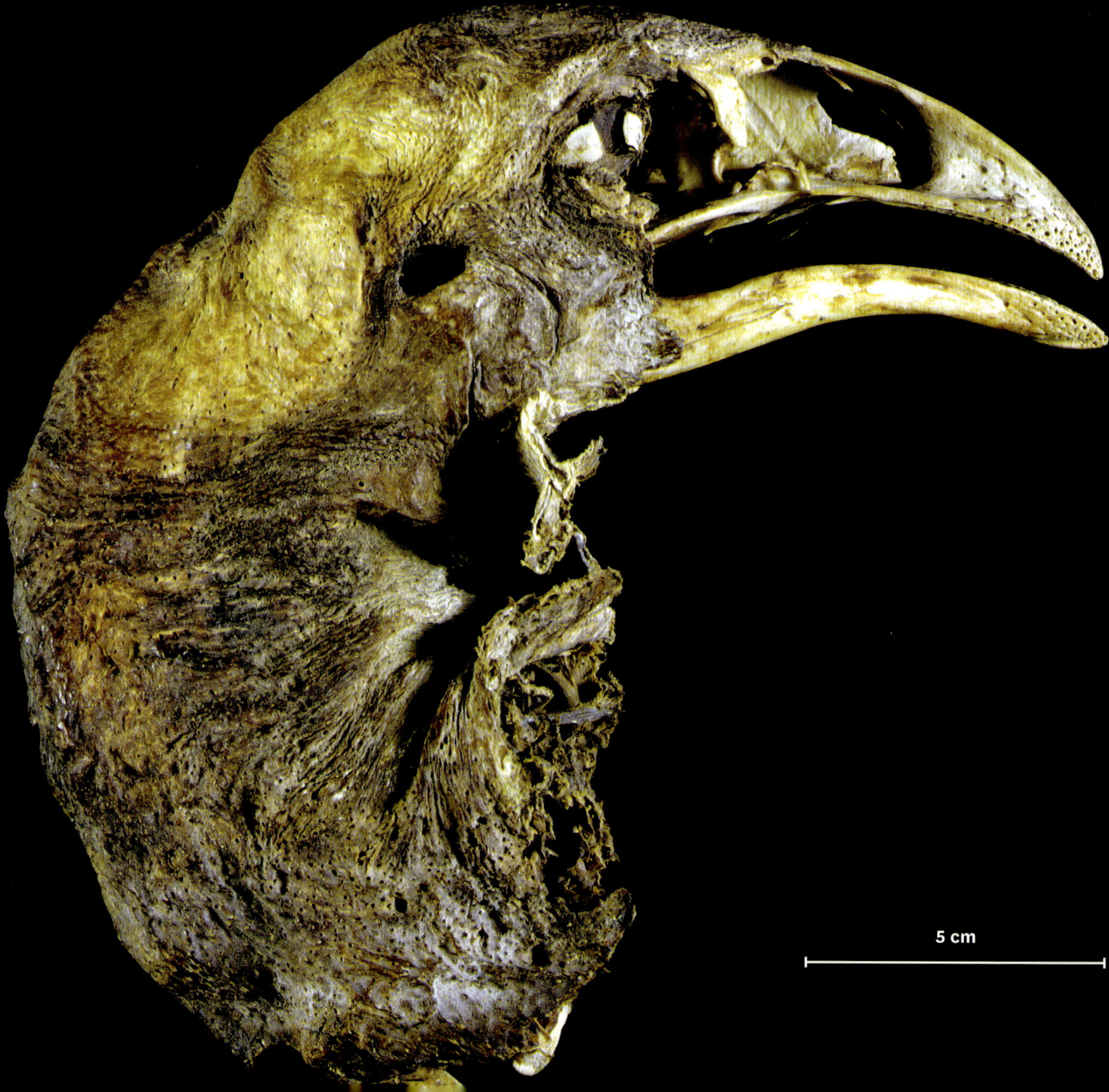
5 cm

Upland moa head

Megalapteryx didinus

Moa are an extraordinary group of birds, being most famous for their giant size. Giant moa could reach up to 3 metres high, making them the tallest known birds to ever live. They are also the only kind of bird to entirely lack wing bones, although their ancestors would have had wings.

Unfortunately, all species rapidly became extinct soon after humans began hunting them for food. What is not well appreciated is that there were nine species of moa that varied greatly in size, shape and habits. All moa were large birds, but some weighed only around 12 kilograms compared with 242 kilograms for the largest moa. This compares with 62 kilograms for the average weight of an adult human.

Perhaps the most unusual of all moa was the upland moa (*Megalapteryx didinus*), the species found in the highest environments in the South Island. It was a fairly slender, athletic species, growing to between 28 and 80 kilograms, with particularly long toes and feathers going right down its legs. Such features would have allowed it to survive well in cold, rocky places. We know quite a lot about the upland moa because it is the species that has been most commonly found mummified. This occurred because upland moa tended to live in drier and cooler environments than other moa species. Such mummified specimens have allowed detailed studies to be carried out on, for example, moa feathers and the muscles of the head. AT

Upland moa (*Megalapteryx didinus*) mummified head, S.400, unknown collector, Cromwell, Central Otago, probably in the 19th century.

Vanuatu petrel

Pterodroma occulta

In January 1927, collectors with the Whitney South Sea Expedition shot six petrels at sea off the Banks Islands in northern Vanuatu. Along with more than 40,000 other birds collected on an extraordinary twelve-year expedition, the petrels form part of the encyclopaedic collection of more than one million bird specimens held at New York's American Museum of Natural History (AMNH). Robert Falla (at one time Director of the Dominion Museum, which is now Te Papa), visited New York and saw the petrels. He considered them to be a small form of the white-naped petrel, a species known to breed only at the Kermadec Islands in northern New Zealand.

In 1997, I saw a similar bird, found dead in northern New South Wales in 1983, in the collection of the Australian Museum in Sydney, and was struck by its small size. After discussing the Sydney bird with Department of Conservation petrel expert Mike Imber, who was familiar with the New York specimens, we decided to name these birds as a new species: *Pterodroma occulta*, the Vanuatu petrel. The AMNH kindly agreed to exchange one of the 1927 birds with Te Papa, so we now hold one of the few specimens in existence.

In 2009, Australian Stephen Totterman, who had spent many years researching the birds of Vanuatu, discovered the breeding site of the petrel: on the top of a remote active volcano on Vanua Lava (one of the Banks Islands). Since then, together we have managed to collect some basic information on their breeding biology, and found that cats were killing them at their nests. It's unclear whether this enigmatic seabird will survive much longer. Sadly, many of the little-known birds on remote Pacific islands face a similarly uncertain future. AT

Vanuatu petrel (*Pterodroma occulta*) study skin, OR.26905, paratype, collected by R Beck, 30 miles east of Mera Lava, Banks Islands, Vanuatu, 27 January 1927.

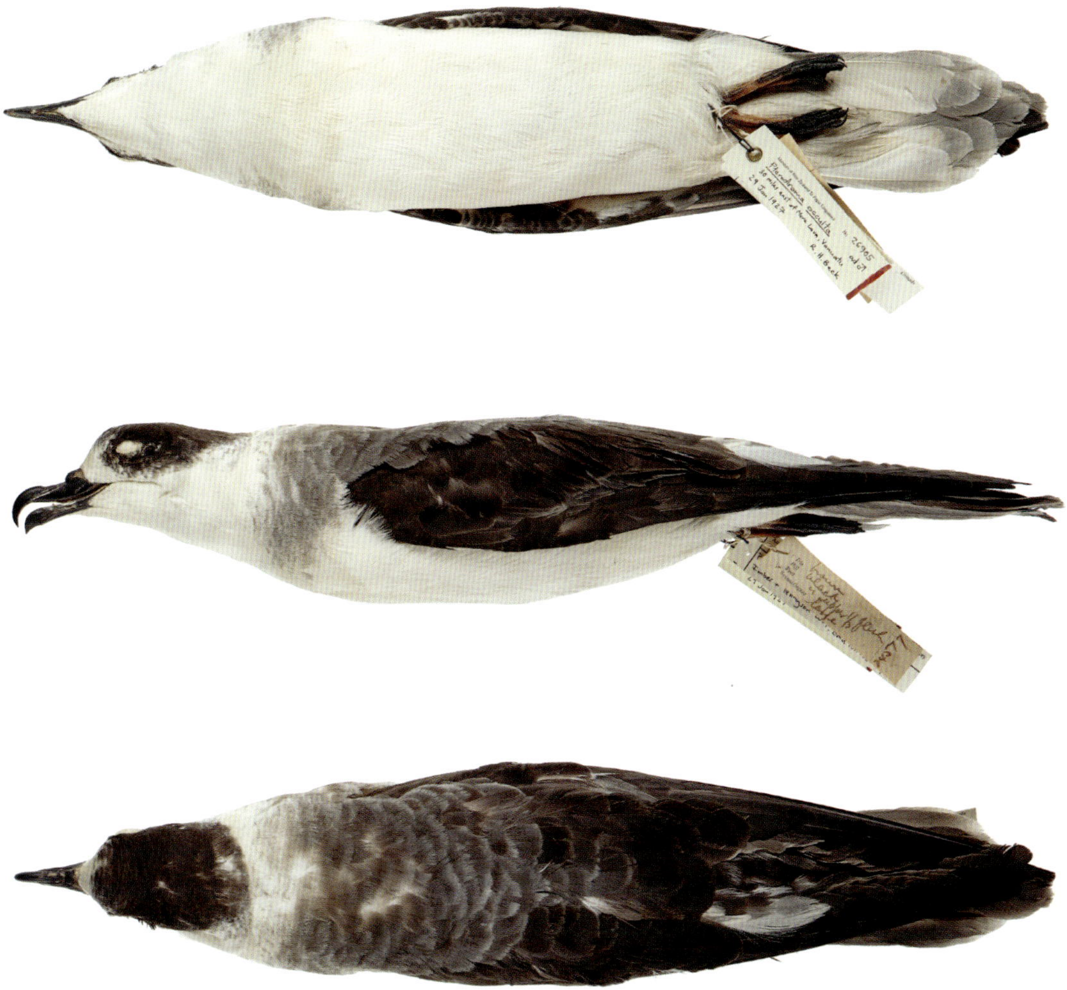

Vegetable caterpillar

Ophiocordyceps robertsii

A taxonomic collection holds as many specimens of flora and fauna as possible. Sometimes it is important to have the different developmental stages, such as larvae or nymphs, pupae and adult insects. It might be that the specimens are in a particularly interesting condition, for example caught during moulting or copulation. In the case of the vegetable caterpillar (āwhato), the specimens collected have been mummified.

The caterpillars of the porina moths or forest ghost moths (*Aoraia* species) crawl through the leaf litter doing their daily business of eating. If one happens to ingest the spores of a certain fungus, *Ophiocordyceps robertsii*, these spores will start to grow in the caterpillar's body.

Normally, the caterpillar digs into the ground at dawn and arises at dusk hungry as ever. Not so the one with *O. robertsii* spores in its belly – it remains underground, still alive but turned into a food source for the fungus. The growing fungus mummifies the caterpillar; eventually, its fruiting body reaches the surface above and produces spores for another hungry caterpillar munching along.

The vegetable caterpillar and its relatives were once much sought after. In Tibet and China, the mummy was consumed as an aphrodisiac; in New Zealand, Māori used to burn the āwhato and mix the crushed charcoal with fat and water to produce ink for moko. This ink was especially dark and also antiseptic, so the moko would heal quickly. It was likewise popular among Pākehā: at the beginning of the 1900s, the Auckland to Rotorua train would slow down so that people could run alongside and sell mummified caterpillars to tourists as curios, for the grand price of 10 shillings each. JK

(A) Vegetable caterpillar (*Ophiocordyceps robertsii*), TMP031944, no collection data.
(B) Porina moth (*Wiseana cervinata*), TMP031943, no collection data.

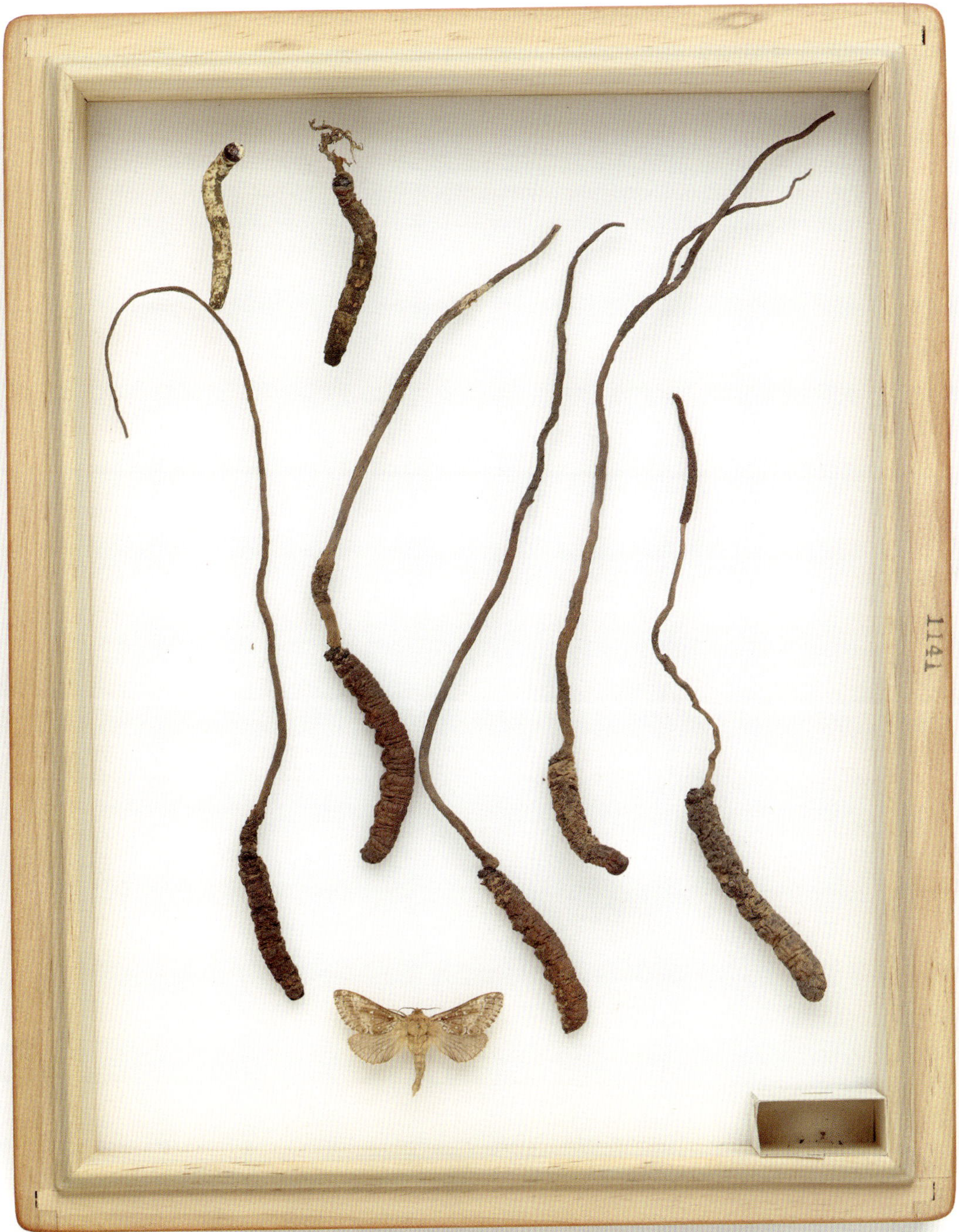
1141

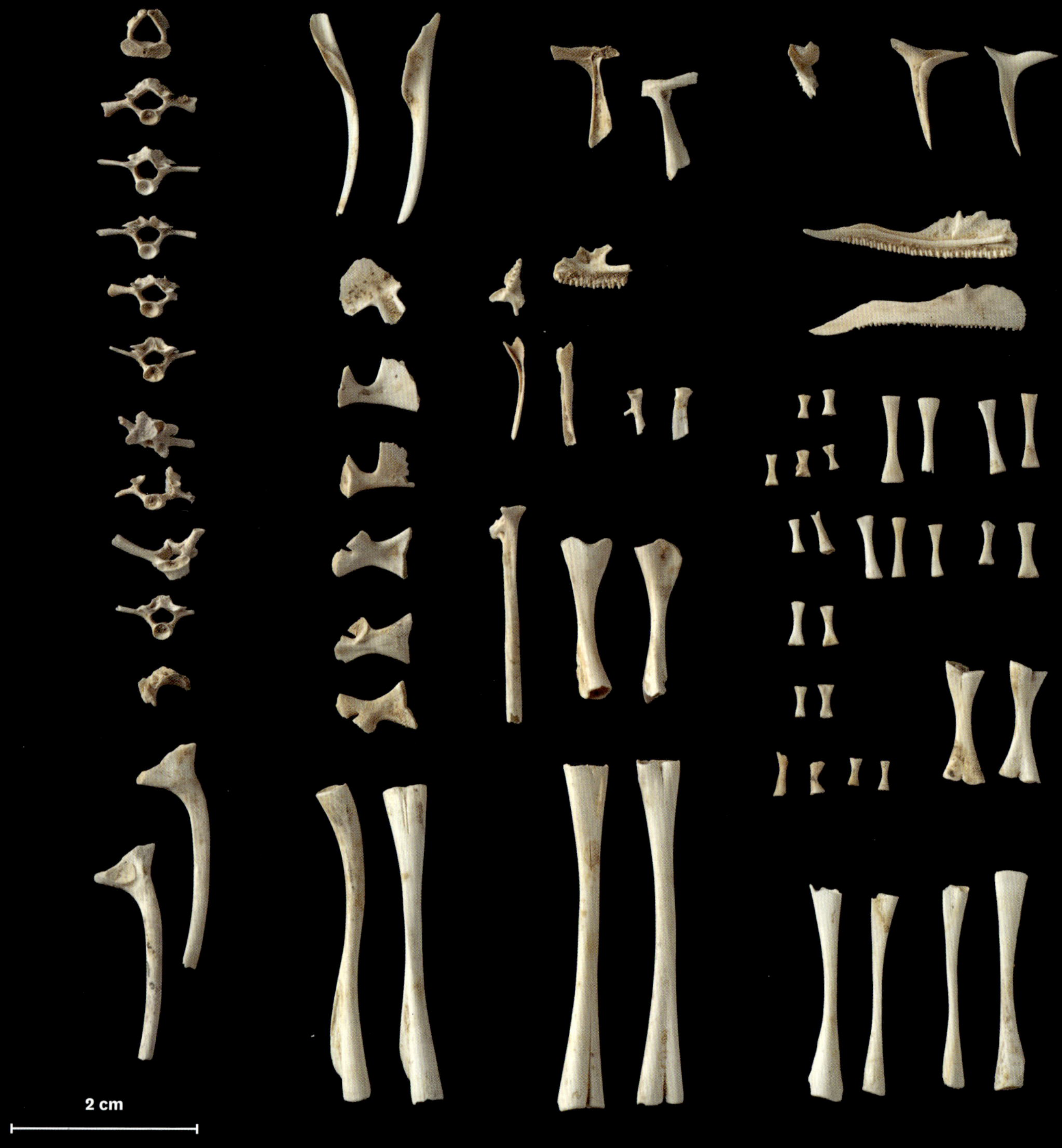
2 cm

Waitomo frog bones

Leiopelma waitomoensis

New Zealand's *Leiopelma* frogs (pepeketua) are thought to have diverged from other frogs around 180 million years ago. They have some unusual 'primitive' features, including extra vertebrae and the remains of tail muscles. Some lay their eggs in areas of moist ground rather than in water, and the tadpole stage of their life cycle occurs inside the large egg. The young hatch as tiny froglets and the male, who carries them on his back, cares for them.

Four species of living *Leiopelma* frog are recognised in New Zealand today, although recent research into their genetics and morphology indicates that the current taxonomy needs revision. All living *Leiopelma* frogs are threatened with extinction, and have greatly reduced distributions compared with pre-human times. Three extinct species of *Leiopelma* have been described from fossil bones, although ongoing research is likely to result in further changes to the number of recognised species. The extinct species are all larger than the living species; the largest, the Waitomo frog (*Leiopelma waitomoensis*) shown here, reached a whopping 10 centimetres in length! Despite its name, the Waitomo frog was not restricted to Waitomo and its bones have been found throughout the North Island (although most remains have been found in caves in the Waitomo district). LS

Waitomo frog (*Leiopelma waitomoensis*), S.023415, holotype, collected by T Worthy, Haggas Hole, Waitomo, March 1983.

Water bear

Phylum Tardigrada

Tardigrades, also known as water bears or moss piglets, are tiny eight-legged creatures but quite distinct from that better-known group of eight-leggers, the arachnids. They are an ancient group with fossils known from the mid-Cambrian, a little over half a billion years ago. The name tardigrade means 'slow-stepper', and these animals were first documented by scientists in the 1770s. The vast majority of tardigrade species are only a fraction of a millimetre long, making them the smallest animals in Te Papa's collection. However, tardigrades have a toughness that belies their size.

Tardigrades can withstand radiation that would kill us very quickly. They can also survive extreme pressures (up to 6000 atmospheres), high temperatures (a few minutes at 150°C) and being frozen (down to −270°C for over two weeks). The European Space Agency has even exposed them to the hard vacuum of space; the tardigrades survived better than we would have.

Tardigrades prefer damp or aquatic environments. Being such hardy creatures, they're found on mountain tops, in ocean depths and just about everywhere in between, but the easiest places to find them are in moss or lichen. While they need moist environments, they have a neat trick should things get too dry. They can shut down, and reactivate again upon exposure to water … even if it's decades later.

About 1200 species of tardigrade are known so far. We know of around 90 species in New Zealand, although a lot more are likely to be discovered. Te Papa has about 2500 specimen lots of tardigrades, including 27 primary types. PS

Tardigrade (*Echiniscus zetotrymus*), AT.000014, length approx. 0.3 mm, collected by D Horning Jr from lichen (*Usnea ciliata*) on rock, Jollybrook River, Lake Sumner area, Canterbury, 18 April 1971.

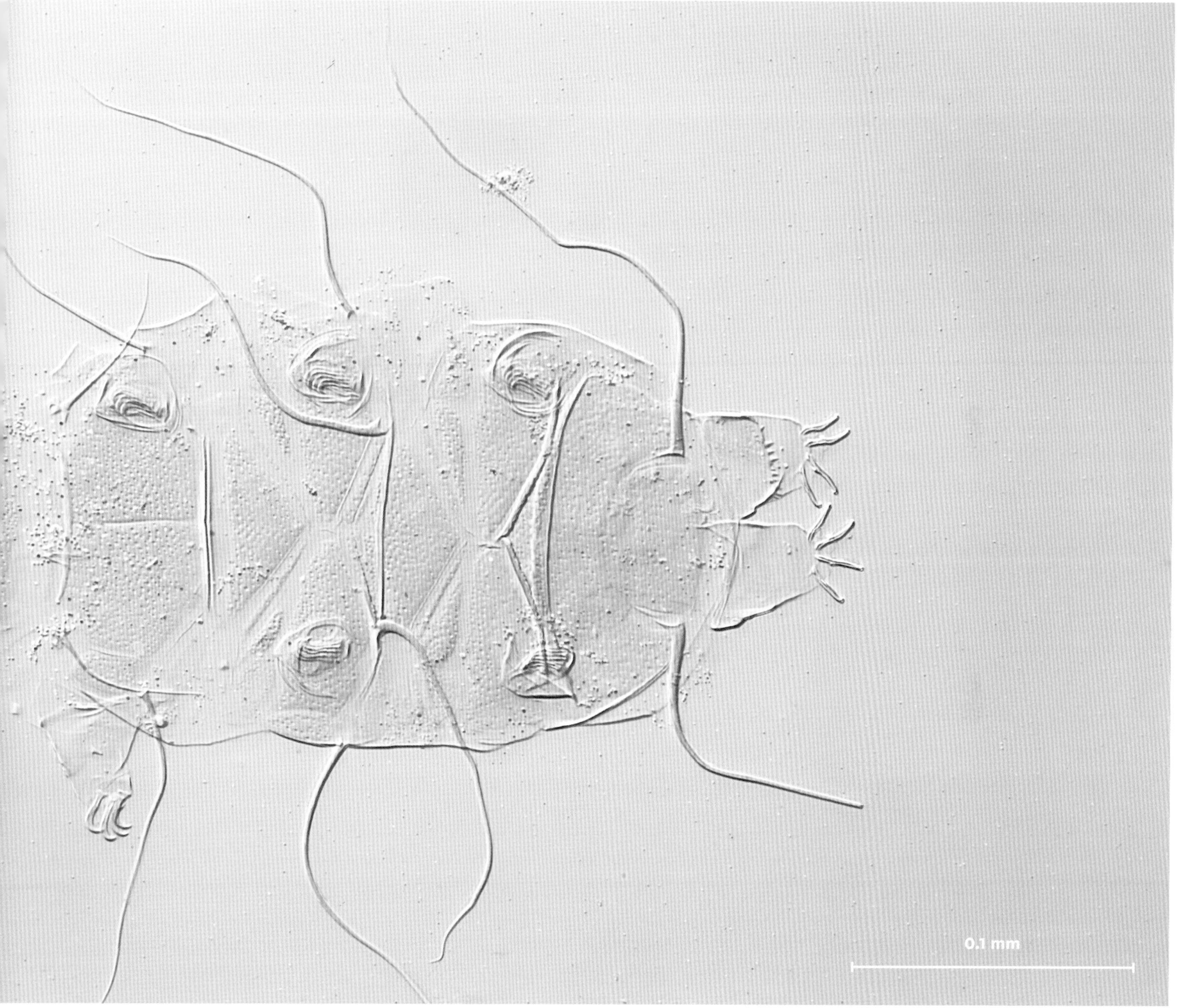
0.1 mm

West Coast bush wētā

Hemideina broughi

The specimen shown here is the holotype female of the West Coast bush wētā, *Hemideina broughi*. A holotype is a specimen that exemplifies and carries the name for the species chosen from the material available when it is first described. It may seem hard to believe, but this magnificent-looking wētā was collected in 1895, making it one of the oldest insect type specimens in Te Papa's collections. It was caught by Mr J Brough around 800 metres above sea level at the Karamea Saddle in the Nelson area. The eminent naturalist Sir Walter Buller described the species and named it after its collector. The Colonial Museum acquired the specimen from Buller a few years later.

This species is rather unusual among wētā. It is nominally part of the genus *Hemideina*, a group that includes well-known species such as the Wellington (*H. crassidens*) and Auckland (*H. thoracica*) tree wētā. However, it lacks several typical *Hemideina* features while possessing a number of characters similar to those seen in giant wētā (*Deinacrida* species). This combination of features means that *H. broughi* is regarded as one of the most primitive of all New Zealand wētā.

These insects are not dangerous, but are more aggressive than their relatives if disturbed. With a body length of more than 80 millimetres, this makes them appear even more intimidating to those with an aversion to wētā! As the common name suggests, the West Coast region is part of their geographical range. They are found in rainforest in the upper north-west of the South Island, with Greymouth marking the southern-most part of their distribution. They like to live in hollows in trees and logs. **PS**

West Coast bush wētā (*Hemideina broughi*), AI.000632, holotype, collected by J Brough, Karamea Saddle, Nelson, 1895.

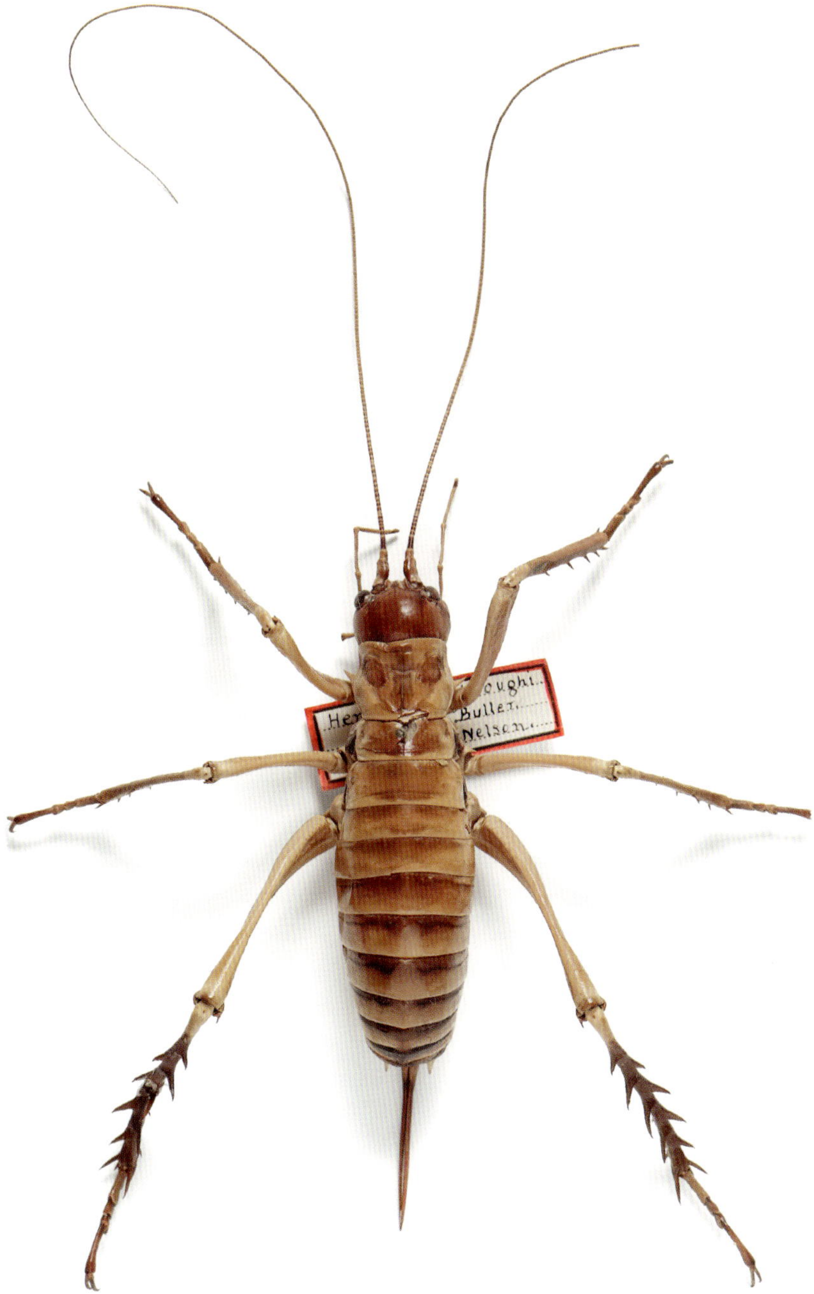
Buller
Nelson

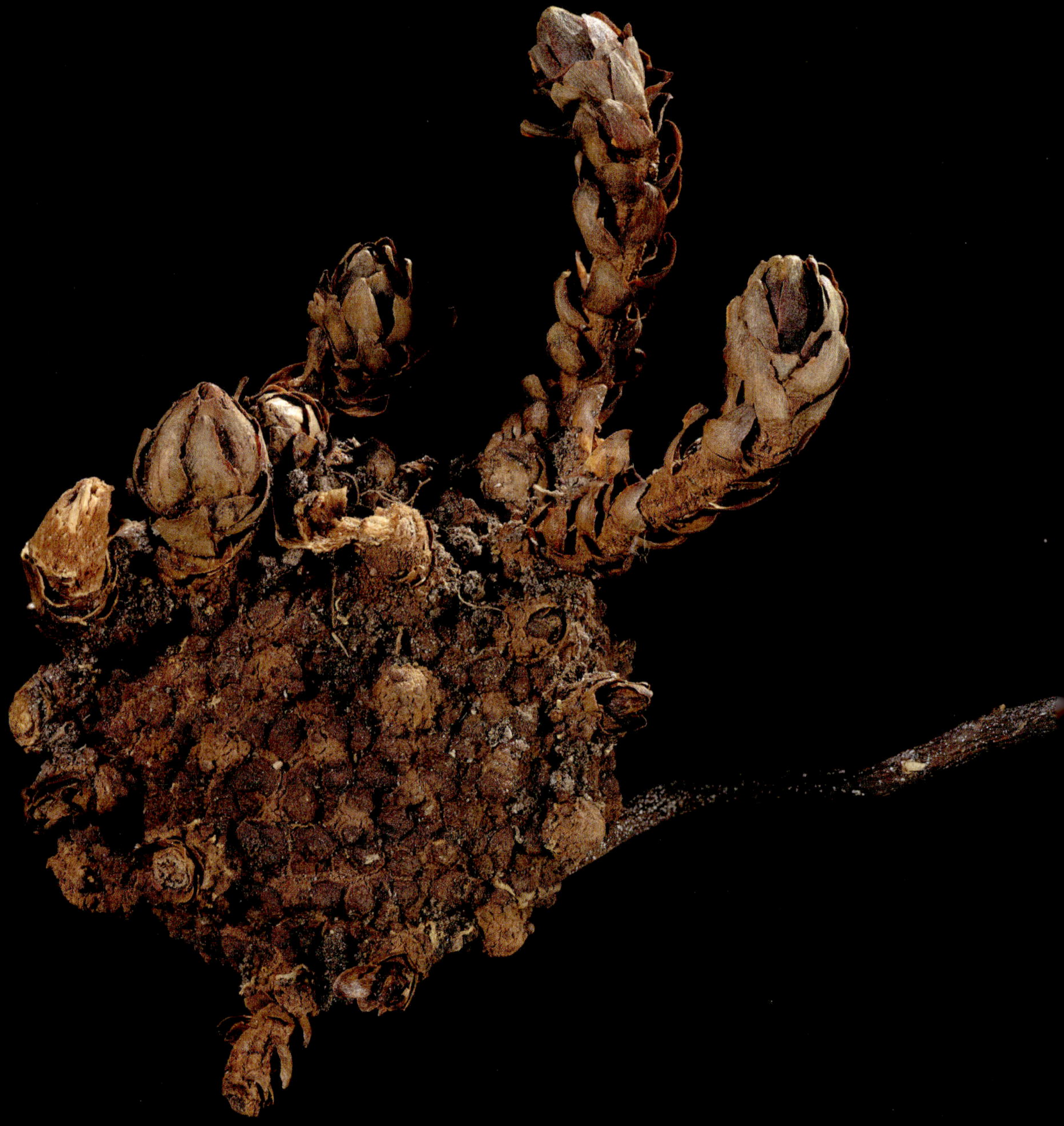

Woodrose

Dactylanthus taylorii

Not many museums can boast of having flowers from the underworld among their collections. Te Papa's herbarium has over thirty specimens, the oldest dating from 1878. *Dactylanthus taylorii*, commonly known as woodrose, is a parasitic plant only found in the North Island of New Zealand. Its Māori names pua o te rēinga (flower of the underworld) or waewae o te atua (fingers of the gods) refer to the subterranean habit of this plant and the finger-like arrangement of its flowers.

Unlike many other plants, *Dactylanthus* is leafless and so does not produce its own food. Instead, it attaches to the roots of nearby shrubs and trees, and steals food from them. The best time to find this unusual plant is when it is flowering. Small clusters of pink-brown flowers emerge among the leaf litter and produce a distinctively musky fruit-scented nectar. Another eerie aspect of *Dactylanthus* is then revealed: the smell attracts short-tailed bats, which come down to the forest floor and crawl around the flowers to drink their nectar and, consequently, pollinate them.

Since its discovery by Reverend Richard Taylor on a rainy night in March 1845, *Dactylanthus* has become very uncommon: it is currently considered a Threatened – Nationally Vulnerable species. Besides the damage caused to the nectar-rich flowers by possums and rats, collectors have greatly contributed to its decline. Plants used to be dug up and boiled until the deformed root of the host, which resembles a wood rose at the point where *Dactylanthus* attaches to it, was exposed. The wood roses were then varnished, mounted and sold as curios. A testimony of this fashion is also represented in our collections. CL

Woodrose (*Dactylanthus taylorii*), SP053917, collected by L Anderson, Mt Egmont, Taranaki, 5 April 1935.

The collectors

A brief history of those who built the collections

Chatham Island snipe specimens purchased from the collector and dealer Sigvard Dannefaerd, showing the original field labels.

Colin Miskelly

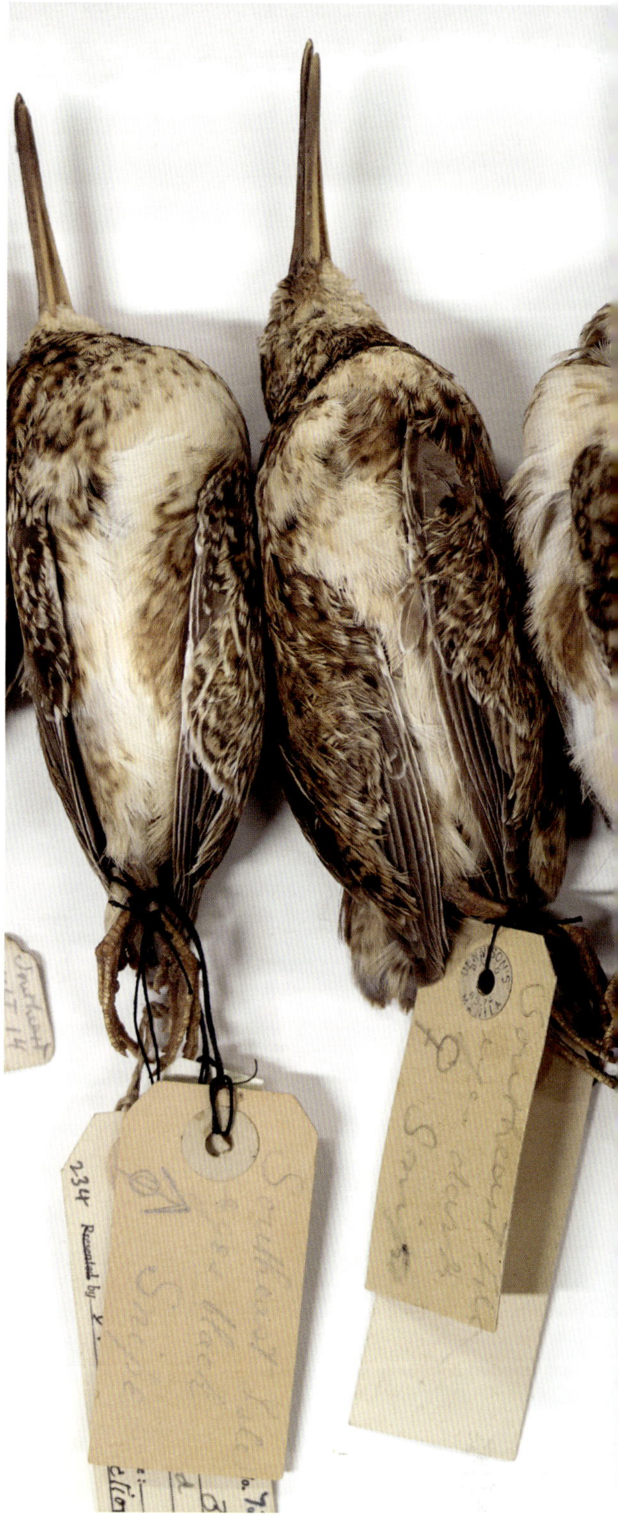

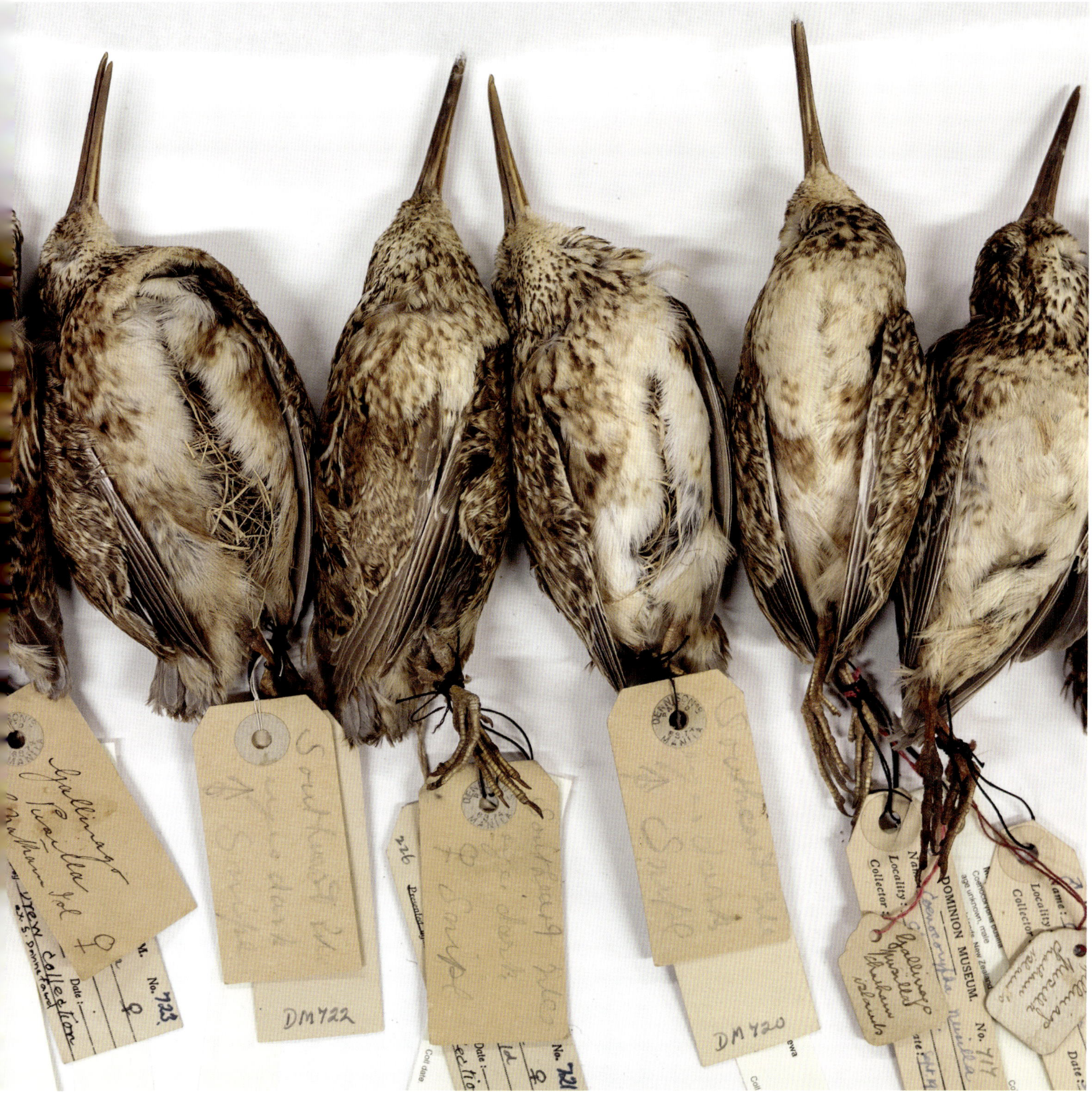

DM722
DM720
DOMINION MUSEUM.
Locality
Collector

Te Papa holds a treasure trove of more than 1.5 million natural history specimens, gathered over the more than 150 years since the museum came into existence. From its humble beginnings as the Colonial Museum in 1865, the museum has grown and diversified as it has shifted location twice and changed its name three times (to the Dominion Museum, then the National Museum of New Zealand, and finally its current name).

Few specimens survive from those earliest years, when founding Director James Hector and his small but loyal team collected scientific specimens during their survey work throughout the country. Although primarily employed as geologists, Hector and his staff had wide interests in natural history, and a mandate to collect and curate any plants, animals, fossils or minerals that attracted their attention.

Before the museum was founded, there was little opportunity for science to be undertaken in New Zealand. Almost all specimens of any significance were sent to the great museums of Europe, where curators and other scientists described thousands of new species in journals and books that few New Zealanders had access to. This started to change from 1868, with the founding of the New Zealand Institute and the annual publication of its *Transactions* by Hector and his team. Home-grown science was greatly enabled, with the new museum providing a natural home for specimens of the hundreds of new species described in the journal's pages.

Many of the earliest collectors were employed in other fields, including as clergymen, run-holders, merchants, civil servants, politicians and lawyers. The study of natural history was both a hobby and a way to network with other like-minded citizens and leaders of society. Notable collectors during the colonial era included the politician Walter Mantell (geology and palaeontology), the surveyor Frank Clarke (fishes), Reverend William Colenso (plants), lecturer and conservator of forests Thomas Kirk (plants) and museum staff members John Buchanan (plants) and Frederick Hutton (zoology).

While the collections grew mainly through donations, exchanges, field collecting and bequests, the museum board or the government occasionally agreed to purchase important collections. In this way, those of the ornithologist and lawyer Walter Buller, the professional collectors and dealers Henry Travers and Sigvard Dannefaerd, and government steamship captain John Bollons were acquired.

Throughout the museum's history, the most prolific contributors to the collections have been the staff, led by Directors who were themselves actively engaged in many branches of natural history. Following Hector's retirement in 1903, the reins were taken up by the ethnologist Augustus Hamilton, followed by the palaeontologist Allan Thomson. While all three men worked across many disciplines, it was Thomson's successor WRB (Reginald) Oliver who left the largest legacy across a diversity of collections, including 13,964 botany specimens, more than 2000 lots of molluscs and 118 specimens of birds.

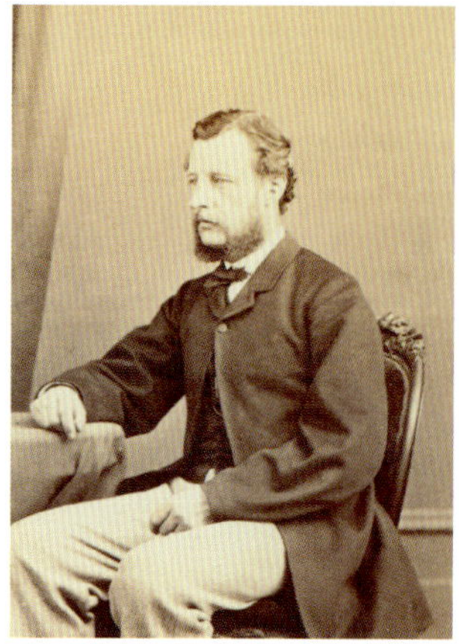

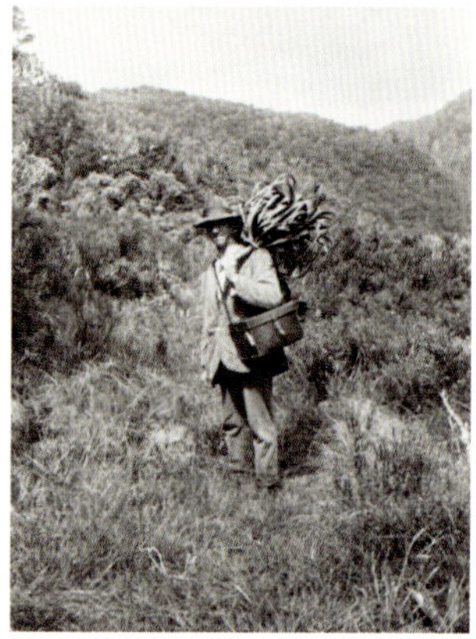

James Hector, c. 1868.

Reginald Oliver collecting plants at Wilmot Pass, Fiordland, March 1927.

Succeeding Directors from 1947 through to the establishment of Te Papa in 1992 had narrower research interests, but continued to play a personal role in the development of the collections. Robert Falla was primarily an ornithologist (acquiring 886 bird specimens), while his successors Dick Dell, John Yaldwyn and Alan Baker worked mainly in various branches of marine zoology, including molluscs, crustaceans, urchins, starfish and marine mammals. Dell was particularly prolific as a collector and describer of marine molluscs, adding more than 3000 lots to the collection, while Yaldwyn was an important contributor to the palaeontology collection, adding 329 lots of bird, mammal, frog and reptile bones.

Besides their significant personal collecting efforts, the mid-twentieth-century Directors influenced the development of the collections through recruiting an expanding team of specialist curators and collection managers. After more than eighty years of making do with cramped accommodation and few staff, the museum enjoyed a growth spurt after World War Two, when able to make full use of the imposing Dominion Museum building in the Wellington suburb of Mt Cook (the building was opened in 1936, but was requisitioned by the military during the war). This expansion in both expertise and collection efforts began under the stewardship of Robert Falla (Director 1947–66) and continued under Dick Dell (1966–80) and John Yaldwyn (1980–89). These four decades could be considered the golden years for natural history at the museum, at least in terms of the rapid development of the biological collections and the leading role that the museum's Directors and natural history curators had in the New Zealand science scene.

Specialist curators and researchers who progressed collection development during 1950–90 included John Salmon and Ron Ordish (entomology), Fred Kinsky and Sandy Bartle (ornithology), Bruce Hamlin and Nancy Adams (botany), Charles McCann (lizards), Jock Moreland, Graham Hardy and Chris Paulin (fishes), Winston Ponder and Frank Climo (molluscs), and Phil Millener and Trevor Worthy (fossil vertebrates). Between 1947 and 1972, Dominion Museum staff organised a series of research expeditions to remote regions of New Zealand (including Fiordland, Stewart Island, the Chatham Islands and the subantarctic islands), resulting in thousands of specimens being added to the collections. Notable among these were numerous fishes, molluscs and other invertebrates from the deep ocean floor, brought to the surface in dredges towed by the MV *Alert* (and later the MV *Acheron*), captained by Alex Black. Many of these were species previously known only as fossils, providing insight into how little was known about the organisms living in the inaccessible abyssal depths surrounding New Zealand.

Exploring the ocean depths is a complex and expensive process, and since the 1970s the museum has benefitted greatly by working collaboratively with other research agencies, including the New Zealand Ministry of Fisheries (Fisheries Research Division, and Fisheries Management Division), Victoria University of

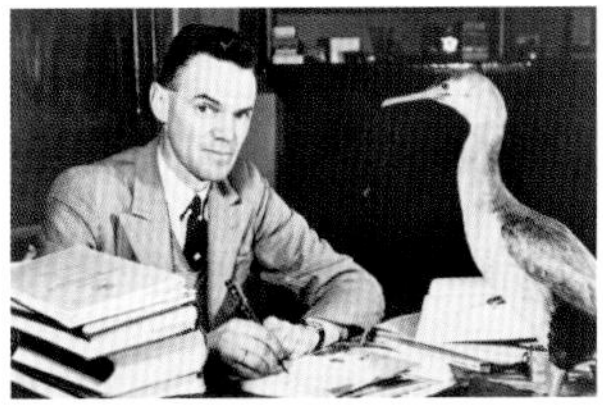

Robert Falla, c. 1935, watched over by a juvenile pied shag.

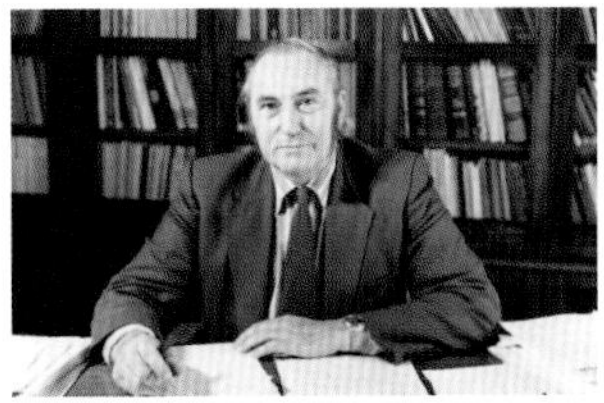

Richard 'Dick' Dell, Director of the National Museum, 1975.

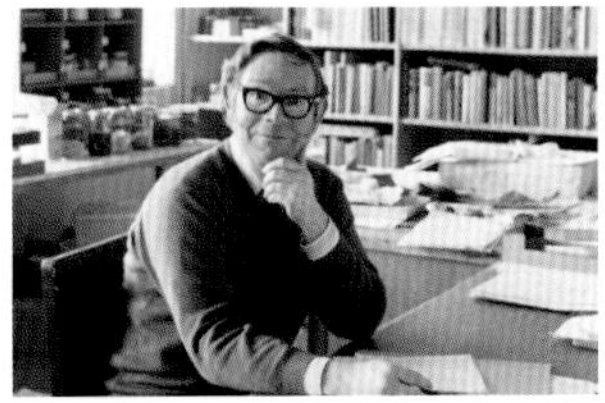

John Yaldwyn, Assistant Director of the National Museum, 1976.

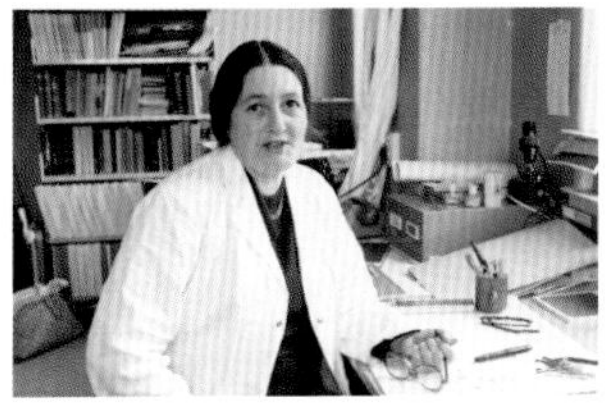

Nancy Adams, Technical Officer (Botany), National Museum.

Wellington, the New Zealand Oceanographic Institute and the National Institute of Water and Atmospheric Research (NIWA). Collections of fishes, seaweeds, molluscs, crustaceans and other marine invertebrates have grown enormously through museum staff collecting tens of thousands of specimens while participating in expeditions based on large research vessels owned or chartered by some of these institutions, including the *Ikatere*, *James Cook*, *WJ Scott*, *Tangaroa* (two different vessels of the same name) and *Kaharoa*. These expeditions explored parts of New Zealand's extensive Exclusive Economic Zone (one of the largest in the world), as well as down to the Ross Sea, and into the tropical and subtropical South Pacific. Many significant specimens of birds, fishes and marine invertebrates were also obtained via the fisheries Scientific Observer Programme, from catches made by commercial deep-sea fishing vessels.

The museum's holdings also grew through the acquisition of major collections from other individuals and agencies. These included teacher and school inspector Donald Petrie's herbarium containing 11,879 lots (around 1800 species) donated in 1922, the Edward Petherick collection (formerly the Newtown Museum) acquired via Wellington City Council in 1929, the Drew Collection of 288 mounted birds purchased in 1939, postal clerk George Hudson's magnificent collection of New Zealand insects (bequeathed in 1946), and wool auctioneer Creagh O'Connor's collections of birds, beetles and snails donated or bequeathed between 1922 and 1952. More-recent acquisitions have included 2635 lizard specimens donated by Tony Whitaker (Ecology Division, Department of Scientific and Industrial Research) in 1977, and the Russel Price collection of more than 21,000 bird bones excavated from Lake Poukawa, Hawke's Bay (acquired from Hastings Cultural Centre in 1982). The museum also received many hundreds of skulls of introduced rodents and stoats, and jaws of larger mammals (including wild deer, goats, sheep and pigs), from Ecology Division and the New Zealand Wildlife Service during the major restructuring of government research and environmental agencies in the 1980s, and over 2000 fishes, including types, donated by NIWA in 1997. Significant bequests and donations since then have included dietician Pauline Mayhill's collection of New Zealand land snails (21,559 lots in 2002), University of Canterbury professor Robert Pilgrim's research collection of bird lice and fleas (4779 slides, including the largest and most diverse collection of flea larvae in the world, donated in 2010), the NIWA Freshwater Fishes collection of nearly 37,000 specimens donated in 2014, and important collections of living and fossil molluscs donated by John Graham, Bruce Hazelwood and David Roscoe between 1990 and 2003.

Attitudes towards collecting specimens for research purposes have changed over the life of the museum, particularly in relation to native birds. Most endemic birds have been legally protected for more than a century, and even vagrant species have received comprehensive protection since 1953. While permits could be sought

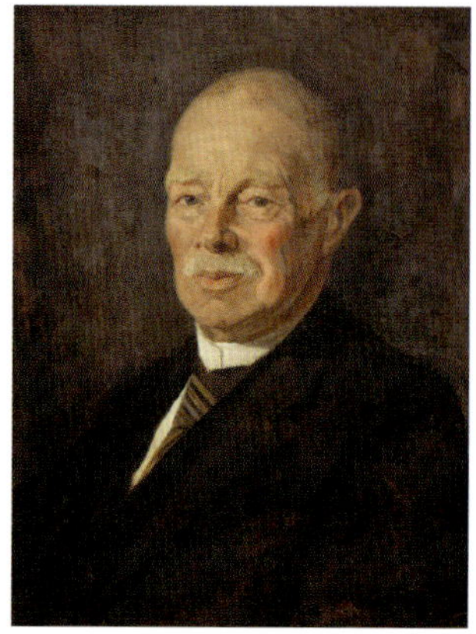

George Vernon Hudson, 1930s.

for scientific collecting, changing societal values and the personal conservation ethics of curators have meant that shooting birds for the collection is an activity of a bygone era. However, the museum continues to acquire bird specimens that are found dead, are euthanised due to injury or illness, or that are accidentally killed. Wildlife Service staff (particularly senior officers Don Merton, Brian Bell and Christopher Robertson) were particularly supportive of the museum, ensuring a steady supply of skins, skeletons and abandoned eggs of some of New Zealand's most endangered bird species found in the course of their work.

Te Papa's collections continue to grow incrementally, especially within the main research areas of current staff. Based on these collections, species new to science – some endemic to New Zealand – continue to be discovered on a weekly basis. Prolific collectors among the staff during the 'Te Papa' era include Bruce Marshall (molluscs), Pat Brownsey, Leon Perrie and Peter Beveridge (mainly ferns, but also moss and liverwort specimens collected at annual Bryophyte Workshops since 1983), Wendy Nelson (seaweeds), Clive Roberts, Andrew Stewart and Carl Struthers (fishes), Ricardo Palma (bird lice), and Alan Tennyson (birds and fossils). Notable among these are the many thousands of fossil bird, reptile, mammal, frog and fish bones plus molluscs extracted from Miocene (sixteen-to nineteen-million year-old) fossil beds at St Bathans in Central Otago by Alan Tennyson and his colleagues. The ongoing descriptions of previously unknown species from these ancient lake-bed deposits continue to transform our understanding of the history and composition of New Zealand's land-based biological heritage.

While the museum's natural environment collections are much more extensive than when the Colonial Museum opened in 1865, there are always exciting new discoveries to be made and significant specimens and collections to be acquired. As curators and collection managers, we are privileged to share Sir James Hector's legacy of scientific enquiry, along with the duty of care to maintain the collections in good condition for future generations of researchers and scholars.

The collection and the field

A brief tour around natural history research at Te Papa

The Te Papa fishes team during the four-week NORFANZ offshore deep-sea sampling expedition of 2003. Scientists from Te Papa, NIWA, Australia and France were on the voyage.

Michael Keith

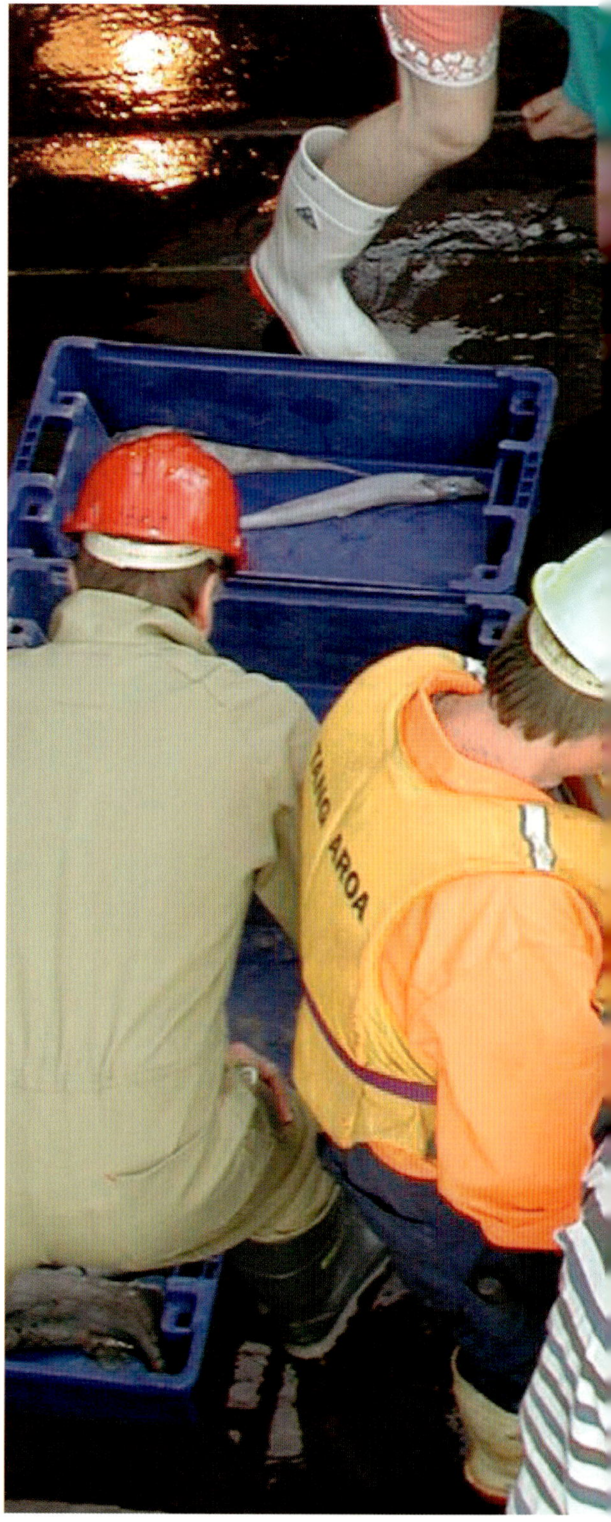

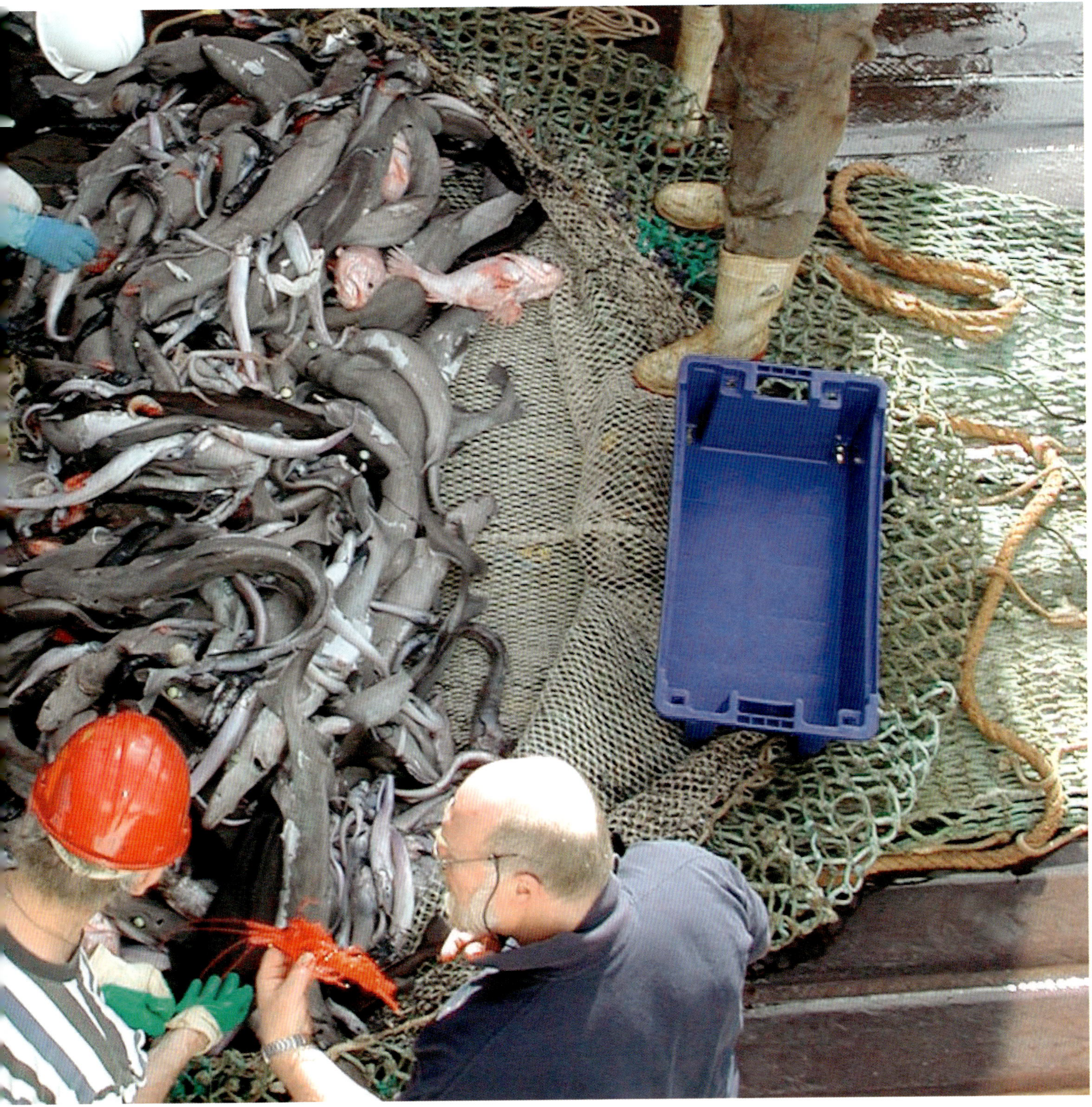

Writing for exhibitions at Te Papa, you occasionally get to work with the museum's scientists, drawing on their specialist knowledge of the environment. You appreciate their expertise, but you also get an inkling of their other world, the gravitational force which holds them in their orbit as you swing by – the museum's Natural History Collection. So writing this piece offered an opportunity to explore that attraction.

Natural history research at Te Papa springs from a relationship between two spheres. One is 'what's in here' – the Collection (with a capital C: the collective of many separate collections gathered from the natural world). The other is 'what's out there' – the field: the natural world itself, and specifically the living things, present and past, of New Zealand's biogeographic region. For the scientists – curators, collection managers, along with researchers brought in on specialist topics – research is a dynamic interaction between the two spheres. Investigating the collection feeds work in the field; fieldwork builds the collection and also feeds a closer look into it.

Built up over some 150 years of such interactivity, the Collection with its myriad specimens (1.5 million and counting) has acquired critical mass for researchers. It has become an invaluable library of material forms – broad, deep and old enough to address questions about the unique characteristics of New Zealand's biodiversity and its changes over time. It is also sufficiently representative of the region to be internationally significant, attracting attention from scientists across the globe.

Susan Waugh manages Te Papa's research programme. 'The research environment here means you explore big-picture questions, like why we have so many species in a particular plant or animal group, and what were the drivers of these species' diversity,' she says. 'You have the capacity to ask historical questions, for example, about the loss of genetic diversity over

Top left: Te Papa photographer Jean-Claude Stahl near Pongaroa, Wairarapa, in 2010.

Below left: Rick Webber and researcher Graham Bird collecting in the intertidal zone at Mana Island as part of the Kapi Mana Bioblitz in March 2011.

Right: Antony Kusabs and Susan Waugh collecting plant specimens at Matiu/Somes Island in Wellington in November 2012.

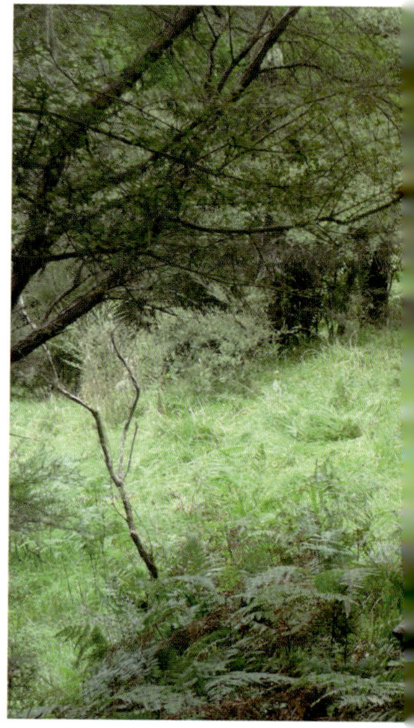

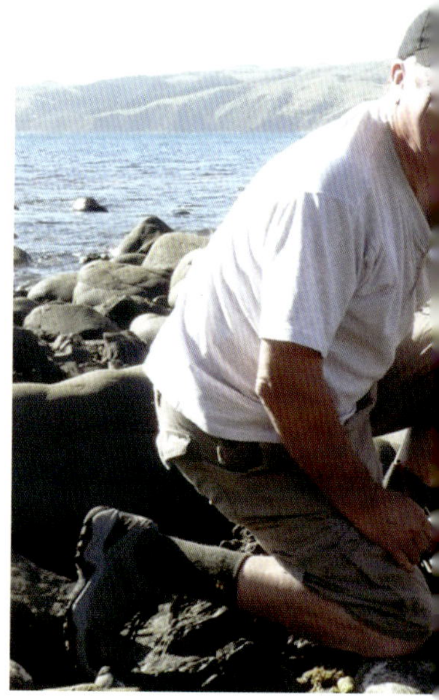

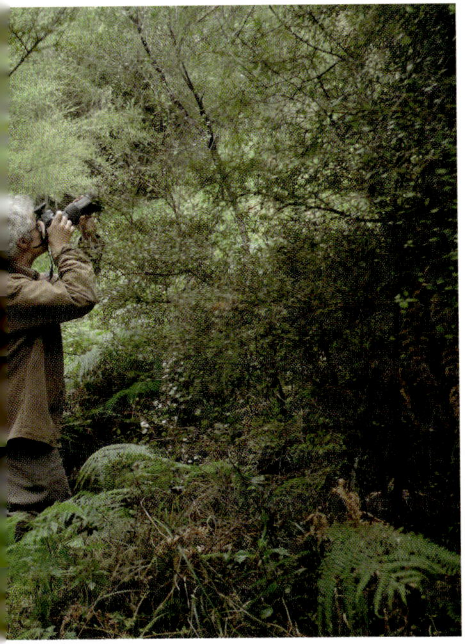

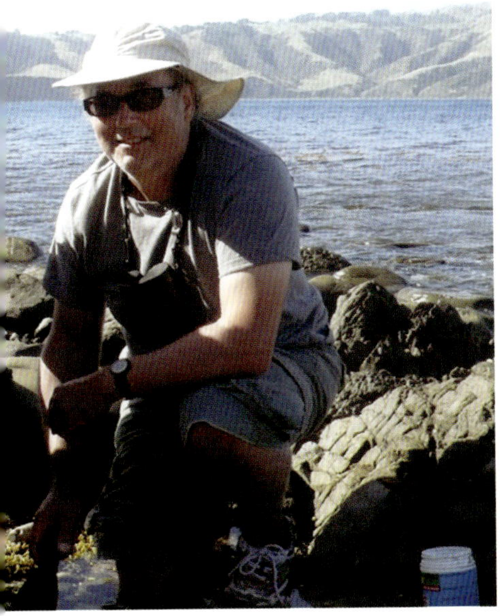

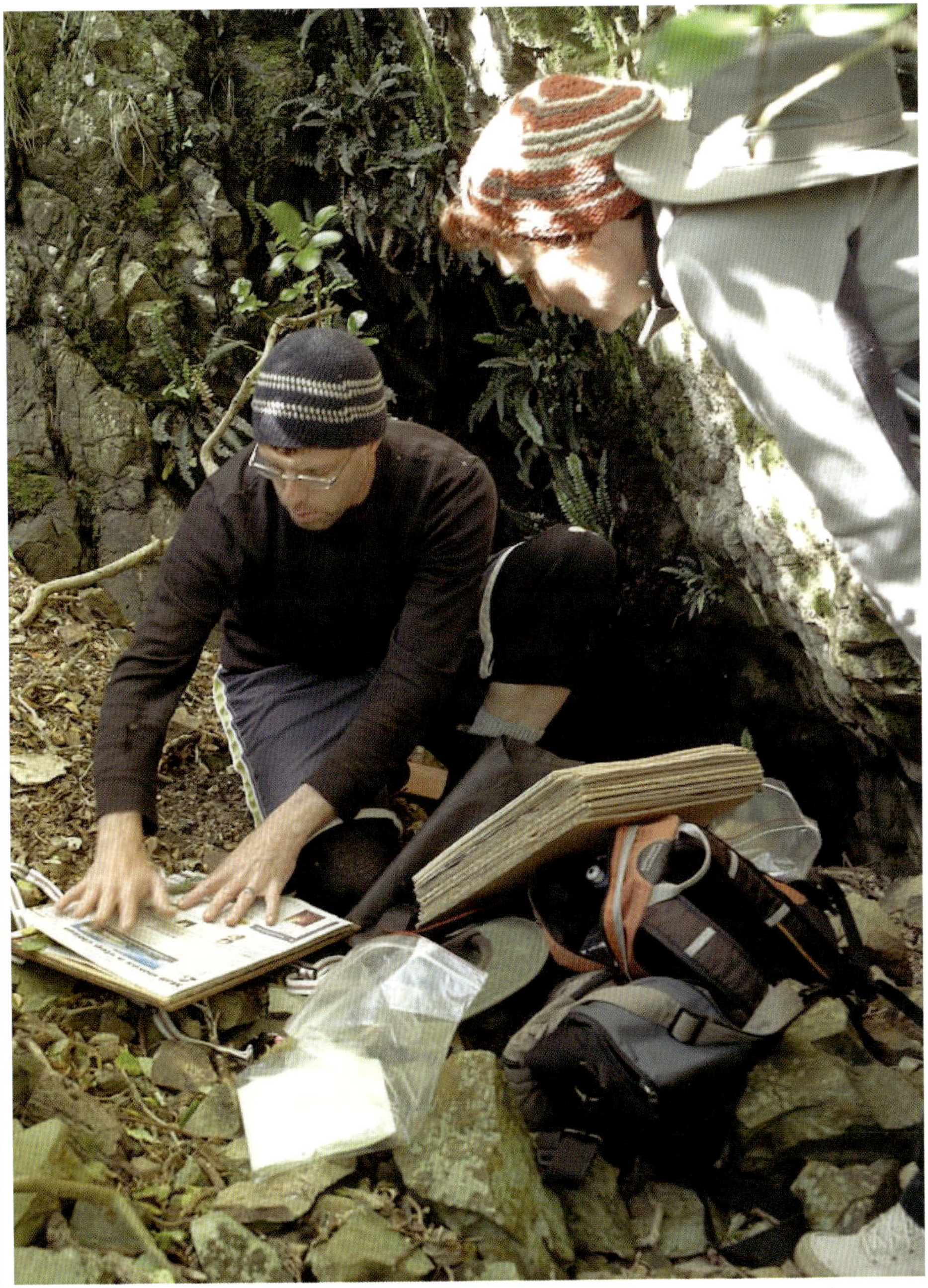

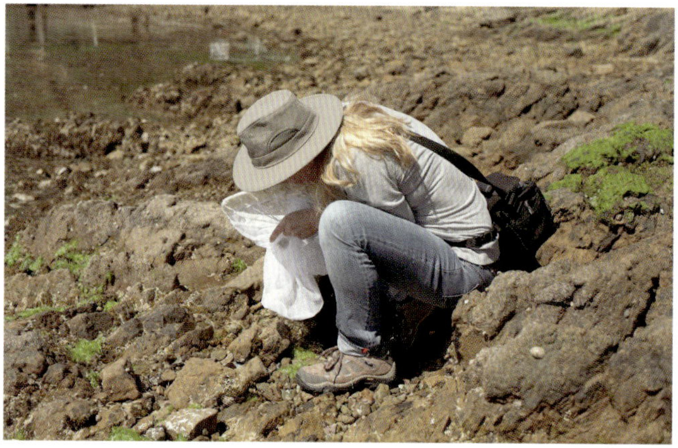
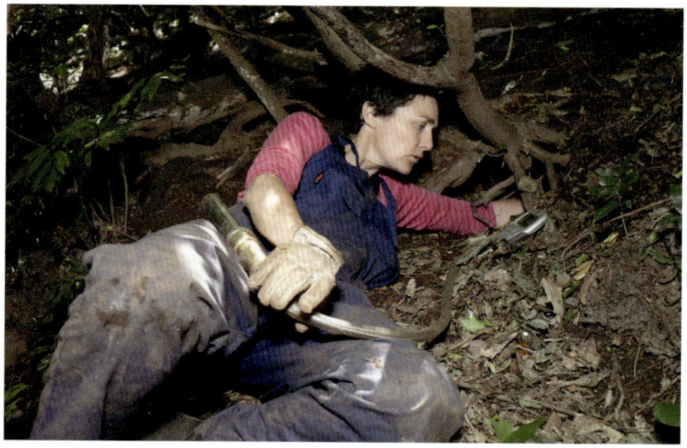
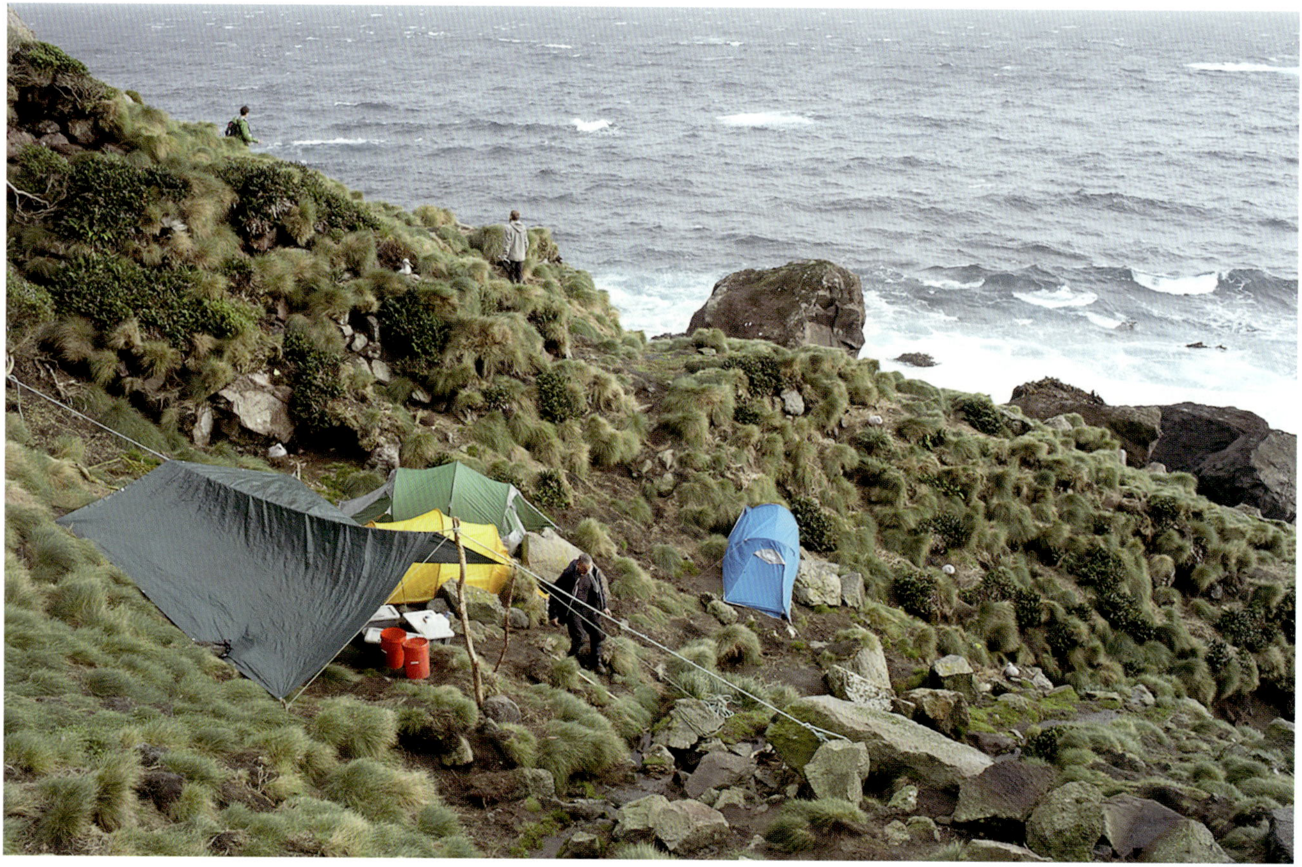

decades. Or studying the drift of breeding cycles in seabird groups as they carve out their niches in changing environments. The Collection is a really rich resource that enables us to explore this history through real specimens, collected over long timeframes, and to make sense of changes in the environment, including human impacts on the New Zealand biota.'

She points out that the Collection can serve as a time-travel device, enhanced by new methods of investigation. 'For instance, we can do isotopic research on feathers of seabird species collected as early as the 1880s and feathers from the same species today. Comparing results can reveal changes in diet over the decades, show how the advent of commercial fishing and changes in climate cycles have affected their feeding. Our collections supplement research done on the living examples of plants and animals found in nature today. They give us a powerful resource to deliver research that is necessary for understanding and conserving the biodiversity of New Zealand. Many of the millions of objects in the collections are unremarkable in themselves. But the totality of the collections makes for a unique toolset with which to explore biodiversity for New Zealand and our region. Along with our expertise and analytical techniques and science training, the Collection represents an important national science resource.'

Botanist Heidi Meudt's research on New Zealand forget-me-nots started off with documentation on specimens in Te Papa's and other New Zealand plant collections. 'Three-quarters of forget-me-not species are threatened,' she says. 'They are difficult to find, and we need information on what species there are and where they are for their conservation. So we mapped out where all the previous specimens had been collected and began a programme of visits to those locations. This summer we went to one place that had not been visited for one hundred years. And we found that species there.'

Top left: Julia Kasper collects specimens on Matiu/Somes Island in 2012.

Top right: Susan Waugh explores the contents of a seabird burrow with a burrow-scope at Titi Island, Marlborough, in 2012.

Below: Colin Miskelly at the Te Papa research team's base on Hautere/Solander Island in May 2016.

Fieldwork brings in new specimens, which also provide material for DNA analysis and microscopy. Fresh material and new techniques enable researchers to explore questions about the plants' evolutionary history, how they got to New Zealand, what their closest relatives are and how they are related to one another.

Core business for natural history collections is preserving, examining, cataloguing, categorising, and documenting specimens. Bruce Marshall is a mollusc specialist. In his time at Te Papa, he has overseen the growth of the mollusc collection from 50,000 to 350,000 lots. He has described and named hundreds of species in his career. Most specimens he's dealing with are less than 10 millimetres long, needing scanning electron microscope imagery to identify. It's painstaking work, and illustrates the ratio of time spent collecting specimens and processing them – in his case, expeditions taking a few weeks produce work lasting many years.

Knowing what species you're dealing with is the essence of research for museum scientists. It's behind all the efforts to name and describe new species and confirm the identity of existing ones. This can be of vital economic consequence, as with the Quota Management System of the NZEEZ, where commercial fisheries are allocated their catch quota by species. Ichthyologist Clive Roberts has led the development of the national fishes collection for almost three decades. One result is the four-volume tome, *The Fishes of New Zealand*, an essential taxonomic handbook for commerce and conservation, and recreational fishers. 'It's a publication that's needed at least once a century,' he says. 'More often if possible!'

The NZEEZ has an area of over 4 million square kilometres, and new species continually turn up at Te Papa from fieldwork and fishing activity. 'The average depth of the EEZ is 2000 metres,' says Clive. 'Our sampled depth so far is 1200 to 1500 metres. So we've probably found only two-thirds of the species out there. Going by the rate of discovery over the last thirty years, there could be about 700 more species to find – work for another three generations of scientists maybe.'

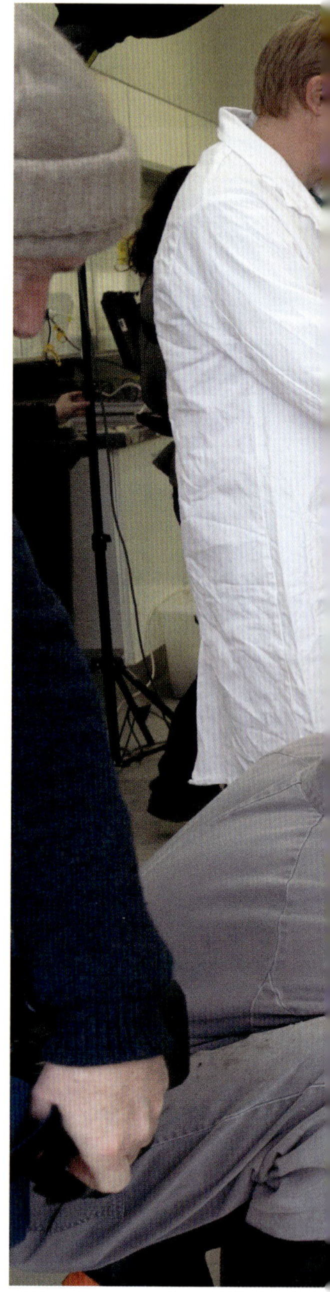

From left: Tom Schultz, Carl Struthers, Jeremy Barker, Matthew Chaplin, Clive Roberts and Andrew Stewart during a dissection of a great white shark at Te Papa in 2011.

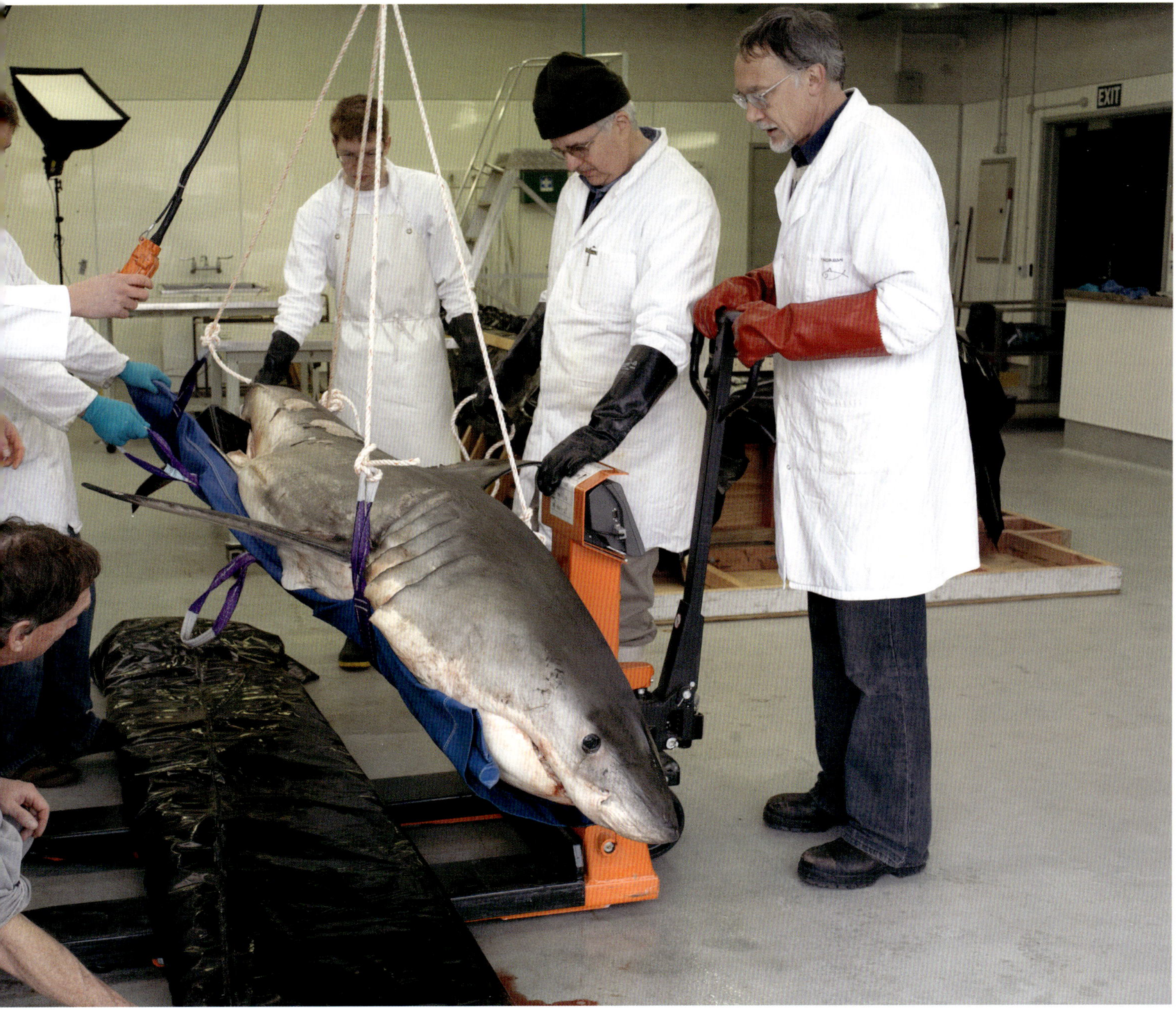
EXIT

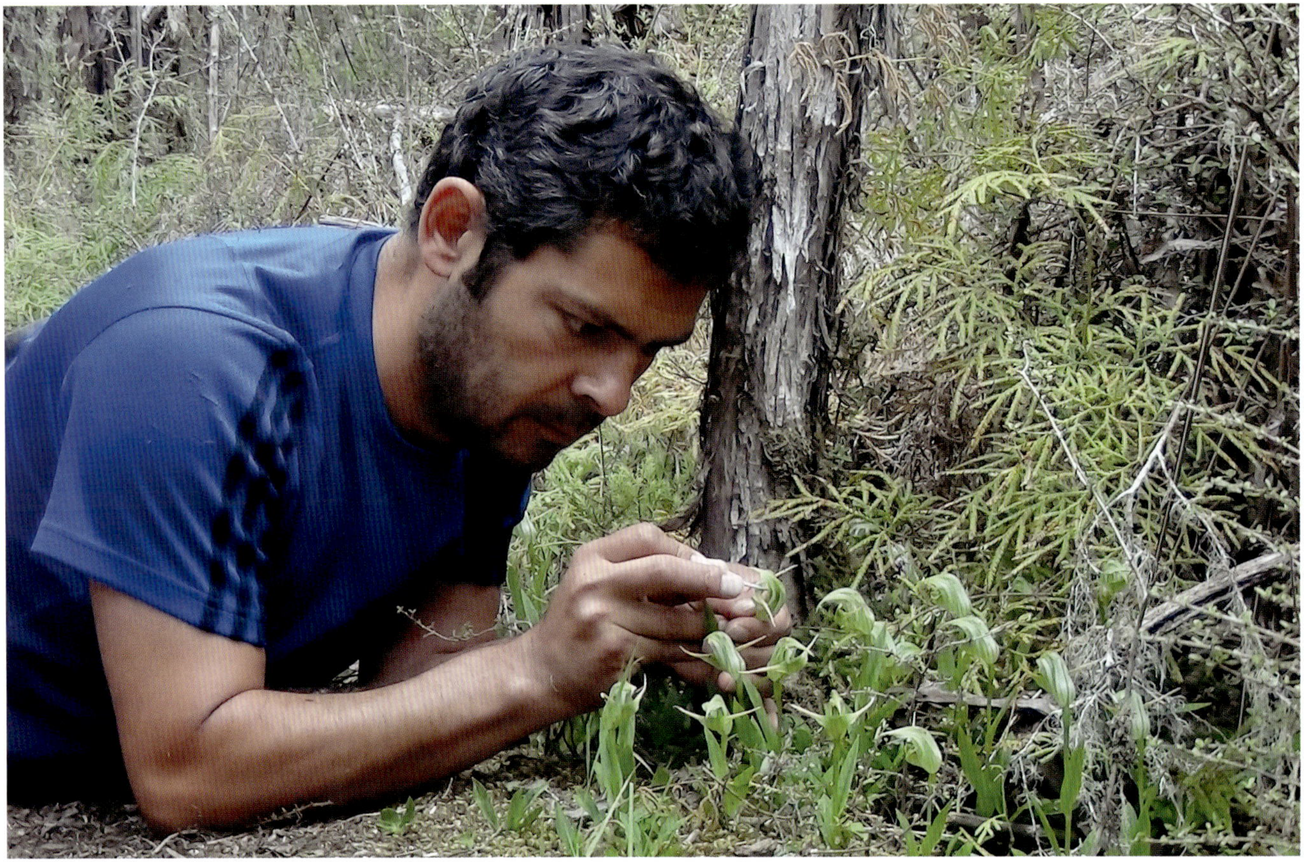

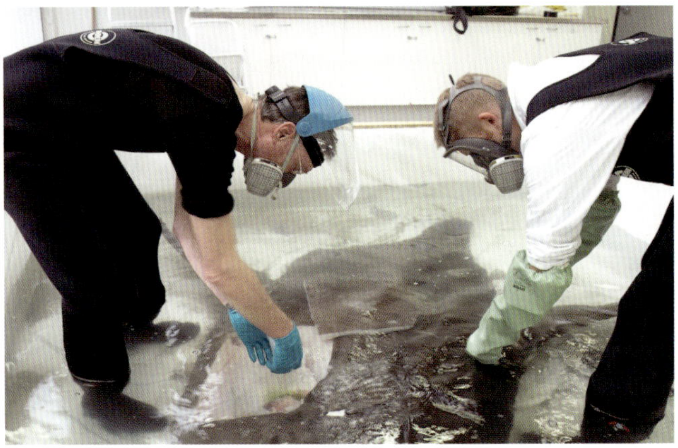

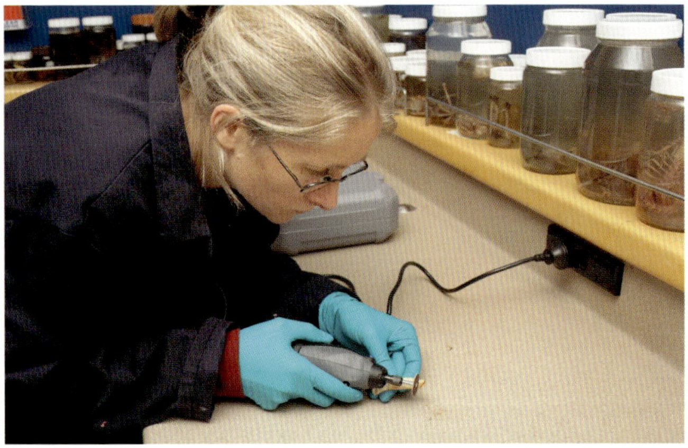

That kind of intergenerational awareness is part of the culture among the scientists in relation to the Collection. Botanist Patrick Brownsey speaks of his sense of 'standing on the shoulders of those who have gone before' as he works to catalogue the fern flora of New Zealand and help to bring the entire botany collection online. Perhaps it fosters the organisation's capacity for long-term thinking that Susan Waugh speaks of.

Palaeontologist and bird specialist Alan Tennyson offers another example of research longevity in the St Bathans fossil project. He and his associates have been returning to the site of this ancient lake bed in Central Otago every summer for eighteen years. The deposit has yielded an extraordinary wealth of land and freshwater fossils from sixteen to nineteen million years ago, unique in this country's palaeontology. 'It continues to produce ground-breaking material,' he says. 'These are the earliest fossils we have of a classic New Zealand fauna – moa, kiwi, tuatara, ducks, pigeons, parrots, even freshwater fish similar to today. But there are also tropical elements, like crocodiles and flamingoes.'

You sense among the scientists a special enthusiasm for being out in the field, discovering what species are out there, whether in the ancient past or the present, and then systematically piecing together their place and relationships in the web of the planet's organic life. Botanist Carlos Lehnebach sums it up as 'trying to understand nature, how things work – are there things out there nobody has yet discovered?' Botanist Leon Perrie gets a buzz out of seeing something he's never seen before: 'a genuinely new species, or new evidence about something previously collected. More broadly,' he says, 'being able to know what's happening in nature. Fit it in boxes. In my early years the world outside was a green blur. Now I know answers to a lot of it, I quite enjoy that.'

Above: Carlos Lehnebach studying green-hood orchids in Nelson Lakes National Park in 2016.

Below left: Tom Schultz and Carl Struthers prepare a sharp-tail sunfish for preservation in 2013.

Below right: Lara Shepherd takes a sample from a bone fragment for DNA testing in Te Papa's genetics laboratory.

For Heidi Meudt, the buzz comes from the challenge of the research process itself: 'From collecting in the field, to describing, to integrating in a paper – spending three years looking at fifteen species, day to day, week to week. And interacting with people in the field – landowners, Department of Conservation. Co-collecting with iwi. Communicating through blogs with the Te Papa audience along the way.'

Entomologist Julia Kasper comes from a background of forensic entomology. She brings an applied approach to her field to add to Te Papa's range of scientific expertise. 'Everything related to human health is of interest to me,' she says, 'particularly bloodsuckers as transmitters of disease.' Her current focus is on mosquitoes, and her previous work was their monitoring and surveillance for biosecurity. Endemic mosquitoes don't transmit disease, but with climate change modifying the environment, exotic species that do could become established here. Her specialty animates work in a part of the Collection that is not well developed. 'It would be good to have a reference collection of exotic species here,' she says, 'particularly to communicate with the public about the risk.' She, like other Te Papa scientists, appreciates the mixture of research and public communication that museum work offers.

Communication and education about marine life has been an enduring source of satisfaction for Te Papa's crustacean specialist Rick Webber during his lifetime of work at the museum. One of his current projects is refining taxonomic descriptions of the larval stages of parasitic crabs that live on mussels. It's research that could contribute to management of a parasite that causes millions of dollars' worth of damage to the mussel-farming industry.

Imaging specimens in the collection is a vital part of their documentation, and Rick early on developed a useful set of skills in scientific illustration to complement his descriptions. Jean-Claude Stahl started out as a seabird specialist. He has now developed his interest in photography into a high degree of artistry

Left: Palaeontologists at work at the fossil dig at St Bathans, Central Otago, in 2013.

Right: Colin Miskelly and Alan Tennyson study Snares crested penguins on Tini Heke/Snares Islands in 2013.

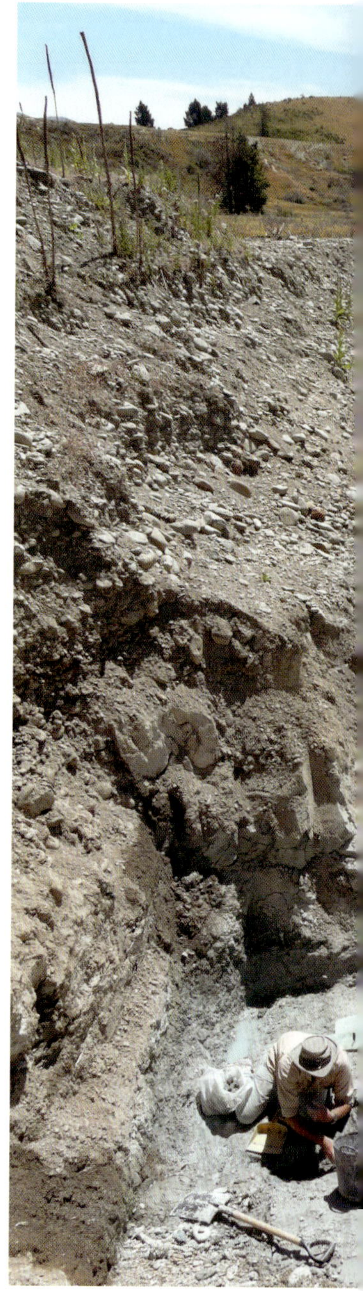

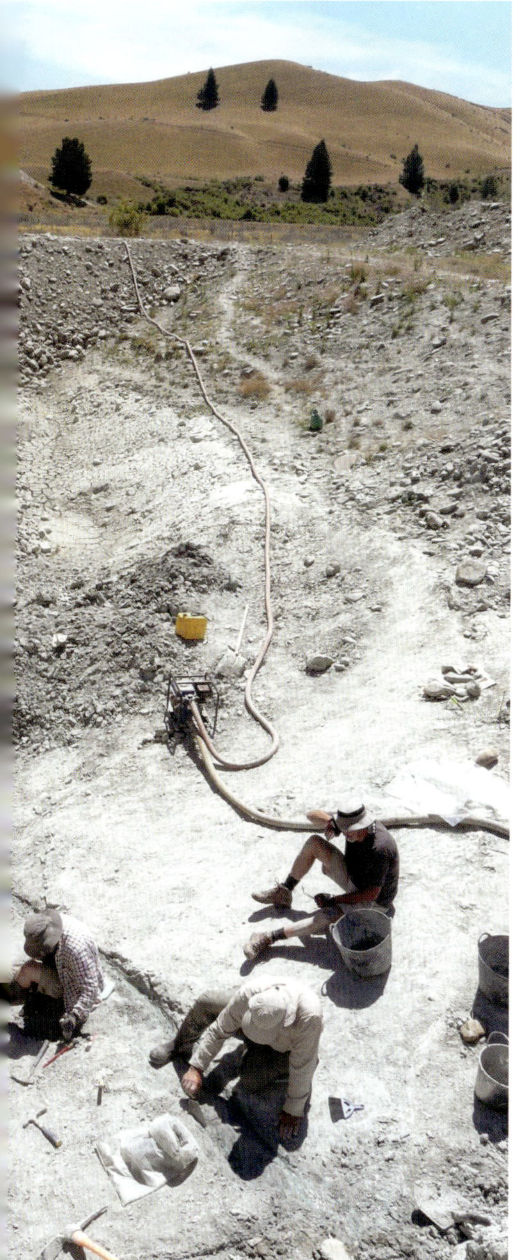
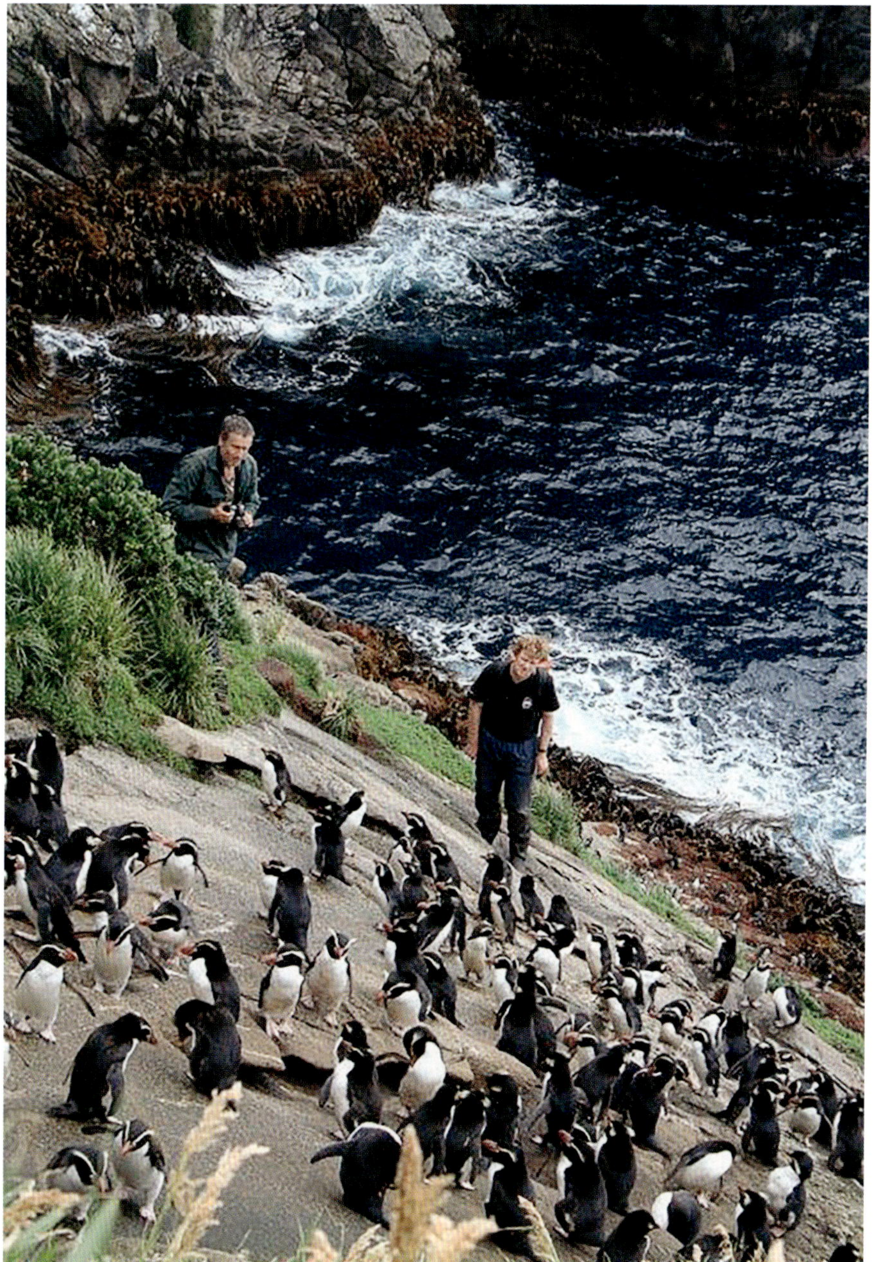

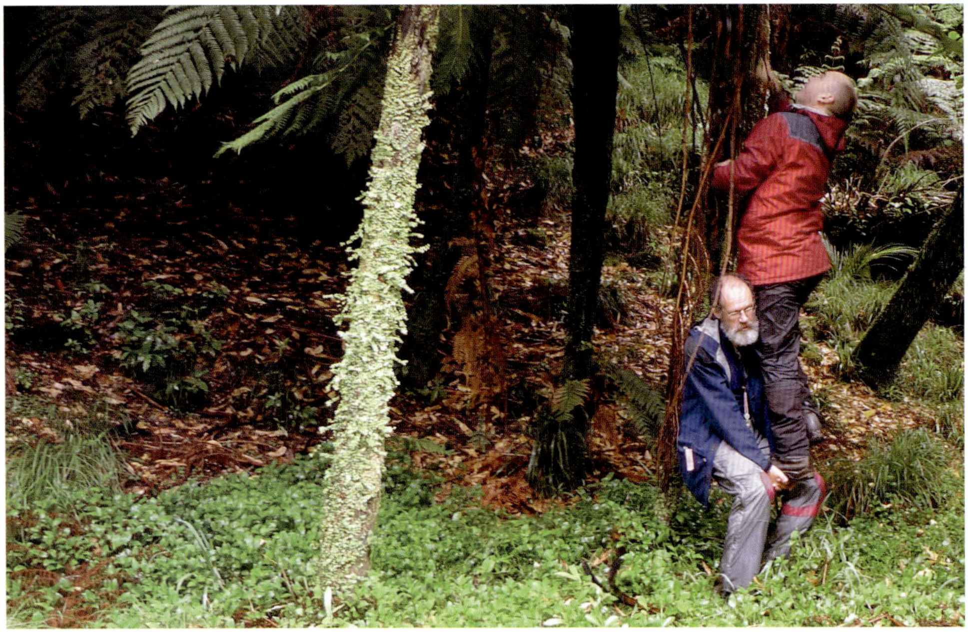
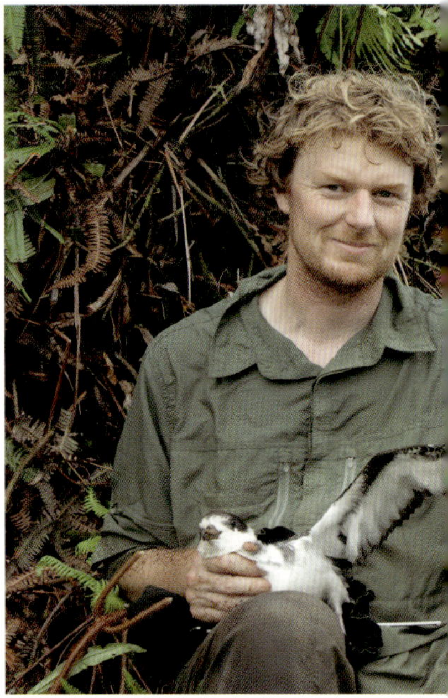
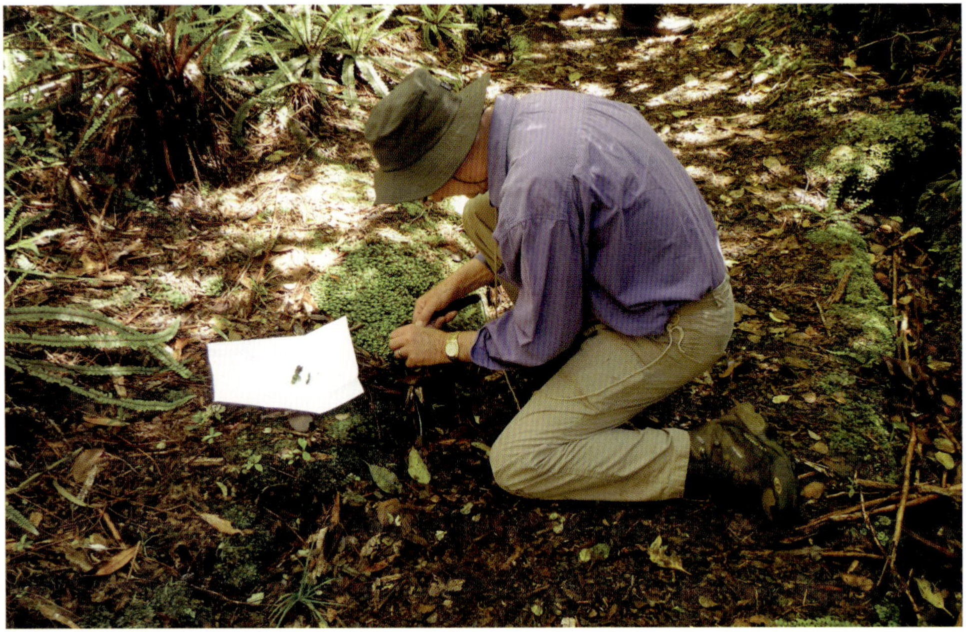
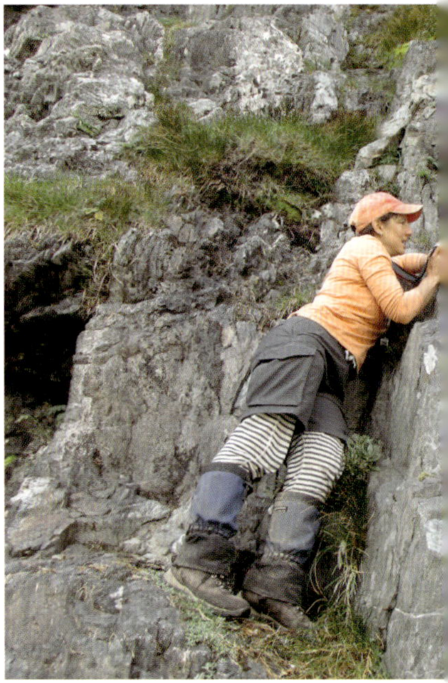

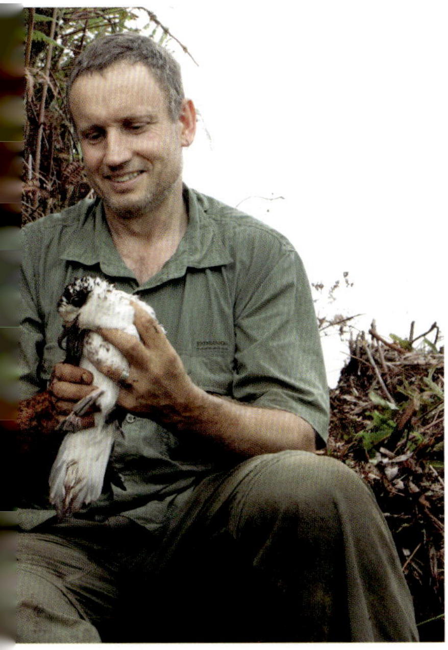

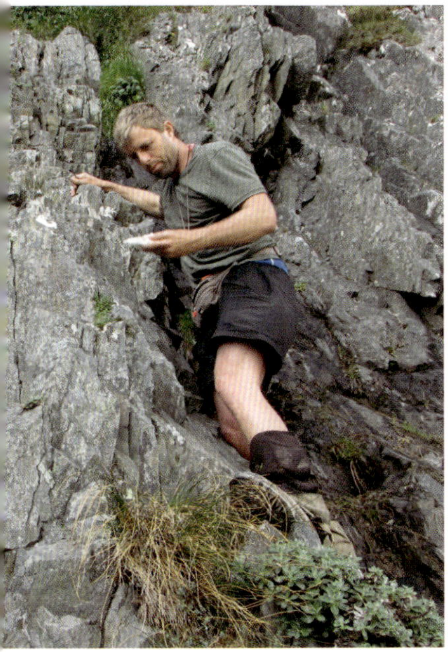

as well as technical accuracy, working across the Collection with plants, vertebrates and invertebrates, as many of the images in this book will attest. Jean-Claude moves between the field and the museum, documenting expeditions and outdoor studies, as well as working with specimens back in the photographic studio.

Work with the bird collection is often dependent on opportunism, with most native species being protected. One hundred years ago, specimens in the collection were often hunted and killed by the collectors. Entomologists, botanists and ichthyologists can still get permission to collect specimens live from the field. But when curator Colin Miskelly plans fieldwork, he does not have that authority and he can never be quite sure what he will find.

'We often seek permission to collect dead birds that have been found,' he says. 'It's hard for many to imagine what value a half-eaten prion corpse from a skua kill has for the collection,' says Colin. 'But as well as the skeleton we now have 10 cubes of tissue added to our library of genetic material for the species. So we can tell if a dead bird of that species that turns up, say, on Paraparaumu Beach, is from the same breeding site.'

These sketches of research activity at Te Papa will perhaps serve to illustrate the dynamic flow the human element provides between the Collection and the field. The Collection is a repository and a springboard that animates the research. The research animates the Collection, ultimately giving voice – and respect – to all the material forms contained within it, and to their relationships with the whole of nature.

Top left: Leon Perrie gets a helping hand from Landcare Research's David Glenny to sample a *Cyathea cunninghamii* tree fern near Pipiriki, during a collecting trip with the Department of Conservation in 2013.

Top right: Alan Tennyson and Colin Miskelly with Vanuatu petrel during an expedition to Vanua Lava in the Banks Islands group, Vanuatu, in 2011.

Below left: Pat Brownsey collecting moss specimens at Tupapakurua Falls, near National Park, central North Island, in 2013.

Below right: Heidi Meudt and Antony Kusabs document forget-me-nots at Hummock Peak, Eyre Mountains, Southland, in 2018.

Authors

PB JD BH JK MK
AK CL BM HM
CM LP DP CR RS
TS LS PS JS AS
CS AT SW RW

Pat Brownsey is Research Fellow Botany at Te Papa, where he has worked since 1977. He has previously been the museum's Curator of Botany and Senior Curator Natural Environment. Pat is an authority on the ferns of New Zealand, Australia and the Pacific, and is now working with Landcare Research to publish an online eFlora of New Zealand. He has co-authored *New Zealand Ferns and Allied Plants* (David Bateman, 1989, 2000), written contributions for the *Flora of Australia* and *Flora of New Zealand*, and published over one hundred scientific papers in refereed journals. PB

Jenn Dalen started at Te Papa in 2003 as Collection Manager for Botany, bringing her background in marine biology and systematics of red algae to work with Te Papa's extensive seaweed collections. In 2013, Jenn's work expanded across the natural history collections and later Te Papa's humanities collections. Now, as Manager Science and Humanities Collections, Jenn manages a dedicated and experienced collection management team to ensure that Te Papa's collections are cared for and accessible for the many ways they are used at the museum. JD

Bridget Hatton began volunteering at the Te Papa herbarium in 2015 while studying towards a Graduate Diploma in Science at Victoria University. She now works part-time as Research Technician Botany, does casual work for other natural history groups and continues her voluntary work databasing Te Papa's lichen collection. BH

Julia Kasper has been Curator Invertebrates at Te Papa since 2017, and is an entomologist specialised in flies. She studies the taxonomy and distribution of lower Diptera in New Zealand and the Pacific Islands, with a strong focus on biosecurity. Her background is in medical and forensic entomology and she also investigates odour-host recognition of flies using chemical ecological methods. Julia has a strong interest in outreach, especially on topics such as medical entomology, olfactory physiology and freshwater ecology. JK

Michael Keith has had a long association with Te Papa and was its Head of Writing from 2002 to 2008, contributing to Te Papa's exhibition texts, marketing, website and other materials for the visiting public. He now runs Shearwater Associates, where he is a writer, editor, textbook developer and publishing and educational project manager. MK

Antony Kusabs is Kaitiaki Taonga Collection Manager for Botany at Te Papa, where he has worked since 2010. Prior to this he was a self-employed ecological consultant. He has extensive field botany expertise through previous employment with the Department of Conservation and Landcare Research. Antony manages Te Papa's botany collection. He enjoys improving the care of collection objects, engaging with researchers and other visitors, field work, and supporting exhibitions. He also likes authoring online narratives related to the botany collection. AK

Carlos Lehnebach has worked at Te Papa as Curator Botany since 2009. His research interests include the classification of flowering plants, their pollination and the conservation of threatened species. Carlos specialises in orchids, an interest that developed in Chile (his home country). He has described several orchid species that are unique to New Zealand and was recently awarded the Dan Hatch Medal for his contributions to the study of New Zealand's orchids. Carlos has primary responsibility for the museum's New Zealand flowering plant collection and the historical collections of foreign flowering plants. CL

Bruce Marshall is Te Papa's Kaitiaki Taonga Collection Manager for Molluscs. He began collecting shells in 1959, aged eleven, in the area between Tauranga and East Cape and concentrating on living and fossil molluscs, while also collecting other fauna. He quickly became known to Charles Fleming and museum researchers Baden Powell, Dick Dell and Winston Ponder. Bruce joined the National Museum in 1976. He edits several journals, has published 144 papers and named 564 new taxa, and received a DSc from Victoria University in 2012. BM

Heidi Meudt is Researcher Botany at Te Papa, where she has worked since 2006. Prior to that, she completed her PhD in botany in 2004 at the University of Texas at Austin, and was a National Science Foundation International Postdoctoral Research Fellow at Massey University. Heidi has authored over thirty scientific publications on the taxonomy and evolution of southern hemisphere plants. Her research integrates data from morphology, DNA, pollen and chromosomes. She was an Alexander von Humboldt Experienced Research Fellow (2012–14) at the University of Oldenburg, Germany, and is a Fellow of the Linnean Society of London. HM

Colin Miskelly is Curator Vertebrates at Te Papa, where he has worked since 2010. He has been involved with wildlife research and conservation since his school days, gaining a PhD in zoology before working for the Department of Conservation for nineteen years as a scientist and manager. Colin has published more than one hundred scientific papers and several books, and is the project manager and editor of the highly popular *New Zealand Birds Online* website. CM

Leon Perrie has been Curator Botany at Te Papa since 2003. Although he dabbles with flowering plants and bryophytes, his research focus is on ferns and lycophytes. He is assisting with the fern chapters for the eFlora of New Zealand. He is also increasingly working in the Pacific, including co-authoring a book about the ferns of the Solomon Islands. Closer to home, he is a regular participant in the activities of the Wellington Botanical Society, and an active contributor to the citizen science website *iNaturalist*. LP

Dean Peterson is former Associate Director Collections, Research and Learning, now Director Strategy and Performance at Te Papa. A graduate of Michigan State University (PhD Physical Chemistry), he spent ten years at the Jet Propulsion Laboratory working in the field of atmospheric research before being appointed to NASA Headquarters as Co-Programme Manager for the Upper Atmosphere Research Program (UARP). He joined Te Papa in 2016 after working for Antarctica New Zealand and National Technology Networks and managing the Marsden Fund (the fundamental research fund for the New Zealand government), the James Cook Fellowships, and the Rutherford Discovery Fellowships. DP

Clive Roberts, Curator Vertebrates, has worked at Te Papa since 1987. Fishes have been a life-long passion, guiding his education and scientific career – collecting, researching and documenting the diversity of New Zealand, South-West Pacific and Southern Ocean fishes. Clive's popular and scientific publications number over 240, and include three co-authored books: *New Zealand Fish: A complete guide* (1989); *The Rockpool Fishes of New Zealand* (1992); and *The Fishes of New Zealand* (Te Papa Press, 2015, vols 1–4), which was awarded the 2016 Whitley Medal for best zoological publication in Australasia. CR

Rodrigo Salvador has been a postdoctoral researcher at Te Papa since 2017. He obtained his Bachelor's and Master's degrees from the University of São Paulo, Brazil, and his doctoral degree from the University of Tübingen, Germany, in 2016. His area of expertise is the molluscs, particularly land and freshwater snails, and his main research interests are systematics, biogeography and palaeozoology. Rodrigo has published over sixty articles in international scientific journals and several pieces of popular science. He is also the co-founder and editor-in-chief of the *Journal of Geek Studies*. RS

Tom Schultz has worked in the Natural Environment/Natural History department at Te Papa since 2002. He is currently Kaitiaki Taonga Collection Manager responsible for several vertebrate groups, working mainly in the bird, mammal, reptile and fossil collections. Tom spent six months working on the articulation of cetacean skeletons for the *Whales/Tohorā* exhibition and has travelled regularly to North America to install and move it between venues. TS

Lara Shepherd has been Te Papa's geneticist since 2011; she manages Te Papa's DNA laboratories, the only such laboratories in a New Zealand museum. Lara has over sixty peer-reviewed scientific publications and was awarded a Rutherford Discovery Fellowship in 2012. Her research uses DNA sequences for identification and to understand the evolution of New Zealand's plants and animals. She has worked on such diverse groups as land hoppers, fungi, ferns, fishes and extinct birds, and has a particular interest in developing non-destructive methods for sampling DNA. LS

Phil Sirvid is Kaitiaki Taonga Collection Manager for Entomology (insects and allies) and has been at Te Papa since 1993. He started early in entomology, rearing Australian gum emperor moths from caterpillars when he was five. At that age Phil was arachnophobic, but he has since developed an inordinate fondness for spiders. He has published a number of papers on these fascinating creatures, spanning topics ranging from taxonomy to medical arachnology. Phil has a PhD from Victoria University, Wellington. He was also one of the team behind the recent *Bug Lab* exhibition. PS

Jean-Claude Stahl is an imaging specialist at Te Papa, where he has worked since 1986. He trained as a biologist and took part in seabird research first on the French subantarctic islands Crozet and Kerguelen, and then on the Auckland, Snares and Solander Islands. He completed a doctorate in animal ecology before settling in New Zealand, and is now involved in digitising the plant and animal specimens of Te Papa's science collections. He specialises in macro and micro photography, and many of his images are published on Te Papa's Collections Online web pages. JS

Andrew Stewart has worked for Te Papa for over thirty-five years and he is now Kaitiaki Taonga Collection Manager for Fishes. He has had a fascination with fishes from before starting school and his speciality areas are the Southern Ocean and deep-sea fishes (especially anglerfishes). Andrew is the joint winner of the 1996 New Zealand Association of Scientists 'Science Communicator Award', and he is co-editor and co-author of the four-volume publication *The Fishes of New Zealand*. AS

Carl Struthers is Research and Technical Officer Fishes at Te Papa, where he has worked since early 2005. His primary research focus includes the taxonomy and phylogeny of the families Sebastidae, Moridae and Aulopidae, but he also has a strong interest in several other New Zealand and Antarctic fish families. Carl is an experienced field worker, and has participated in many fish-collecting expeditions using scuba in coastal waters and blue-water expeditions to remote locations, both within and outside New Zealand waters (including the Kermadec Islands, Auckland Islands, Fiji, Tuvalu, New Caledonia, Tonga and Japan). Carl is also one of the editors and a contributing author of *The Fishes of New Zealand*. CS

Alan Tennyson is Curator Vertebrates at Te Papa, where he has worked since 1996. Prior to that he worked in the Department of Conservation's Threatened Species Unit and as a marine campaigner at the Forest & Bird Protection Society. His work has mainly focused on seabird conservation and research on New Zealand's extinct birds. He is the author of more than 130 scientific publications and co-authored the books the *Checklist of the Birds of New Zealand, Norfolk and Maquarie Islands, and the Ross Dependency, Antarctica* in 2010 and *Extinct Birds of New Zealand* in 2006, which was a 2007 Montana New Zealand Book Awards finalist in the Environment Category. AT

Susan Waugh is the Head of Science at Te Papa, where she has worked as manager of the Natural History team since 2011. Susan maintains an active research interest in seabird biology, conducting some of the first satellite-tracking studies of albatrosses carried out in New Zealand and more recently working to improve the conservation status of threatened endemic New Zealand seabirds and reduce the impact of fisheries on these birds. Susan studied zoology and biology, initially at Canterbury University and then at Cambridge University, England, and completed her PhD at Rennes I University in France in 1998. SW

Rick Webber is Curator Invertebrates at Te Papa, focusing on marine species and specialising in Crustacea. Since starting work at the National Museum over 35 years ago, lobsters, crabs and shrimps have been his main interest, particularly their larval development, with land hopper taxonomy a recent addition to his studies. Rick takes particular pleasure in producing taxonomic illustrations. His publications include descriptions of New Zealand spider crab larvae (1981) and a major contribution to Crustacea in the *New Zealand Inventory of Biodiversity* (2010). RW

Image credits

All images by Te Papa photographers Jean-Claude Stahl, Maarten Holl, Michael Hall, Norman Heke, Mike O'Neill and Kate Whitley unless otherwise credited here.

A—B **22** (Aorangi skink) photograph by Raymond Coory; **41** (Blobfish) photograph by Kerryn Parkinson © NORFANZ Founding Parties.

12–13 photograph by Jane Ussher

C—E **62** (Deep-sea anglerfish) photograph by Carl Struthers; **64** (Deep-sea molluscs) photographs (SEMs) by Bruce Marshall

44–45 photograph by Jane Ussher

F—G **82** (Flaming flagfin) photograph by Carl Struthers; **89** (Forget-me-not pollen) photographs (SEMs) by Niki Murray

72–73 photograph by Jane Ussher

H—M **111** (Hagfish) photograph by Carl Struthers; **115** (Hoodwinker sunfish) photograph by Carl Struthers; **141** (Male-eating bivalve) photograph by Raymond Coory; **142** (Moa egg) nā Te Rūnanga o Kaikoura i tautoko; **144** (Moss-dwelling forget-me-not) photograph by Phil Garnock-Jones

104–105 photograph by Jane Ussher

N—P

157 (Nelson cave spider) photograph by Raymond Coory; **163** (New Zealand harvestman) photograph by Raymond Coory; **167** (New Zealand spider orchid) photograph by Pam Shearer; **177** (Pseudoscorpions) photographs by Raymond Coory

146–147 photograph by Jane Ussher

R—S

178–179 photograph by Jane Ussher

212–213 photograph by Jane Ussher

T—Z **219** (Tateid snails) photographs by Raymond Coory; **224** (Tū Hononga) photographs by Pat Stodart

The collectors **244** (top) photograph by James Wrigglesworth; **244** (bottom) photograph by JT Salmon; **245** (second from top, third from top, and bottom) photograph by Trevor Ulyatt; **246** painting by Mary Tripe

242–243 photograph by Jane Ussher

The collection and the field

251 (top left) photograph by Leon Perrie; **256** (top) photograph by Andreas J Zeller; **259** (left) photograph by Alan Tennyson; **259** (right) photograph by Antony Kusabs; **260** (top left) photograph by Eru Te Huia; **260** (top right) photograph by Stephen Totterman; **260** (bottom left) photograph by Leon Perrie; **260** (bottom right) photograph by Brian Rance

Index

Page numbers in bold refer to illustrations.

D

E

F

G

H

I

First published in New Zealand in 2019 by Te Papa Press
PO Box 467, Wellington, New Zealand
www.tepapapress.co.nz

A catalogue record is available from the National Library of New Zealand

ISBN 978-0-9941460-5-2

Digital imaging by Jeremy Glyde. Additional digital imaging by Sjoerd Langeveld (pages 12–13, 44–45, 72–73, 104–105, 146–147, 180–181, 212–213, 242–243).

Designed and typeset at Strategy Creative
Printed in China by Everbest Printing Co Ltd

Front cover image: moa egg (Te Papa, ME012748)
Back cover images: (clockwise from top left) kākāpō (Te Papa, OR025019); moss-dwelling forget-me-not (Te Papa, SP104478); blobfish (Australian Museum, I.42771-001); freshwater limpet (Te Papa, M325933)